THE PETERSON FIELD GUIDE SERIES®

A Field Guide to Geology

Eastern North America

DAVID C. ROBERTS

Illustrated by

W. GRANT HODSDON

Sponsored by the N
the National
and the Roger To.,

D0914015

HOUGHTON MIFFLIN COMPANY

Boston   New York

1996

For information about permission to reproduce selections from
this book, write to Permissions, 215 Park Avenue South,
New York, NY 10003

PETERSON FIELD GUIDES and PETERSON FIELD GUIDE SERIES
are registered trademarks of Houghton Mifflin Company.

For information about this and other Houghton Mifflin trade and
reference books and multimedia products, visit The Bookstore at
Houghton Mifflin on the World Wide Web at
http://www.hmco.com/trade/.

Library of Congress Cataloging-in-Publication Data
Roberts, David C.
A field guide to geology. Eastern North America / David C. Roberts ; illustrated by W. Grant Hodsdon.
p. cm. -- (The Peterson field guide series; 47)
ISBN 0-395-66326-1 (cloth). -- ISBN 0-395-66325-3 (pbk.)
1. Geology--East (U.S.) 2. Geology--Canada, Eastern. I. Title. II. Series.
QE78.3.R62 1996   96-5715
557—dc20   CIP

Printed in the United States of America

VB 10 9 8 7 6 5 4 3 2 1

EDITOR'S NOTE

The landscape of eastern North America is legendarily diverse. It comprises mountains of various sizes, vast plains, and rocky coastlines. Most of us have great curiosity but little expertise in how these landforms came to be. A car trip spurs more wondering, as we pass through structures exposed to view in road cuts. In his *Field Guide to the Geology of Eastern North America*, David Roberts has given us a guide to interpreting these rocks and landforms. He helps us find answers to our common questions: What kind of rock is that? How old is it? How did it get there? Roberts gives us not only the big picture — how rocks form, what makes mountains — but a host of little ones. Each of his Highway Sections covers a specific route in eastern North America that illustrates the what, why, and when of the area's geology, from the gold veins of the Canadian Shield, to the fossils and sandstones that are evidence of a warm, shallow sea in the interior of the continent, to the massive folding, faulting, and general upheaval of the Appalachians.

In 1934, my first *Field Guide to the Birds* was published. It was designed so that live birds could be readily identified at a distance by their field marks without resorting to the "bird in hand" characters that early ornithologists relied on. The Peterson System, as it is now called, is based primarily on patternistic drawings with arrows that pinpoint the key field marks. The Peterson System is generally incompatible with geology, of course; no two formations are exactly alike, even those formed in the same way. But this book is one of a new generation of Field Guides that helps the reader understand not only what is out there, but why it is there. As we read the *Field Guide to Geology*, we learn to recognize the structures found around us. First we notice the basic differences between rock types; we learn to spot the fine-grained texture of a rock formed deep within the earth, the stratified look of a rock that hardened from layer upon layer of loose sediments, and the tortuous folds of a rock transformed by intense heat and pressure. Roberts shows us how to tell ripple marks left by a stream from those left by an ocean, how to tell where a glacier stopped, and where to look for valuable minerals.

To stand virtually smack in the middle of the continent and find ripple marks left by waves of an ancient ocean, or to recognize the abundant evidence of the glaciers that once covered the Northeast, is like having the power to look back in time. *Geologic* time, that scale of millions and billions of years that is otherwise beyond our perception. With experience, and with this excellent Field Guide to show us the way, we can learn to see the world around us with new understanding.

Roger Tory Peterson

CONTENTS

Editor's Note   v

1 How to Use This Book   **1**
2 Some Geologic Background   **7**
   Igneous Rocks   7
   Sedimentary Rocks   16
   Metamorphic Rocks   25
   Geologic Formations   32
   Geologic Time and the Evolution of Plate Tectonics   45
3 The Canadian Shield   **57**
   Superior Province Highway Sections   67
   Southern Province Highway Sections   82
   Grenville Province Highway Sections   102
4 The Stable Interior   **113**
   Allegheny Plateau Highway Sections   134
   Cumberland Plateau Highway Sections   143
   Northeast Corridor Highway Sections   155
   Cincinnati Arch Highway Sections   167
   Michigan Basin Highway Sections   176
   Illinois Basin Highway Sections   184
   Ozark Dome Highway Sections   193
   Western Tier Highway Sections   211
5 The Appalachian Province   **219**
   Central Appalachian Highway Sections   250
   Ouachita Mountains Highway Sections   267
   Western Windows Highway Sections   274
   New England Highway Sections   296
   Atlantic Provinces and Southern Quebec
      Highway Sections   317

6   **The Coastal Plain**   **329**
      Coastal Plain Highway Sections   360

Appendix A: The Origins of Names in the Geologic Time
            Chart (see front Endpaper)   371
Appendix B: Radioactive Dating   375
References   379
Index   385

A FIELD GUIDE TO GEOLOGY
Eastern North America

... in the vast hieroglyphic record which our globe composes, page lies beneath page, and inscription covers over inscription, — it presents to the student a theme so vast and multifarious that it might seem . . . he [must] despair of ever being able effectually to grapple with it. "But," to borrow from one of the most ingenious of our Scottish metaphysicians, "in this, as in other instances in which nature has given us difficulties with which to cope, she has not left us to be wholly overcome. If," says Dr. Thomas Brown, ". . . she has placed us in a labyrinth, she has at the same time furnished us with a clue which may guide us, not indeed, through all its dark and intricate windings, but through those broad paths which conduct us into day. The single power by which we discover resemblance or relation in general, is a sufficient aid to us in the perplexity or confusion of our first attempts at arrangement."

— Hugh Miller, *The Testimony of the Rocks*, 1852

1

HOW TO USE THIS BOOK

This book will help you to read the testimony of the rocks, to understand their origins and histories. Earth's crust is truly a prodigious book, testifying to over three billion years of geologic history, including not only the history of life and cycles of changing climates, but also numerous cataclysmic crustal events that are almost beyond human comprehension. It is a story that challenges our minds and our imaginations.

This book will help you to answer questions about landscapes, rock types, rock structures, and the evolution through billions of years of the eastern United States and southeastern Canada. It will help you to become familiar with the physiognomy of eastern North America, and guide you in reading clues that geologists have used to interpret what happened there in the past. You can learn to see what others don't notice and appreciate what to them is meaningless. You will look at the world with different eyes.

Your local road cuts and rock outcrops (places where the bedrock is exposed) may look pretty much the same every morning when you get up, even over the course of an entire lifetime. But the land is changing today, as it always has. Millions of years ago, the rocks you see today may have been way below the surface. In the future, they may disappear, destroyed by the slow but mighty processes of weathering and erosion that disintegrate them and carry their remains away. Over hundreds of millions of years, mountain ranges are created and then destroyed. Even whole continents change. The one we are living on is but a piece of an earlier great supercontinent that broke apart and whose fragments will eventually rejoin.

It may seem difficult to interpret the great variety of rock types to be found, but if you think in terms of resemblances and relationships, you can bring order out of apparent chaos. Certain observations suggest the origins of certain kinds of rocks. Mud left by a flooding river can become compacted and cemented into shale. Sand in a beach, dune, or riverbed can become sandstone if its grains are cemented together by mineral matter. The lava from a volcano cools into the rock called basalt, while the molten material that doesn't reach the surface may cool into another rock, gabbro, with large, angular mineral grains, and the ash from a volcanic eruption may turn into a third type of rock called tuff. When the mineral-filled water from a warm spring evaporates, the minerals left behind form travertine.

When you examine rocks, think in terms of their geologic history and origins. Shale was originally mud, and sandstone was sand. Basalt and tuff prove the existence of past volcanos. The rock gabbro cooled and formed below Earth's surface. Where there is travertine, there were warm springs millions of years ago.

Other observations show how rocks break down and disappear. Flagstones in a pavement will crack and disintegrate. Pieces flake off stone walls of buildings. Smaller rock fragments fall from the faces of cliffs and road cuts. Old marble statues and gravestones gradually lose detail. Those are all examples of the weathering that is constantly attacking rocks. You can see how water moves weathered fragments from one place to another in sand or mud washed down from a bare spot onto a sidewalk or driveway during a heavy rainstorm. And of course, we are all familiar with the muddy water of many rivers as they carry the products of erosion from all over their drainage basins to points farther downstream.

The instability of Earth's crust is evidenced in many ways. Earthquakes are caused by movement along major cracks in the crust called faults. There are cracks or fault zones in some outcrops where the rock has broken, and masses have moved up, down, or horizontally in relation to each other. Large masses of solid rock have cracked without moving because of stresses in the crust, as large sections of the crust move and rotate (the movement is described in the section on the plate tectonics theory in chapter 2). Such cracks are visible in many outcrops. In some places, rock layers have clearly been tilted or folded by horizontal pressure, another result of movement in the crust. Volcanic eruptions are spectacular evidence of the fact that there is molten material below the solid crust. The eruptions are caused by movements and stresses in the crust that cause major cracks, through which the molten material reaches the surface.

Rocks show that what is happening today also happened in

the past. You can understand what they are telling us if you observe carefully and make deductions from your observations. Surprisingly good geological detective work can be done without an encyclopedic knowledge of geology. The more you see and understand, the more exciting it gets. Become familiar with some background information, tune up your imagination, and you are ready to go.

■HOW THIS BOOK IS ORGANIZED

Chapter 2 contains the background information you will need to help you interpret what you observe and to put your observations into the framework of the continent's history. It includes an introduction to sediments — rock and mineral fragments that settle from water, wind, and melting glaciers — and to the three basic types of rock: igneous, sedimentary, and metamorphic. This chapter will help you understand the origins of rocks and major rock structures, as well as geologic phenomena such as the work of glaciers and the formation of caves.

The rest of the book, chapters 3 through 6, is organized according to geologic regions, or physiographic provinces, of the continent (see the map on the back endpapers of this book). Each province has a particular combination of landscape, rock structure, and history. After a general description of the geology of the province, each of these chapters describes selected areas in more detail.

The Highway Sections of chapters 3–6 describe the bedrocks and structures that can be seen along certain highways in the region. In some areas there are few or no outcrops or road cuts that expose the bedrock, while in others there will be more interesting outcrops and road cuts, more than can possibly be described in a book this size.

■READING THE TESTIMONY OF THE ROCKS

Once you know a little about the origins of rocks, you can start to recreate the past. A piece of sandstone in your hand may have come from the bed of a river that flowed from a great mountain range now long gone. The limestone you drove by yesterday may be a mass of tiny broken fragments of animals, now extinct, that lived in a warm, tropical sea. The granite on the facade of a building cooled slowly from a molten mass far below Earth's surface.

Much of North America was covered by glaciers during the last ice age, which ended about 12,000 years ago. With a little practice, it is easy to recognize country that has been glaciated and to identify the boundary between glaciated and unglaciated regions.

This book will help you visualize the land of each physiographic province and how it was formed. The map on the back endpapers suggests the history of rocks, organisms, and geologic processes extending over billions of years. Each province has its own story.

One province, the Stable Interior, consists of almost horizontal layers of rock that originated on the floors of oceans containing early primitive fishes, or on land masses so old that their animals were very different from those of today. The Appalachian Province rocks were folded and broken and changed into other rock types by the heat and pressure caused by great sections of Earth's crust colliding with each other. They contain granite and other related rocks that formed during the crustal movements. The Canadian Shield contains massive fragments of ocean basins so old that their waves carried only the simplest of organisms and broke on shores devoid of animals or plants. The rocks of the ocean basins were crumpled and changed by crustal collisions and surrounded partly by granite that formed far below Earth's surface and partly by granite that was changed by the high heat and pressure of the collisions. The Coastal Plain is a series of geologically recent ocean bottom rocks and unconsolidated sediments, with fossils of recognizably modern animals that exhibits a history of shorelines moving back and forth as sea level rose and fell.

The time chart on the front endpapers shows the succession of geologic eras, periods, and epochs. Familiarity with the chart will enhance your feeling for the long continuum of geologic time. No human can truly comprehend a time period of a million years, much less a billion years. But we can live with *relative* ages. We can imagine that the Ice Age occurred relatively recently during the Quaternary Period, while the Grenvillian Orogeny (crustal collision) happened much, much earlier, late in the Proterozoic Eon. The dinosaurs lived in between, during the Mesozoic Era, which was closer to the Ice Age than to the Grenvillian collision.

■ A WORD ABOUT MAPS

In a pocket-sized field guide, the maps are necessarily small and some detail has to be sacrificed. Figure 1 is a good example. The map shows Interstate 20 where it traverses the edge of the Coastal Plain between Weatherford and Abilene, Texas. The Coastal Plain rocks were deposited on top of older Stable Interior ones. Because of differences in the rate of erosion, the actual border between the younger and older rocks is very irregular, but on a small map like Figure 1, the border between younger and older rocks is a simple curved line.

Maps also often fail to show geologic details. For example, one small to medium-sized hill can have several varieties of bedrock, which are impossible to show on a small map. In the exposures you study, you may find rock types that are not shown on the maps.

Even on more detailed maps, the boundaries may be educated guesses, because the actual boundaries are hidden by topsoil or glacial debris with a plant cover (or buildings, concrete, and asphalt). Moreover, a younger geologic map of a given area may differ from an older one because of new information, and maps may differ because different geologists have interpreted ages or rock types differently.

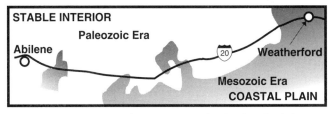

Figure 1. Interstate 20 between Weatherford and Abilene in Texas nine times crosses the boundary between Stable Interior rocks from the Paleozoic Era (white) and younger Coastal Plain rocks from the Mesozoic Era (shaded) that were deposited on top of them. The boundary's irregularity is due to recent erosion. On a small map of a larger area, the boundary would be drawn as a simple curved line, similar to the line of the highway, without the detail seen in this more accurate map. The chart on the front endpapers shows where the Paleozoic and Mesozoic eras fit into the geologic time scale.

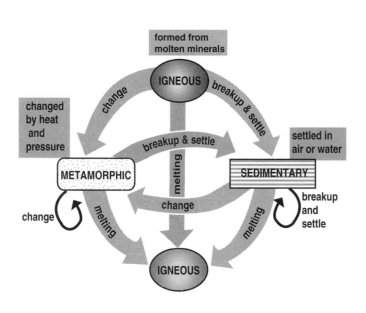

2

SOME GEOLOGIC BACKGROUND

Rocks are classified in three basic categories, according to how they formed: igneous, sedimentary, and metamorphic. Putting a rock in the right category is the first step in identifying it—and it is often all that you really need to know. Igneous rocks formed from molten mineral matter that exists deep below Earth's surface. Some cooled below the surface, and some cooled on the surface. Sedimentary rocks consist of sediments, or grains such as those that make mud, sand, and gravel. Metamorphic rocks are those that have been changed in texture and composition by great heat and pressure in Earth's crust.

There are a great many different kinds of rock, and you can expect to find some in your travels that you cannot identify. Sometimes even the experts are puzzled by outcrops and have to take samples to their laboratories. However, you will probably be able to recognize most of the rock types you encounter if you can identify the most common rocks of the three categories.

Rocks are aggregations of mineral grains, so familiarity with some of the more common minerals will help in the process of identification. A few key minerals are described here; see Frederick H. Pough's *A Field Guide to Rocks and Minerals* for more details on mineral identification. A hand lens will help you find mineral grains, minute structures, and small fossils or fossil fragments. You may wish to buy a rock or mason's hammer to produce fresh, unweathered surfaces; weathered surfaces can be misleading.

■ IGNEOUS ROCKS

Igneous rocks originated as molten minerals below Earth's crust. Common igneous rocks include granite and diorite,

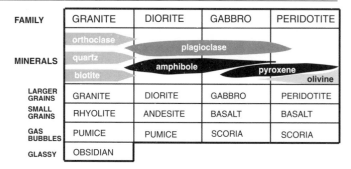

Figure 2. The families of igneous rocks.

which cooled below the surface, and rhyolite and basalt, which cooled on the surface as volcanics (Figure 2). Igneous rocks form most of Earth's crust.

Igneous rocks vary in color from almost white through yellowish, pinkish, brownish, and gray to black, depending on what minerals they contain. Lighter colored varieties contain large amounts of the minerals quartz and orthoclase feldspar. Dark varieties contain high percentages of dark to black minerals, such as amphibole and pyroxene, as well as some kinds of plagioclase feldspar. Any of these rock types can contain minerals other than the primary ones listed.

◆Intrusive Igneous Rocks

When molten mineral matter cools below the surface, it forms intrusive igneous rocks, so called because the molten material has intruded, or moved into, the crust. The rock is a solid mass of readily visible, irregular to squarish or angular mineral grains that have grown and cooled tightly together. Figure 3A is a sketch of this igneous texture. Some intrusive igneous rocks show cavities left by former gas bubbles.

Within the Earth's crust, the molten minerals are insulated from the atmosphere, and therefore cool slowly. Slow cooling gives the mineral grains time to grow large enough to be visible without magnification. The actual grain size varies with the cooling conditions; the faster the mixture cools, the smaller the grains are.

Intrusive igneous rocks lack the obvious layers that charac-

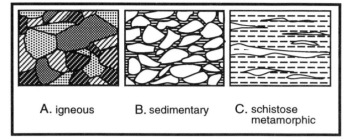

Figure 3. Simplified sketches of rock textures. A represents an intrusive igneous rock such as granodiorite, made up of four minerals. B represents quartz grains in a sandstone, surrounded by cement; in an actual rock, the grains may be more closely packed. The texture in C has been changed by heat and pressure.

terize sedimentary and volcanic rocks, although more or less parallel major cracks, or joints, may make them appear to be layered. The particular grainy texture of the rock, though, will show you that it is not sedimentary or volcanic.

● Granite

Granite is the most common intrusive igneous rock in the continental crust (Plates 4, 31). It is made up primarily of the minerals quartz, orthoclase feldspar, and biotite. Granite varies considerably in color, from more or less whitish through pinkish and yellowish to gray, according to the color of the orthoclase. The quartz is grayish and translucent and fractures during weathering (or under a hammer blow) with an irregular surface (Plate 41). It reflects sunlight like broken glass. The orthoclase breaks with smooth plane surfaces, a type of break that is called cleavage (Figure 4). Orthoclase feldspar cleaves in two directions at right angles to each other; in other directions it has an irregular fracture (Plate 41). If the conditions are right, you will see sunlight reflecting from cleavage planes of the feldspar. The third major mineral of granite is biotite, or black mica, which cleaves in only one plane. In other directions, it breaks irregularly. Thus, biotite occurs as black reflective flakes, or as "books" of flakes (Plate 41). Some granites contain muscovite, or white mica, a mineral that cleaves like biotite but has a grayish to whitish translucent appearance.

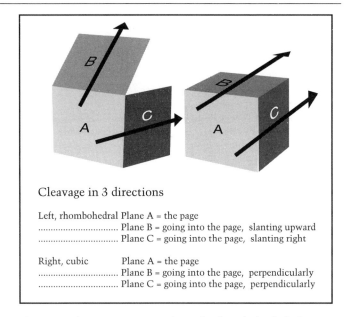

Figure 4. Cleavage in minerals. Left, rhombohedral cleavage, as in the mineral calcite, with cleavage in three directions not at right angles; right, cubic cleavage, as in the mineral galena, with cleavage in three directions at right angles.

● Diorite

Diorite is typically speckled black and white, more or less half and half; it consists of amphibole, which is black, and a white variety of plagioclase feldspar. The elongated amphibole grains cleave in two directions that are not at right angles. In small grains, the cleavage gives their broken surfaces a splintery appearance unlike the flat flakes of granite's biotite. Plagioclase cleaves in two planes that are almost, but not quite, at right angles (94°). White plagioclase is so similar to white orthoclase that it is easy to confuse them. However, if you have the right plagioclase-containing rock sample, and hold it in the light just right, you can see on some cleavage surfaces tiny parallel lines placed close together. Aside from the feldspar, you can distinguish true diorite from granite by diorite's higher percentage of the black mineral and lack of quartz.

● Gabbro

Gabbro is distinguishable from granite and diorite by its dark gray or black color. It consists of dark plagioclase, amphibole, and pyroxene, another black mineral. Pyroxene looks splintery, like amphibole; it, too, cleaves in two planes, but the planes in pyroxene are almost at right angles, and the grains typically are not elongated as are those of amphibole.

● Peridotite

Most peridotite consists of pyroxene and olivine. Typically, olivine is transparent to translucent green, but there are also light gray or brown versions. It breaks irregularly and has a sparkly reflection. The presence of olivine distinguishes peridotite from gabbro.

Gabbro and peridotite are relatively rare in outcrops because they typically form not under continents but in the crust under the oceans, below sediments and lava flows. We see rocks of the ocean crust only where they have been pushed up onto the continent by forces such as continental collisions, and even then, the lower crust is seldom exposed.

● More Igneous Rock Types

Intrusive igneous rock types grade into one another, as shown in Figure 2. For instance, granodiorite is similar to granite but has a significant infusion of plagioclase and amphibole; gabbrodiorite is similar to gabbro but contains a lot of amphibole. Syenite looks rather like granite but has little or no quartz, consisting essentially of orthoclase and biotite. Syenite that contains a significant amount of plagioclase as well as orthoclase is called monzonite. Table I summarizes the differences between granite and its close relatives.

Table II shows the mineral composition of some dark-colored granular rocks, including three types that you might confuse with gabbro and peridotite. Amphibolite is a metamorphic rock (Plate 44). The others are intrusive igneous rocks. Diabase is a relatively fine-grained rock in which small, more or less rectangular grains of plagioclase are surrounded by larger grains of pyroxene. Anorthosite may be dark. It can be white or black, depending on which variety of plagioclase it has (Plate 7).

Table I

	Granite	Granodiorite	Syenite	Monzonite
Orthoclase	x	x	x	x
Plagioclase		x		x
Quartz	x	x		
Biotite	x	x	x	x

Table II

	Gabbro	Diabase	Amphibolite	Peridotite	Anorthosite
Plagioclase	x	x	x		x
Amphibole	x		x		
Pyroxene	x	x		x	
Olivine				x	

Table I. This chart shows how granite and granitelike rocks differ in their mineral compositions.
Table II. The mineral compositions of dark-colored granular rocks that might be confused with one another. Amphibolite is actually metamorphic, but could be mistaken for an igneous rock; the others are intrusive igneous.

◆The Origin of Igneous Rocks

Planet Earth is made up of layers. The crust, consisting of solid rock, extends down about 15 to 35 miles (24–60 km) below continent surfaces and about 3 to 6 miles (5–10 km) below the ocean floor. The second layer, where igneous rocks originate, is called the mantle. It continues down to about 60 to 120 miles (100–200 km). The upper part of the mantle consists of solid mineral matter that is hotter and denser than that of the crust, and the lower part of the mantle is a very thick material, about the consistency of tar, that can flow slowly. Below the mantle is a liquid layer, and below that is the solid core.

Masses of igneous rock form as a result of stresses in the crust caused by squeezing and distention as sections of the crust collide, separate, and grind past each other with infinite slowness but implacable force. The crust cracks, and pieces of it move past one another along cracks called faults.

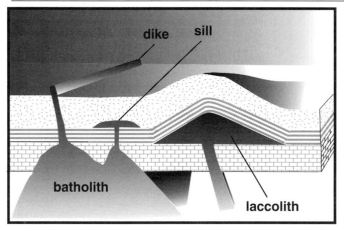

Figure 5. The forms of intrusive igneous rock bodies. The largest is a *batholith,* an irregular mass of solidified mineral matter, by definition over 40 miles (70 km) in known diameter. A similar mass with a smaller diameter is a *stock.* Where molten material has been forced up through a crack in enclosing rock, it hardens into a *dike,* a sheet of igneous rock. If the intruding material forces itself between two layers of sedimentary rock, it forms a *sill.* A thick sill with the overlying rock pushed up into a dome is a *laccolith.*

Figures 14 and 15 show the kinds of faults. Faults that extend into the mantle tap the material there; a crack in the solid rock above permits mantle material to become a "mush" of hot liquid mineral matter with crystals or solid mineral grains forming in it. Minerals melt at temperatures ranging from about 1110°F (600°C) to 1560°F (850°C). The crustal stresses force or float the mush up into the crust.

Molten material that reaches the surface flows or explodes out of a volcano or a long fissure in the crust. If it remains below the surface, it congeals into bodies of the shapes indicated in Figure 5. Many batholiths and stocks form along fault zones and are shaped like thick sheets; others cool as short, more or less circular "cakes," flat cones, flattened mushrooms, discs, or balloons. Some have round protuberances extending upward from their upper surfaces.

The whole process happens very slowly. Geologists estimate the material in a batholith may move upward at a rate of about 6 feet (2 m) per year, so it takes a few thousand to a million years for it to reach its highest level. Most of the cooling takes place at the final level, and estimates of cooling rates range from 50°F (10°C) to 1020°F (550°C) per million years, with the rate decreasing as time goes by. Thus, cooling takes a few million years.

◆Extrusive Igneous Rocks

When molten mineral matter reaches the surface, it is called lava. Rocks formed when lava cools are called extrusive igneous rocks. Lava cools faster on the surface, so there is not enough time for the mineral grains to grow large. The extrusive rocks rhyolite, andesite, and basalt (Figure 2; Plates 8, 24, 28) have mineral grains barely visible or invisible to the unaided eye. Obsidian, or volcanic glass, has no visible grains, even under magnification.

Each intrusive igneous rock has an extrusive equivalent with essentially the same mineral composition but with much smaller grains. Granite-rhyolite and gabbro-basalt are two such pairs (Figure 2).

The most viscous lavas are the rhyolitic ones, which have small, short-lived flows. The most fluid, basaltic lavas, can travel long distances. In Hawaii, basaltic flows from Kilauea have flowed 10 or more miles (16 km) from the crater into the ocean.

Some lavas contain gases such as water vapor, gaseous carbon compounds, sulfur, hydrogen, chlorine, and fluorine. Gas bubbles that do not escape from the cooling lava before it solidifies leave pockets in the subsequent rock. Pumice, from viscous rhyolitic lava, is a rock so frothy with gas pockets that it will float. The basaltic equivalent is scoria, which is darker.

Extrusive igneous rocks harden from lava flows, and periodic eruptions may result in rock layers that may resemble the layers of sedimentary rocks. However, a hand lens will reveal the finely grained igneous mineral texture. If the color is a reddish or gray orthoclase one and you can see quartz grains, the rock is probably rhyolite. If the rock is black, it is basalt. Some extrusive igneous rocks have a speckled texture that indicates andesite.

In contrast, sedimentary rocks such as shale and fine-grained sandstone consist of tiny quartz grains with few or no other mineral grains, except for mica flakes in some cases. On weathering, igneous rock breaks into irregular angular chunks; most shale weathers into smaller chips. Sandstone also weathers into chunks, but its sandy texture distinguishes it. Some basaltic flows crack, when they cool, into vertical hexagonal or pseudohexagonal columns, a phenomenon called columnar jointing that is a proof of solidified lava.

Ash that explodes out of a volcano is made up of tiny, jagged, glassy mineral shards. After the shards fall to the ground, they may be cemented together by mineral matter to form a sedimentary rock called tuff. Many old lava flows are interbedded with layers of tuff (and other sedimentary rocks). Tuffs vary considerably in appearance and may be difficult to identify. Some outcrops of tuff look like shale; others resemble lava rock.

Under certain conditions, grains or crystals of one mineral

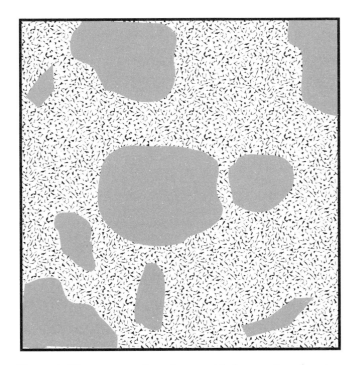

Figure 6. The texture of porphyry, with large mineral grains in a finer-grained groundmass.

in a deep molten mass will solidify and grow relatively large, while the other mineral matter is still liquid. After a volcanic eruption, when the liquid portion cools, its minerals will cool into smaller grains. The resulting rock will have large grains or crystals in a finer grained matrix. Such rock is called a porphyry, and the larger components are phenocrysts (Figure 6).

Although lava extrudes most spectacularly from volcanoes, the testimony of the rocks is that by far the greatest bulk of lava has issued from long cracks, or fissures, in Earth's crust. Fissure flows occurred both on continents and on the ocean floor in the past; they still occur today in many places on the ocean floor.

■ SEDIMENTARY ROCKS

Sediments are masses or deposits of mineral grains or other particles that settle out of water or air. Sand and silt settle to the bottom of bodies of water. Rock salt and gypsum are left behind when ocean water evaporates. Wind-blown dust, sand, and volcanic ash fall out of the air onto the ground. Boulders, pebbles, sand, and silt melt out of a glacier's ice.

A prime characteristic of sedimentary rocks is their distinct layers, formed as sediments accumulated in level and parallel "beds." Intrusive igneous rock and many kinds of metamorphic rock do not show layers. Sedimentary rocks may have cracks along the planes between layers (the bedding planes). Weathering along the cracks will emphasize the layering of the rock.

Most sedimentary rocks consist of particles cemented together by mineral matter such as quartz, calcite (calcium carbonate), or hematite (iron oxide) that precipitates out of ground water. The solid precipitate surrounds the particles and holds them together. With magnification, you can see that the particles of some sedimentary rocks are not as closely packed as those in igneous rocks (Figure 3B).

Conglomerate, sandstone, shale, and limestone are common sedimentary rocks. Sedimentary rocks cover the igneous rocks of Earth's crust in many regions. Almost all of the bedrock in the Stable Interior and the Coastal Plain is sedimentary, as is much of the Appalachian Province bedrock.

Figure 7 shows the grain sizes for different categories of sediments. The largest particles—boulders, cobbles, and pebbles, are moved by flooding rivers or streams, or by storm waves at seashores. Quiet waters allow the smallest grains to settle as silt and clay.

"Sand" refers strictly to particle size. There are many kinds of sand, differing in composition. The sand most familiar to many of us consists almost entirely of quartz grains. A "clean" quartz sand is 90 percent or more quartz. A "dirty" quartz sand has more than 10 percent of other mineral grains or silt grains. Some dirty quartz sands, including those that melt out of glaciers, have sand-size fragments of very fine-grained rocks as well as mineral grains.

Some sands consist primarily of other minerals. The sand at New Mexico's White Sands National Monument is grains of gypsum. Olivine sand occurs in Hawaii, along with sand that consists of basalt fragments. Some sand in Florida is

Sediment Grain Sizes		
NAME OF PARTICLE	DIAMETER	
	mm	*inches*
Boulder	greater than 256	greater than 10
Cobble	64–256	2.5–10
Pebble	4–64	0.15–2.5
Granule	2–4	0.07–0.15
Sand	1/16–2	0.0025–0.07
Silt	1/256–1/16	0.00015–0.0025
Clay	less than 1/256	less than 0.00015

Figure 7. Sediment grain sizes.

composed mostly of sand-size fragments of small ocean animal shells and skeletons.

●Conglomerates

Among the most easily identifiable sedimentary rocks is conglomerate, a solidified gravel (Plates 28, 32, 37) usually formed of pebbles and sand cemented together. Many conglomerates formed in the channels of ancient rivers. A torrential downpour in a mountain range can quickly fill the streams with water that rushes down toward the nearest lowland. The force of the water rolls and bounces pebbles, cobbles, and sometimes even boulders downstream, along with smaller particles. The force of the currents determines how large the rocks are and how far they are carried. In general, the water carries smaller particles farther, but some sand gets trapped among the larger particles. Most of the sand and smaller grains are carried farther, out onto the lowland. The pebbles in these conglomerates are typically rounded from knocking together and from abrasion by sand in the water.

Gravel can also form on lowlands where weathering breaks the bedrock into chunks. At the seashore, wave action can break and abrade rocks to produce gravel. Marine fossils in the resulting conglomerate or in closely associated sandstone are field marks of marine conglomerate.

Breccias are conglomerates with angular pebbles, unworn by water action (Plate 5). Volcanic breccia consists of angular rock fragments embedded in a matrix of volcanic ash or lava (Plate 43), and limestone breccia contains angular chunks of limestone.

●Sandstones

Typical sandstone consists of 90 percent or more sand-sized quartz grains, cemented together by quartz or calcite. Plates 12, 15, 21, 22, 30, 32, 34, 36, and 42 show samples and outcrops of sandstone. Arkose is a sandstone with more than 10 percent feldspar grains. Graywacke is a sandstone with a greater than 10 percent admixture of feldspar and other grains besides quartz. Quartz sand is characteristic of river channels, flood plains, and near-shore lake and ocean deposits. It can be carried by currents too weak to move pebbles and cobbles. When rivers rise over their banks, they deposit sand on their flood plains as well as in their channels. Where a river flows into a lake or ocean, any sand it carries drops to the bottom (possibly to be moved farther by waves and currents).

Wind-blown sand forms large deposits in some regions, and some sandstones, particularly in the Rocky Mountain region, originated as sand dunes.

The texture of sandstone grades from coarse to fine, with fine-grained varieties grading into shales. Sandstones are gray, tan, yellow, and even green, depending upon the content of minerals other than quartz. Those that contain large amounts of hematite are various shades of red to purple (Plate 42).

Most sandstones formed in distinct layers at least a few inches thick, and they tend to weather into angular chunks along bedding planes. Some sandstones have much thicker layers, measured in tens of yards or more.

Sand moved by a water current forms small ridges called ripple marks at right angles to the direction of current flow. The sand grains move up a gradual upcurrent slope and drop down a steeper downcurrent slope on each ridge. Current ripple marks are preserved in some sandstone layers and may be exposed by weathering (Plate 46). They tell you that you are examining either an ancient riverbed or an ocean or lake bottom along which there was a constant current; they will even tell which direction the current moved.

The ripple marks seen at many beaches, formed by waves washing back and forth across the sand, are oscillation ripple marks. On these, the inland- and offshore-facing slopes have the same pitch. If you see large areas of oscillation ripple marks, you know you are seeing an ancient beach.

A delta is a deposit of sand and silt where a river flows into a lake or the ocean (Figures 8B and 148). The delta grows larger as sand slides down its outer margin into deeper water, forming inclined layers. A horizontal layer is deposited where the river flows along the top of the growing delta. The inclined layers make a regular pattern in resulting sandstones called delta cross bedding (Plates 12, 34).

Current cross bedding (Figure 8A) is another feature of some river channel sands and sandstones. Irregularities in the bed of a river can promote the development of small delta-like structures, with thin, inclined layers dipping in the direction of water flow. As time goes by, low places are filled in, the current shifts direction, and new small deltas grow in other places. The result is localized small slanting layers within a major layer of sand or sandstone.

Graywacke is a "dirty" sandstone that contains a lot more than quartz grains. It is made up of angular, sand-sized grains of quartz and feldspar, plus sand- to granule-sized rock fragments, all in a matrix of clay and silt. In most cases, the rock fragments are of extrusive igneous rocks and of the metamorphic rocks slate and phyllite. Graywackes come in vari-

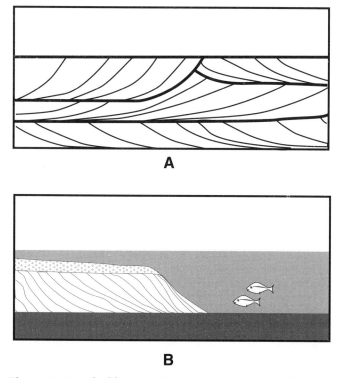

Figure 8. Crossbedding. A shows current cross bedding, as formed in a river channel where the current has shifted in direction; B shows a section of a delta, with cross bedding occurring as sand slides down the outer edge. A horizontal layer of sediment at the top of the delta covers the upper edges of the earlier cross beds.

ous shades of gray, including yellow and green tints. Fossils in these rocks indicate an oceanic origin for most graywackes, which are usually associated with lavas and tuffs. Graywacke is found where there has been major crumpling in Earth's crust and presumably represents a miscellaneous mixture of weathered material dumped quickly into a nearby sea from a volcanic area.

● **Shale**

Silt and clay particles are so small that they are easily carried by a current that is too sluggish to move even sand. When silt and clay are cemented by calcite or quartz, they form the rock shale. Plates 12, 14, 15, 21, 22, 28, 32, and 36 show shale in outcrops.

After a river floods, a large amount of silt and clay will settle out of the still and evaporating flood waters, so silt and clay (and shale) are typical flood plain deposits, as is sand (and sandstone).

The silt and clay that enters lakes and oceans can be carried great distances before settling to the bottom. Because of their lightness, silt and clay tend to settle farther from shore than the sand does.

The clay minerals in shale come from the weathering of feldspars and micas. Some shales also contain tiny mica flakes. In typical shale, the individual grains are all but invisible without magnification, but some varieties are sandy and grade into fine-grained sandstones.

In the strict sense, the word "shale" is limited to a rock type that exhibits obvious bedding planes. In other varieties of shaly rocks, called mudstone or siltstone by the experts, the layers blend gradually into each other, and there are no distinct bedding planes, although the bedding may be indicated by different colored layers. The term "shale" includes these varieties in the following chapters of this book. Like sandstones, shales come in several colors, from cream to jet black, including shades of green, brownish orange, and purple (Plates 42, 48).

Most shales are easy to recognize in outcrops because of their characteristic thin bedding, which causes them to weather into chips and thin blocks that eventually turn into mud. Some shales and all mudstones and siltstones weather directly into mud, so their weathered surfaces are covered with dried mud, with relatively hard rock directly below.

● Redbeds

Shales and sandstones in shades of red and purple are called redbeds. The color comes from hematite, an iron mineral that occurs as microscopic grains in shale and mudstone, and as a coating on sand grains in sandstone. The hematite presumably forms in a relatively warm climate that has distinct wet and dry seasons. The seasonal climate promotes the combination of an iron mineral in the sediments with atmospheric oxygen to form the iron oxide hematite. Most redbeds apparently formed on dry land, but there are marine redbeds that may have been formed from red mud being washed into the ocean.

● Limestone and Dolomite

Limestone is a kind of rock made up of 90 percent or more calcium carbonate, or calcite; dolomite is a similar rock composed mainly of the mineral dolomite (Plates 12–15, 23). Dolomite and most limestone originated in ocean regions that were beyond zones of silt and sand deposition (Figure 9). There are three varieties of limestone: clastic limestone, chemical limestone, and organic limestone. They differ in composition and origin, but all are made up mostly of calcite. Calcite is a mostly colorless and translucent mineral that cleaves in three planes that are not at right angles to each other. Broken calcite masses show parallelogram-shaped faces (see Figure 4). That characteristic cleavage will be visible in some samples of limestone and will help you identify the rock. Also, in sandstones with calcite cement, you can see light reflecting off the calcite's cleavage surfaces in between the quartz grains.

Clastic, or fragmental, limestone is made up of sand- or silt-sized calcareous fragments from the shells and skeletons of marine animals. The material was broken up by wave or current action in shallow oceans or near the shores of deeper ones. The fragments are cemented together by pure calcite or by a limey mud containing calcite. Clastic limestone is a common ocean deposit, formed in relatively shallow water, where most of the shellfish, corals, sea urchins, and other animals with calcareous hard parts live. The clastic rock from Indiana called "Bedford limestone" is a very popular building stone. If you find a building sheathed with it or another similar stone, you will be able to see the animal fragments.

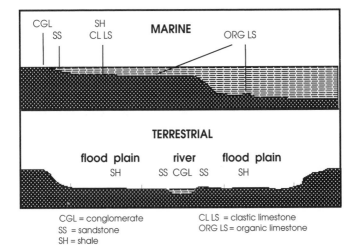

CGL = conglomerate   CL LS = clastic limestone
SS = sandstone   ORG LS = organic limestone
SH = shale

Figure 9. Sedimentary rocks are indicators of land and ocean-bottom environments.

Chemical limestone is formed by organisms, such as some algae, that excrete calcite as part of their metabolic activity. Over a long time, the calcite builds up into layers that become rock. Freshwater limestones and some marine limestones are of this variety. They form in quiet waters where there are no strong currents to disperse the calcite.

Organic, or pelagic, limestone (chalk) is formed when microscopic marine organisms die, and their calcareous skeletons fall to the bottom and become cemented along with fragments of other organisms in a calcareous mud (Plate 35). This type of limestone is relatively rare, and it appears to have formed in quiet-water environments where there were few bottom-living animals, either in deep oceans or along continental shelves. Chalk is white to light yellowish, is porous and very fine-grained, and you can write on a chalkboard with it. It differs from other sedimentary rocks in that it is hard but not brittle, and you can cut it with a pocket knife. Organic limestones occur in North America as the chalks of Kansas and the Gulf Coastal Plain, which are similar to the chalk in the famous white cliffs of Dover, England.

The colors of limestones vary from off-white to black. In between are many shades of yellow, brown, and gray. The gray to black colors come from organic matter; the other colors are caused by iron or other minerals.

Layers of most limestones tend to be thicker than those of shale, and they weather into angular blocky chunks, not into chips as shale does (Plate 12). Sandstone also weathers into chunks, but the limestone chunks are typically somewhat more cubic than those of sandstone. Another distinctive characteristic of limestone is the way it weathers. Rain and melting snow dissolve the calcite in the rock over a long period of time, leaving the surface smooth, with gently rounded corners, edges, and irregularities. Chalk is different from other limestones in that it typically disintegrates instead of weathering into chunks.

Dolomite, or dolostone, is a rock that looks like limestone but is composed of the mineral dolomite instead of calcite. Calcite is a compound of calcium, carbon, and oxygen; dolomite has those three elements plus magnesium. If you find a rock that looks like limestone but has cavities that contain pearly or pinkish crystals with curved surfaces, both the crystals and the rock are probably dolomite. The grains in dolomite rock show cleavage surfaces like those of calcite (Figure 4). Dolomite rock may form from limestone when calcite is somehow replaced chemically with dolomite.

● Chert

In many limestones and dolomites, an obscure chemical process has replaced portions of the rock with chert, a very hard and dense material made of exceedingly fine grains of silica, which is silicon dioxide, like quartz (Plate 13). The silica has migrated to certain points and other minerals have migrated away. Chert is opaque and varies in color from almost white through reddish and greenish to gray and black, depending on its other mineral and carbon content. It typically breaks with splintery or curved surfaces.

Chert forms irregular blobby nodules (Plate 13) and "lenses" that resist weathering. The host rock dissolves and cracks away from around them. The flint from which people once made arrow and spear heads, knives, scrapers, and hand axes, is a variety of chert, sometimes with added black, reddish, or yellowish patterns and bands from iron and other impurities (Plate 45).

Chert also occurs in layers that originally formed in quiet ocean basins. Chert layers originated in siliceous masses of skeletons of microscopic organisms and pieces of sponges. The silica dissolved in the water; where the water became overloaded with silica, the silica precipitated out to form layers of chert. If the water was muddy, the silica migrated away from the silt and clay to form layers of chert interbedded with shale.

● Evaporites

Halite, anhydrite, and gypsum are rock types called evaporites because they consist of minerals precipitated from ocean water as it evaporates. Halite is made of sodium chloride, or common table salt. Anhydrite is a form of calcium sulfate, and gypsum is anhydrite with water molecules attached to its molecules. These rock types are evidence of environments in which shallow ocean embayments dried up and left the minerals behind.

■ METAMORPHIC ROCKS

In Earth's surprisingly active crust, some bodies of rock are subjected to slow, continuing, strong horizontal pressure. Most, if not all, of the pressure is caused by movement of crustal sections, or plates. Over a long period of time, rock may be folded or pushed downward into a zone of the crust where the heat is so great that the rock first becomes plastic and then, as depth and heat increase, the rock melts. The pressure and heat cause the constituent minerals to flow and, in some cases, to chemically recombine into other minerals. If the molten material later cools, the new rock will be different from the original rock—it will have metamorphosed. Metamorphic rocks are widespread in New England, southeastern Canada, the Canadian Shield, and the Piedmont Zone of the Appalachian Province. Schist, gneiss, slate, and marble are well-known metamorphic rocks.

● Argillite and Slate

When shale is subjected to great pressure and then heat, it turns into argillite and then slate. Argillite is shale that has become harder, more compact, and more brittle than typical shale. Its layers tend to be thicker, and it breaks into angular chunks instead of the thinner chips of shale. As the pressure and temperature increase, the argillite turns into slate.

Slate, like shale and argillite, is a rock whose grains are mostly invisible to the unaided eye. Its primary characteristic is that it cleaves with flat surfaces as some minerals do, with cleavage planes that look like bedding planes but aren't. As pressure further compacts the argillite, higher temperature causes some clay minerals to chemically metamorphose into flakes of the minerals muscovite and chlorite. As they grow, the flakes orient themselves with their flat surfaces mostly parallel to each other and perpendicular to the direction of pressure. The tiny flakes cause cleavage zones in the rock (Plate 27), which can be at many different angles to the bedding planes, as Figure 10 shows. The angle between cleavage planes and bedding planes depends upon what part of the fold you are looking at. The parallel cleavage planes of slate are what make it so good for roof shingles. Slate comes in gray, reddish, and greenish colors, due to reddish iron oxide (hematite), greenish chlorite, and carbon.

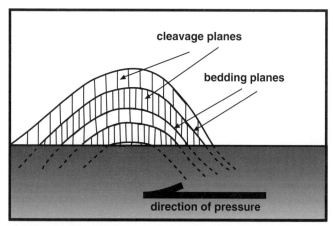

Figure 10. Slaty cleavage. A fold in slate layers, showing cleavage planes perpendicular to the horizontal pressure, not to the bedding planes.

● Phyllite and Schist

If pressure and heat continue to build up, minerals in slate begin to combine into somewhat larger grains to form phyllite and to recombine into readily visible grains of several minerals in schist. In phyllite, clay minerals have continued to change into mica; phyllite is more of a mass of minute mica flakes, with some new orthoclase and the original quartz. It may still cleave as slate does (Plate 27). As in slate, the individual mica flakes cannot be seen without magnification, but they give the rock a shinier appearance than the dull surfaces of argillite and slate show.

If the heat increases further, the mica, orthoclase, and quartz molecules combine into larger masses to form a variety of schist that consists of streaks or bands of easily visible black (biotite) mica flakes with streaks and blobs of quartz and orthoclase.

The word "schist" refers to the rock's texture (Figure 3C). All schists have mostly flaky or fibrous minerals, and the minerals have flowed in response to the heat and pressure (Plates 24, 45). Some igneous as well as sedimentary rocks become schist, and schists have a variety of minerals. Typical biotite schist is described above; other varieties have a

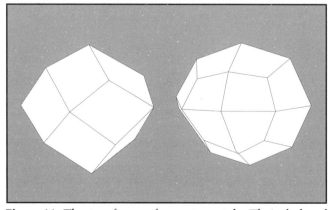

Figure 11. The two forms of garnet crystals. Their dark red color makes them easy to recognize in metamorphic and some igneous rocks.

fine-grained layered look. In either case, the minerals gathered into separate masses and slowly flowed, with the bands and streaks spreading out at right angles to the direction of pressure. Schist easily splits into slabs because of its flaky mineral grains. There are no true bedding planes in schists, as there are in sedimentary rocks.

The darker igneous rocks, such as diorite and gabbro, become chlorite schist and hornblende (a variety of amphibole) schist. Chlorite schist is mostly a mass of more or less parallel chlorite flakes. Hornblende schist is a mass of splintery hornblende grains.

Garnet (Figure 11, Plates 24, 44) occurs as dark red crystals in some bodies of schist. Garnet results from the recombination of minor constituents in the original rock. The presence of garnet crystals is thus a good indication that the rock is metamorphic.

● Quartzite and Metagraywacke

Quartzite and metagraywacke are layered rocks that evolve from sandstone and graywacke because of great heat and pressure (Plate 26). In each case, the minerals melt and then recombine upon cooling.

When sandstone turns into quartzite, its original quartz composition does not change, but the recrystallization gives the rock a more compact texture with quartz grains closely packed like those of igneous rocks (Figure 3). Quartzite is a dense, hard rock that breaks *through* the grains, while typical sandstone breaks *around* the quartz grains. The colors of quartzites resemble those of sandstones.

Metagraywacke is the metamorphic equivalent of graywacke, or dirty sandstone—in other words, a quartzite with a significant addition of minerals other than quartz.

● Marble

Heat and pressure metamorphose limestone and dolomite into marble; the major minerals, calcite and dolomite, melt and then recrystallize as the rocks later cool (Plate 26). Pure marble is whitish and may look like sandstone or quartzite at first blush. The original layering of the limestone is preserved, and the rock has a sugary texture. However, you can see tiny cleavage planes on the calcite grains, and therefore the mineral cannot be quartz. Other minerals, or "impurities" in the marble, carried over or recombined from the limestone or dolomite, form patterns that make "impure" marble an attractive decorative rock when it is cut and polished. For instance, iron minerals make reddish or yellowish streaks, and carbon from organic matter adds black.

● Gneiss

Granite, its close relatives, and other rocks metamorphose into gneiss (pronounced "nice"). Gneiss, like schist, is identified by its texture rather than by its mineral composition. Typical gneiss is a coarse, grainy, more or less banded rock (Plates 5, 7, 25, 44). As in schist, the minerals have flowed, but unlike schist, most of the minerals are not flaky. The banding comes from the separation of different minerals into dark and light layers.

The most common gneiss is granite gneiss. Its mineral composition is the same as that of the original granite, which it superficially resembles. However, the gneiss grains are strung out in a streaked fashion, indicating that the minerals melted and flowed. The texture is thus similar to that of schist but in many cases is less well defined and coarser. There is a relatively common intermediate stage of gneissic granite, characterized by incipient streaking of the granite minerals. The minerals of many gneisses changed during metamorphism from those of a variety of original rocks. Figure 12 shows how remnants of sedimentary rocks can become part of a granite gneiss.

● Granulite

Granulite is a fine-grained metamorphic rock that has evolved from sedimentary, igneous, or lower-grade metamorphic rocks. Some typical granulites consist of plagioclase, amphibole, pyroxene, and garnet. Others have significant amounts of quartz and orthoclase, to approach the composition of granite. Many bodies of granulite are banded. The fine and regular grain size, the banding, and the mineral composition distinguish it from other rock types. Granulite presumably originated deep in Earth's crust, where a large section of the crust broke, and one part of it moved against another. The granulite formed in the zone of grinding movement.

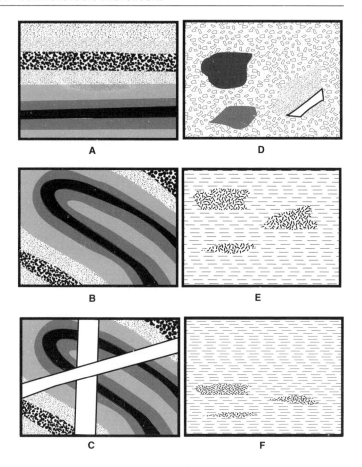

Figure 12. Bands forming in a body of gneiss. A, sandstone and shale below the surface. B, sandstone and shale are thrown into tight overturned folds by horizontal pressure (see Figure 13). C, dikes cut the older rocks. D, parts of the original formation that broke up during an intrusion of granite. If the sedimentary rocks didn't metamorphose earlier, they have by now. E, under continuing pressure and heat, the granite has metamorphosed to gneiss, and the former sedimentary rocks have continued to change. F, with further heat and pressure, the former sediments become streaks as the minerals in the gneiss slowly flow.

● Migmatite

A typical small rock outcrop or road cut, say the size of a trailer truck, shows only one kind of rock, or maybe two related kinds, such as shale and sandstone. But some outcrops show what looks like a mixture of different metamorphic and igneous rock textures, called migmatite. The pattern varies considerably. Some migmatites have layers of granite mixed with layers of schist and gneiss (Plate 23); others have blocklike masses of schist and gneiss in many shapes surrounded by granitic rock. Migmatite, like granulite, is the result of intense metamorphism, and its distinguishing characteristic is its intimate mixture of igneous and metamorphic textures.

● Amphibolite

As its name implies, amphibolite has grains of amphibole minerals as its primary constituents (Plate 44). Typical samples of the rock consist of the amphibole mineral hornblende, mixed in with plagioclase feldspar and perhaps other minerals. You can tell it is metamorphic by the parallel arrangement of the amphibole needles and elongate prisms, and also by the segregation of minerals into layers, giving the appearance of flowing. Amphibolites metamorphose from gabbro, diabase, basalt, or various schists.

● Greenstone and Greenschist

Greenstone, a hard, compact, very fine-grained rock, is generally considered to be a metamorphosed form of basalt (Plate 44). Its greenish color is due to one or more of the green minerals chlorite, epidote, and actinolite. Greenschist is a similar rock with a schistose texture and in many cases, quartz, orthoclase, and muscovite along with the green minerals.

■ STUDYING ROCK OUTCROPS

Exposed rocks can be found not only at road cuts but in natural outcrops along valley edges and walls; in creek and riverbeds, at waterfalls, and at dam sites; where roads cross rivers and creeks, and where rivers cut through escarpments. Many parks and nature preserves have outcrops. Rocks are exposed along rocky seashores and the margins of some lakes. Old strip pits, mine dumps, and abandoned quarries are sources of bedrock. In some places, bedrock shows in roadside ditches and borrow pits (Plate 27). In areas that the glaciers did not reach, old stone walls show local bedrock (Plate 27). In glaciated regions, most walls contain mixtures of rock types transported by the moving ice.

If you wish to take rock and mineral samples or fossils home with you, be aware of how dangerous it is to collect along highways. In fact, prospecting rocks along state-owned turnpikes is not permitted. And if you wish to collect on private property, make sure you have permission of the owner.

Some sedimentary rocks contain fossils. Certain ethical guidelines should be observed when collecting fossils. If a rock exposure is along a highway, the fossils should not be collected but left for others to enjoy. However, fossils that are loose and in danger of deteriorating may be collected, if you treat them with respect, keep a record of where they were found, locality information with them, and donate them to a school or other organization if you no longer need or want them.

It is strictly illegal to collect fossils of animals with backbones on government property, and even if you have permission to collect on private land, you should be careful not to diminish the scientific value of a specimen. Any potential significant finds should be taken to a museum or university for examination.

■ GEOLOGIC FORMATIONS

A geologic formation is a body of rock or related different rocks of the same geologic age whose characteristics can be recognized and mapped over a relatively large area. A thumbnail definition says a formation can be mapped over an area at least the size of an average American county. In actual practice, formations vary in size as counties do, and some are bigger than any county.

Almost all formations take their names from the geographic areas where they were first studied seriously or from areas of outstanding outcrops. For instance, the Conway

Granite is named for the town of Conway, New Hampshire, and the La Malbaie Gneiss outcrops at the town of La Malbaie, Québec. The Fort Union Formation of Montana, North Dakota, and Wyoming, consisting of terrestrial mudstone, sandstone, and freshwater limestone, is named for a late-19th-century fort in Montana. The terrestrial redbeds of the Duchesne River Formation are named for a river in Utah, and the Palliser Formation, consisting of marine limestone and dolomite, is named for the Palliser Range of the Canadian Rockies.

■ FOLDING AND FAULTING

Major horizontal pressures in Earth's crust have moved and modified bodies of rock. In many places around the world, thick masses of rock have wrinkled and cracked under the pressures.

◆ Folds

The wrinkling of rock masses is most obvious where layers of sedimentary rock that were originally horizontal have been forced into accordion-like folds. In some areas, the first folds have been further folded by a second and even a third episode of pressure. The four kinds of folds—anticline, syncline, anticlinorium, and synclinorium—are shown in Figure 13.

Small anticlines and synclines, or small parts of larger ones, are evident in some road cuts and rock outcrops (Plates 21–23), but you will never see a whole large one. Large simple folds are as much as 10 to 15 miles (16–24 km) wide and 10 times that long. The process of folding is so slow that the upper part of a large anticline will weather and erode as the structure rises. What you see in many cases is rock layers in the limbs of the folds angling upward toward each other. Similarly, the bottoms of large synclines are hidden below the surface; you observe the layers in eroded limbs dipping downward toward each other. Plate 17 shows folds in the Appalachian Mountains, where the upper portions of anticlines have eroded into valleys.

Even if the limbs of an anticline or a syncline have been

eroded flat, you can recognize the structure if you know the ages of the rocks. In a series of sedimentary rock layers that have not been turned upside down, the layers at the bottom will be the oldest, and the layers at the top will be the youngest. If the series is folded into an anticline and eroded flat, the oldest rocks will be on the inside and the youngest on the flanks of the fold. If the series is folded into a syncline, the opposite will occur; the oldest rocks will be on the flanks and the youngest on the inside. You can visualize this principle if you fold a magazine into an anticline and a syncline and imagine the high points eroded off. If you start with the front cover up, it represents the youngest layer; the back cover is the oldest layer.

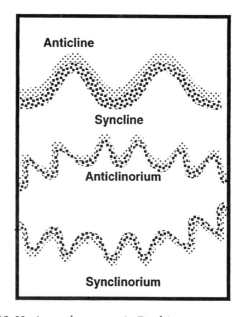

Figure 13. Horizontal pressure in Earth's crust causes folds in sedimentary rock layers. An *anticline* is an upfold, a *syncline* is a downfold. A series of folds arched upward is an *anticlinorium*, a depressed series is a *synclinorium*. Continuing pressure can overturn a fold, as illustrated in Figure 12B, in which an anticline has been overturned toward the left by pressure from the right.

◆ Faults

When sections of Earth's crust are subjected to intense pressure, the crust cracks as well as folds. Faults are cracks along which rock masses have moved. The kinds of faults are illustrated in Figure 14 (see also Plates 32 and 45). Squeezing of the crust causes reverse, thrust, and strike-slip faults. Stretching of the crust leads to normal faults. Imagine an ant standing on the foot wall of the fault, with the hanging wall over its back. If the hanging wall moves downward, the result is a normal fault. If the hanging wall moves upward, the result is a reverse fault. A thrust fault is a low-angle reverse fault. If the movement is horizontal, the fault is a strike-slip or transcurrent fault. The word "strike" in this case refers to a direction of movement parallel to Earth's surface. The familiar San Andreas fault zone in California is a strike-slip fault zone; land to the west of it is creeping northward in relation to the land to the east.

Faults come in small, medium, and large sizes. The largest ones are thrust faults along which major pieces of Earth's crust have overridden others. Most of the big ones in eastern North America have been detected in the Canadian Shield and the Appalachian Province. Thrust faults usually come in groups, as shown in Figure 15.

When one large rock mass moves over another in a thrust fault, the fault itself is filled with rock fragments and pulverized rock dust. When this ground-up material becomes cemented into a solid rock, it is called mylonite. Mylonite consists of thin layers of rock dust, often as thin as $1/32"$ (1 mm) or less, with occasional pebbles and granules (Plate 43). There may be some metamorphism of the minerals, making the rock resemble schist, but mylonite has fine laminations that do not occur in schist. Some mylonites closely resemble finely laminated shale, but the mylonite is a hard rock with no bedding planes, and it weathers into angular chunks, unlike shale.

■ KARST TOPOGRAPHY AND CAVE FORMATION

Caves are typically created when ground water dissolves limestone, dolomite, gypsum, and rock salt. The most common and extensive caves are formed in limestone. Carbon dioxide from the air dissolves in ground water, and some of it

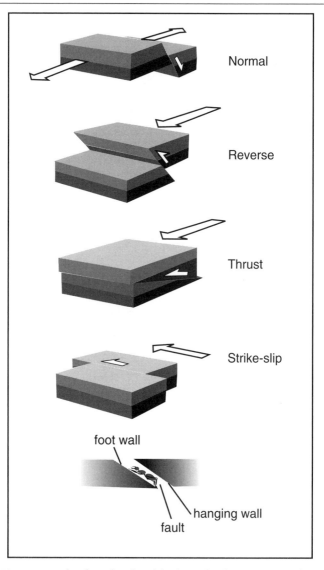

Figure 14. The four kinds of faults. The large arrows show the direction of pressure. Normal faults are caused by stretching in Earth's crust; the others are caused by compression.

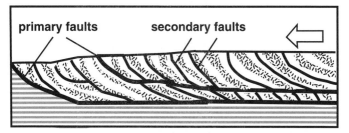

Figure 15. Cross section of a thrust fault system. Heavy lines are faults. At lower right, rock from below has been pulled up and folded over. Pressure was from right to left. Blocks of bedrock have been pushed against and on top of each other.

combines with the water to form carbonic acid. The acidic water creeps along cracks of joints and bedding planes (Plate 16), dissolving the rock and carrying away the dissolved material until a cave forms (Figure 16). As much as $1/32''$ (1 mm) of limestone can be dissolved from a surface each year.

As water seeps below the surface along joints, it can enlarge openings in the rock to form sinkholes. Sinkholes are variously shaped as bowls, steep-sided funnels, and vertical cylinders, partly filled with rock debris. Many sinkholes originate when the roofs of caves collapse. If a cultivated field or a pasture in limestone country has an island of bushes and trees in it, the island may be a sinkhole. In some cases, streams flow into well-developed sinkholes and disappear, continuing for long distances underground. After thousands of years, the waterways through the rock become passages in caves.

Land surfaces with sinkholes and disappearing streams are called karst topography, from the Karst, or Kras, Plateau in western Slovenia. Karst topography can be found throughout the world.

◼ UNCONFORMITIES AND EROSION

Because of the disruptions of weathering and crustal movements, nowhere on Earth is there a continuous series of rocks from oldest to youngest. Unconformities are gaps in

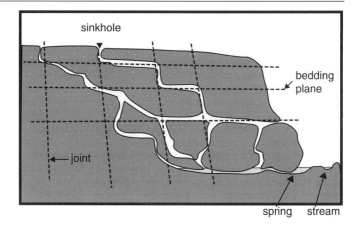

Figure 16. Caves form when water dissolves limestone and dolomite along joints and bedding planes. Only a few of the many joints and bedding planes are shown.

the rock record, each gap representing hundreds to millions of years' worth of missing rocks. The unconformity is an eroded surface or one that had nothing deposited on it for a significant amount of time, between two formations. Because of the time gap, the two formations do not "conform" to each other, even if they contain the same kinds of rocks (Figures 17, 41).

Among the most easily recognizable unconformities are those formed when layers of sedimentary rock are tilted, eroded to a horizontal surface, and then covered by more layers of sedimentary rock (Figure 17; Plate 4). The unconformity is at the zone where the tilted and horizontal layers meet. In the case illustrated in Figure 17, after the pressure stopped, the tilted layers were reduced to a relatively flat surface over another long period of time. Then conditions changed, and gravel was deposited on the eroded surface. The gravel ultimately turned into conglomerate rock, resting on the unconformity, which in this case is called an angular unconformity.

Other obvious unconformities involve greatly different kinds of rocks. For example, in some places you can find sedimentary rocks such as conglomerate lying directly on gran-

Figure 17. An unconformity on the north shore of Chaleur Bay, on Québec's Gaspé Peninsula. Folded Silurian limestone and shale have been eroded and covered by nearly horizontal conglomerate of much later Carboniferous age.

ite, as in Figure 41. Since granite formed below Earth's surface, it must have been somehow uplifted by movements in the crust and then denuded by weathering and erosion. All that must have taken a long time; eventually the eroded surface was covered by gravel that became conglomerate. The unconformity along the surface of the granite represents millions of years. Figure 52 shows a more complex unconformity. The sandstone, shale, and limestone in the upper right originally extended farther to the northwest, and were deposited upon an eroded surface of folded marble and masses of other rocks.

Some unconformities are highly irregular erosion surfaces, even where both formations are sedimentary and horizontal. Where vertical cliffs and road cuts show cross sections of the formations, you can see an erosion surface that curves up and down, outlining significant hills and valleys.

In some unconformities, the rock layers are parallel and there is no obvious erosion surface between two formations, but fossils indicate significantly different ages of the rocks above and below the unconformity.

■ THE EFFECTS OF GLACIERS

A glacier is a slowly flowing mass of snow-covered ice. Glaciers form when more snow accumulates each year than is lost through melting each summer. While snow accumulates on the glacier's surface, the snow below the surface turns into ice.

There are two kinds of glaciers, valley and continental. Valley glaciers flow down mountain valleys. Today's continental glaciers are the ice caps of Greenland and Antarctica. In continental glaciers, the ice in the center is so thick that it flows outwards in all directions, right over lesser mountains and around higher peaks. The greatest thickness of the Greenland glacier is over 10,000 feet (about 3,000 m); its average thickness is approximately half of that. In Antarctica, the thickest part of the glacier is about 14,000 feet (4,300 m). The average thickness is about 5,800 feet (1,800 m).

A valley glacier flows down the mountain until it reaches an elevation at which the climate is warm enough for summer melting to equal winter snow accumulation. A continental glacier will grow outward until it either reaches the ocean or a latitude at which the climate is warm enough to check its progress.

Glaciers greatly modify the country over which they flow. Valley glaciers carry loose rock away from the valley walls, and rock fragments frozen into the ice abrade the walls. This plucking and scouring changes the cross section of a mountain valley from V-shaped to U-shaped and carves the valley's head, where the glacier originates, into a bowl-like shape called a cirque (Plate 40). A valley glacier also carries a heavy load of weathered rock material that accumulates on the ice where the top of the glacier meets the valley wall.

Glaciers trap and transport any loose object they pass over and smooth the bedrock below them. If the bedrock is moderately hard, like granite or limestone, pebbles and cobbles frozen into the bottom of the glacier will cause long parallel scratches, or striae, in the bedrock (Plate 46). A harder surface does not become scratched; a softer surface is ground down but does not show the striae. One of the best evidences of a past glacier is a polished, striated pavement of bedrock. Pebbles and cobbles caught in the glacier may also become striated with scratches in various directions as they turn and roll in the moving ice.

● Glacial Deposition

The ice of a glacier contains clay, dust, and rock and mineral fragments from silt to boulder size. When a glacier melts away, all this miscellaneous material is deposited as a sediment called till. Boulders, cobbles, and pebbles in till that are not from local bedrock are called erratics (Plate 40).

Glacial till may superficially resemble gravel deposited by running water in that both consist largely of pebbles and cobbles. Gravel is typically made up of rounded, relatively tightly packed pebbles and cobbles along with some granules and sand (Figure 18 and Plate 37). Till differs in that it has a more extensive matrix of sandy, silty, or clayey material, with a scattering of granules, pebbles, and cobbles, some rounded but others angular (Figure 18; Plate 37). Till left by a continental glacier typically has fragments of many rock types, some that have been carried long distances. Valley glaciers, being smaller and restricted in locale, have less varied rock fragments. Since moving ice can carry larger rocks than moving water can, boulders are more common in till than in gravel.

At the end of a valley glacier or at the land margins of a continental glacier is a ridge of till called a terminal moraine. If climatic conditions do not change, the glacier neither advances nor retreats; it acts like a conveyor belt, transporting more material each year to the same ridge, where till melts out of the ice during the summer thaw. When the cli-

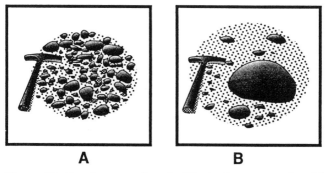

A **B**

Figure 18. Ice Age gravel and till. A shows gravel, with closely packed pebbles and a matrix of sand; B shows till, with a relatively higher percentage of matrix (sand or silt) and a greater variety of grain sizes, from granules to boulders.

mate warms and the glacier melts back, the moraine is left behind to mark the glacier's former edge (Plate 38).

Meltwater flowing through till sorts it into silt, sand, and gravel, which it deposits as glacial outwash (Plate 37). When the glacier has disappeared, the paths of its meltwater streams and rivers will be indicated by outwash.

An esker is a narrow, in some cases steep, sinuous ridge of gravel deposited by a river that once ran on or in a glacier (Plates 38, 39). Glaciers typically contain large cracks, and it is not unusual for a river of summer meltwater to flow into a crack and eventually form a channel inside or at the base of the ice. Whether it flows inside the glacier, at its base, or on its surface, the river takes sand and silt away from the pebbles and cobbles in the ice and leaves the larger material in its bed. When the glacier melts, the river's gravel bed settles to the ground (or remains there). Where the riverbed was thickest, the resulting esker is highest (Figure 19).

As the edge of a glacier melts back, blocks of ice often remain behind. Outwash from the glacier may then bury an ice block. When the block finally melts, it will leave a depression, or kettle, in the outwash (Figure 20). If the depression is deep enough, groundwater will seep into it and form a kettle lake or pond. Most of the lakes and ponds in North America are in the part of the continent that was covered by continental glaciers, and many of them are kettle lakes.

Drumlins are elongated, smooth hills of till, gravel, and sand created by continental glaciers. The elongation is in the

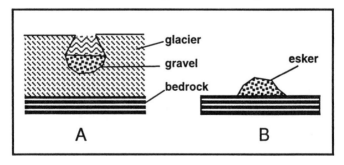

Figure 19. An esker is a ridge formed from the gravel that accumulated in the bed of a stream that ran through or at the base of a glacier. A shows a section of the original stream in the glacier; B shows a section of the esker that is left after the ice melts.

direction of ice flow. In general, when viewed from above, drumlins vary in shape from oval near the glacial margin to long and narrow away from the glacier's edge. Some have irregular shapes. It is not known exactly how they formed. Drumlins occur in large numbers from Nova Scotia through Massachusetts, in central New York, and in Wisconsin and Minnesota. Where they developed, the moving ice somehow molded the landscape of older till and outwash into drumlins behind its edge. Many drumlins are steepest at the ends that face the direction from which the ice came and fall gradually to ground level at the other ends.

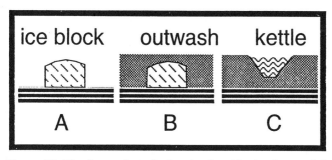

Figure 20. The formation of a kettle. A, a block of ice is left as a glacier retreats; B, the block is covered with sand and gravel outwash deposited by meltwater streams; C, the depression left when the block melts fills with groundwater to become a pond or lake.

◆The Ice Age

Large areas of North America and Eurasia show the evidence of glaciers—till, outwash, striated pavements, moraines, eskers, drumlins, kettles, cirques, and U-shaped valleys. About 2.3 million years ago, the climate of the Northern Hemisphere began to cool enough so that eventually mountain valley glaciers grew larger and joined continental glaciers that were forming. Four different times, great ice caps grew and coalesced with extensive ice fields in the mountain ranges, and then melted away as the climate grew temporarily warmer.

The last continental glaciers reached their maximum extension about 20,000 years ago. By 10,000 years ago, the

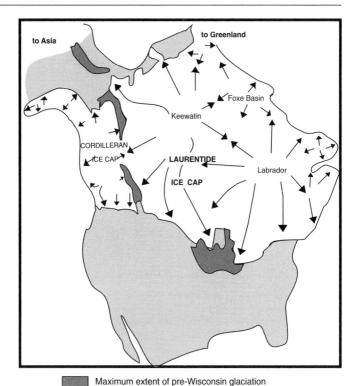

Maximum extent of pre-Wisconsin glaciation

Figure 21. The extent of Ice Age ice caps. The white portion shows the ice cap during the last (Wisconsin) glacial stage and the centers from which it grew. The dark areas were covered by earlier ice caps but not by the Wisconsin ice.

North American glaciers had retreated to central, north-central, and northeastern Canada. In the West, there were still large ice fields and valley glaciers in the Canadian Rockies. By 5,000 years ago, the continental glaciers had disappeared from North America and Eurasia, while the valley glaciers continued to advance until a hundred years ago. Now, they are mostly retreating.

The time of advancing and retreating glaciers, from approximately 2 million years ago up to about 10,000 years ago, is officially the Pleistocene Epoch of the Quaternary Period, but is familiarly known as the Ice Age. There were several other major ice ages in the distant geologic past, long before the Pleistocene one, which left their marks in striated pavements and a rock type made of solidified till, called tillite (Plate 43).

Figure 21 shows North America's continental glaciers at the peak of the fourth and last glacial stage. The borders of the continent look different because sea level was 490 feet (150 m) lower than it is today; the missing water was frozen in the glaciers. Note the land bridge to Asia. The arrows show directions in which ice flowed from four major and three minor growth centers. (Valley glaciers and mountain ice fields in the northwestern United States mountains are not shown.) Note also the area mostly in Wisconsin that was not covered by any of the four Laurentide glaciers.

The continental glaciers of the Ice Age were thick enough to move right over small mountains, and the ice was so heavy that it forced the underlying bedrock downward. In some places, Earth's crust is still readjusting after the melting of the ice.

■ GEOLOGIC TIME AND THE EVOLUTION OF PLATE TECTONICS

In the 1700s, a few scientists became aware that some rocks are much older than others, and they began to develop a time chart to categorize rocks by their ages. By the end of the 19th century, geologists had agreed on the major divisions of time now called eras, and on the subdivisions called periods (see the chart on the front endpapers). The subdivisions were made mainly on the basis of fossils and evidence for geologic upheavals, such as folding, faulting, and metamorphism.

The fossils showed that organisms have changed considerably through geologic time. Fossils from Paleozoic Era rocks are of very ancient ocean animals, much different from those of today, including primitive-looking fishes and relatively simple and primitive land plants. The Mesozoic Era that followed was characterized by a variety of dominant reptiles, including dinosaurs, accompanied by fishes, ocean shellfish, and plants that were more modern-looking. Finally, the Cenozoic Era brought mammals, modern plants, and modern ocean animals.

The periods have been further subdivided into epochs, but most of the epoch names are useful only to specialists. Geologists have also combined eras into even larger categories called eons. The rocks older than the Paleozoic Era are now considered to belong to the Archeozoic and Proterozoic Eons, which collectively are called "Precambrian" (the Cambrian being the earliest period of the Paleozoic Era).

The dating of some rocks by radioactive minerals, outlined in Appendix B, has allowed geologists to assign reasonably accurate ages to the time segments on the geologic time chart. The dates on the time chart are approximate, subject to debate and revision as geologists refine their techniques and discover new details.

◆The Origin of Mountains

Understanding how mountains form is key to understanding the geological history of the Earth. Since the beginning of the Paleozoic Era, the margins of what was to become North America and the neighboring oceanic crust have been very active geologically. For example, the Appalachian Mountains were formed by horizontal pressure from the southeast at the end of the Paleozoic Era. The pressure of moving sections of Earth's crust caused a great thickness of originally horizontal sedimentary rocks to be thrown into large and small folds and, in many places, spectacularly faulted.

On the other side of the continent, the rocks that became the Rocky Mountains were shoved into the continental interior from the west at the end of the Mesozoic Era. They too were drastically folded and faulted. Along the west coast, rocks have been more recently broken and crumpled; in some places, pieces from deep in the oceanic crust have come to rest high up on dry land. Precambrian rocks were similarly folded and faulted before the Paleozoic Era began and show abundant evidence of mountain-making that occurred in the dimmest recesses of the geologic past. Much of the folding and faulting was accompanied by the intrusion of igneous rock into the upper part of the crust and at least some metamorphism of preexisting rocks.

Events of mountain-making, with folding, faulting, intrusion, and metamorphism, are called orogenies. In North America, three eastern and at least four western orogenies occurred after Precambrian times.

◆Pangaea

Observant students of geography have long noticed how neatly the American eastern coast fits against the western coast of Africa and Europe. As early as 1838 it was proposed that the continents may have been joined in the past and later torn apart. The first formal hypothesis of what has been called the continental drift theory was published in 1912. It postulated that a supercontinent, named Pangaea (*pan-JEE-uh*), started to break up in the Triassic Period, and that its pieces have slowly separated since then to become the modern continents. This concept was based on some convincing observations.

For example, the edges of the continental shelves fit even better than the coastlines. Moreover, similar rocks of the same age occur on both sides of the South Atlantic Ocean.

Figure 22. *Mesosaurus*, a 3-foot Late Pennsylvanian to Early Permian reptile. Its skeleton and the rocks in which its fossils are found indicate it lived in water near the seashore and ate fish. It may have had webbed feet, and it probably spent most of its time in the water, but it was not the sort of animal that could have swum across the Atlantic Ocean. The presence of its fossil remains in both Brazil and Africa is evidence that the continents were once joined.

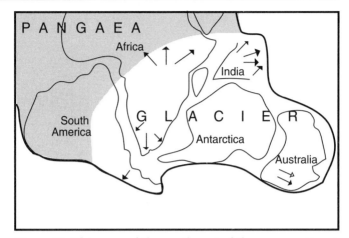

Figure 23. This map of the south part of the supercontinent Pangaea shows how glacial striae now on different continents could have been left by a single ice cap. The striations are evidence that the continents were once joined.

One particular series of rocks in eastern Brazil occupies a half-moon-shaped area that approximates the state of Bahia. The rock series is abruptly cut off at the coast. The same series outcrops in a strip of the same width along the west coast of central Africa, in the area of Gabon and southern Cameroon. If you put geologic maps of South America and Africa together, it is evident that the American outcrop is the tip of one arm of a much larger outcrop area in Africa. It seems certain that the rock series has been torn apart.

Continental drift theorists also pointed to fossils of a small extinct reptile called *Mesosaurus,* which are found only in Pennsylvanian and Permian age rocks in South Africa and eastern South America (Figure 22). The animals apparently lived in fresh water or saltwater near the shore, ate fish, and had short legs and skeletons not specialized for long-distance swimming. It seems unlikely that they could have either swum the ocean or walked from one continent to the other via the Arctic. It is much more logical to assume that the rocks carrying their fossils were once on the same continent but have separated.

Fossils of a characteristic land plant occur in identical form in Permian and Triassic rocks of South America, Africa, Antarctica, and Australia. It is difficult to imagine how the plants could have traversed the oceans, but one can easily imagine them as inhabitants of widespread forests of a single supercontinent. Fossils of some animals with backbones other than *Mesosaurus* tell similar stories.

Rocks, fossils, and continental shelves indicate that Argentina, South Africa, India, Australia, and Antarctica could have been joined in the past. So does evidence of ancient glaciers. Tillites (solidified glacial till) from Pennsylvanian age continental glaciers outcrop in Argentina, Brazil, South Africa, central Africa, central India, and several places in Australia. In some places, the tillites can be seen to rest on striated pavements. If you map the directions of the striae today, there does not seem to be an overall pattern. However, if you arrange the continents as in Figure 23, the striae make a pattern that strongly suggests the presence of a single ice cap.

◆The Moving Sea Floor

Studies of the sea floor made after World War II revealed clear evidence of "continental drift." Researchers discovered a network of volcanic ridges on the sea floor (Figure 24). One, the Mid-Atlantic Ridge, traverses the Atlantic Ocean north to south. At the center of the ridge is a zone of fissures

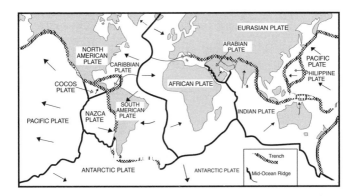

Figure 24. The plates of Earth's crust, bordered by trenches and midocean ridges. Note that the northeastern part of the Pacific Plate is moving northwestward, carrying part of California with it. The edge of the plate in California is the well-known San Andreas Fault zone.

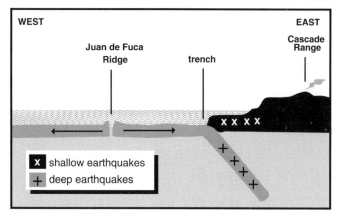

Figure 25. A section of Earth's crust on the northwest coast of the United States. To the left is the Pacific Ocean, to the right is the state of Washington. New crust is forming at the Juan de Fuca Ridge and pushing in both directions. The continent is overriding the eastward-moving crust.

through which lava flows, adding to the oceanic crust. On each side of the fissure zone, the ocean floor is moving away at a rate of about .75 to 1.75 inches (1 to 4.5 cm) per year. This sea-floor spreading shows that the continents along the Atlantic Ocean are separating. The same process is separating other continents.

As scientists discovered the ridges, they found in other areas long, narrow, and very deep submarine canyons, or trenches (Figure 24). The trenches are zones of faults. There, accompanied by earthquakes and volcanic activity, ocean crust on one side of the trench is forced down under the crust on the other side, where it melts.

The ridges and trenches are the boundaries of eight major and several smaller sections of Earth's crust called plates (Figure 24). In essence, the crust is born at the midocean ridges and moves slowly to the trenches, where it is engulfed. The continents are not actually drifting, they are being carried along by the moving crust of the plates.

For instance, off the coast of Oregon, Washington, and British Columbia is a midocean ridge, the Juan de Fuca Ridge, named for a 16th-century Spanish explorer (Figure 25). Sea-floor spreading occurs at the ridge, with crust moving east

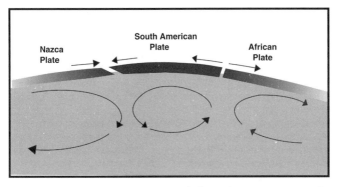

Figure 26. Convection currents below the crust move the rock in the plates.

and west. The eastward-moving crust travels at a speed of about 1.5 inches (about 4 cm) a year toward a trench near the coast, where it is forced under the edge of North America. The area where the crust is forced under the continental edge is prone to earthquakes, and the Cascade Mountains there are volcanoes.

The movement within the plates is explained by a theory of convection. According to the theory, the hottest mineral mixture beneath the crust rises under the midocean ridges, and some flows out as lava there. Most of the material just below the crust moves to one side or the other and cools off when it is a good distance from the ridges. As it cools, it sinks back down, to be warmed up again. The rising and sinking results in continuous currents. These convection currents nudge the older crust of the plates along, and the continents ride on the plates (Figure 26). The super plate that contained the supercontinent Pangaea may have split into smaller plates because of that process. Figure 27 shows a modern concept of Pangaea's breakup.

◆The Plate Tectonics Theory

Studies of mountain origins, the discovery of Pangaea, and research on the moving sea floor together led to what is now known as the plate tectonics theory—the ultimate frame-

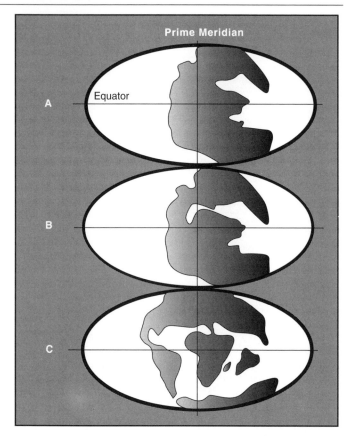

Figure 27. An early restoration of the breakup of the super-continent Pangaea. A shows Pangaea as it was 200 million years ago. B shows the supercontinent 160 million years ago; Eurasia is more deeply separated from Africa. C shows the situation 80 million years ago. North and South America have separated from Africa. India is moving up to collide with Asia. North America is still joined to Europe, and Australia is still joined to Antarctica.

OROGENY	COLLISION OF SUBCONTINENTS	RIFTING OF SUPERCONTINENT
		NOW
Appalachian	500 - 250 M Y	
		850 M Y
Grenvillian	1.1 B Y	
		1.5 B Y
Hudsonian	1.6 B Y	
		2.2 B Y
Kenoran	2.6 B Y	
		2.8 B Y
Laurentian	3.0 B Y	

Figure 28. A simplified version of the plate tectonic cycle. Approximate dates of orogenies on the left alternate with approximate dates of continental splitting on the right (MY = million years ago; BY = billion years ago). We are now in the latest rifting phase.

work into which observations of geologic history fit. The plate tectonics theory states that Earth's crust is composed of plates, bordered by midocean ridges and trenches, in which portions of the crust are moving under the influence of convection currents below. The theory explains what you can see in the rocks today and what you can predict about the   future.

The supercontinent Pangaea was formed when several subcontinents crunched into what was to become North America and Greenland. Ancient mountains rose from a collision as early as the Ordovician Period, with folding, faulting, intrusion, and metamorphism that was named the Taconic Orogeny. More violent activity occurred later in Devonian-Mississippian times (the Acadian Orogeny) and in Pennsylvanian-Permian times (the Alleghenian Orogeny). The same three orogenies, with different names, were recognized long ago in Europe, and evidence of the orogenies was later found in northern Africa.

The Taconic, Acadian, and Alleghenian orogenies were caused by pieces of Earth's crust slowly crashing (geologists sometimes use the term "docking") against the core of the

North American continent from the east, southeast, and south, one after another, during the Paleozoic Era.

Later, as Pangaea began to break up in the Triassic Period, pieces of crust assaulted the west coast of the continent, which was itself moving westward because of sea-floor spreading in the Atlantic. Major mountain-making activity occurred periodically in the West from the Triassic Period into the Tertiary Period. Today, earthquakes and volcanic activity show that the crust there is still unstable.

Older rocks show evidence of a continuing cycle of super-continents breaking apart and reforming. The Paleozoic orogenies occurred roughly between 500 and 250 million years ago, during an era of continental collisions that might be called a super orogeny, the Appalachian Orogeny. The Appalachian Orogeny was followed by rifting as Pangaea broke up. Precambrian rocks show that alternating continental rifting and collision has gone on for well over 2 billion years (Figure 28).

We are now in the latest part of a rifting phase. Unless the forces in Earth's crust change, the movement will reverse a few million years from now, and the continents will begin a new collision course.

3

THE CANADIAN SHIELD

The Canadian Shield is a vast region of hardwood and evergreen forests, myriad pristine lakes, and bedrock that predates the Cambrian Period. It makes up most of Canada east of the Great Plains and Rocky Mountains, with extensions down into Minnesota, Wisconsin, Upper Michigan, and New York. The surface of Earth's crust there is so old and tortured it is like a patchwork quilt of sedimentary, igneous, and metamorphic rocks that at first blush seems to have no logical order. The rock masses are eroded remnants of incredibly ancient geologic structures engulfed by igneous intrusions and crossed by sills and dikes (see Figure 4). Nothing is less than half a billion years old. In the rest of eastern North America, most rock ages are calculated in *millions* of years; in the Shield, the ages are in *billions* of years.

The rocks exposed on the Shield are the ancient rocks that are covered by younger formations on most of our continent. Since they formed before the Cambrian Period, they are often called Precambrian in age, and they make up the so-called basement complex. They are remnants of several different orogenies, when mountains slowly rose as island chains and subcontinents collided, and rocks were folded, faulted, intruded, and metamorphosed (Figure 28). In between these mountain-building episodes were times of rifting, during which supercontinents split and river and ocean waters deposited sediments, while volcanoes spewed forth lava and ash. After each cycle of collision and rifting, some of the resulting rocks were modified by the next orogeny. And at all times, erosion was working to destroy what orogenies had built.

■ THE CONTINENTAL NUCLEUS

The Canadian Shield is the continental nucleus because the rocks generally become younger as you move farther away from the Shield. The basement rocks around it are mostly covered by Paleozoic formations in all directions but the northeast (where they extend to Greenland). Farther away, the Paleozoic formations are covered by more recent Mesozoic and Cenozoic ones.

Virtually every kind of rock can be found in the Canadian Shield. Sedimentary and volcanic rocks represent the sediments, volcanoes, and fissure flows of Precambrian continents, subcontinents, island chains, and the ocean basins in between. Metamorphic equivalents of those sedimentary and volcanic rocks show how they were affected by orogenies. Intrusive igneous rocks (such as granite) formed during crustal movements and were later metamorphosed (such as granite to granite gneiss). During the orogenies, layered formations were folded and faulted, so rock layers occur at all angles in outcrops and road cuts.

◆ The Lay of the Land

The center of the Canadian Shield is a depression marked by Hudson Bay and the country to the west of it that is covered by Paleozoic sedimentary rocks. The Paleozoic rocks are eroded remnants of a series that formerly was continuous with those of the Central Lowland to the south. Movements in the crust after the early part of the Paleozoic Era uplifted the regions around the central portion, and the Paleozoic covering was eroded away to expose the underlying older Precambrian rocks.

Around the central depression is a country of low hills and lakes, with the bedrock mostly covered by a thin veneer of till and outwash from the continental glaciers of the Ice Age. Along the south and southeast margins as far east as Labrador, the bedrock was raised somewhat higher than it was raised in the rest of the Shield. There, the raised crust has been eroded into low mountains, including the Laurentians of Québec and the Adirondacks of New York. The crests of the mountains are generally around 3,000 feet (900 m) in elevation. In the rest of the Shield, the relief is in most cases no more than 300 feet (90 m) (Plate 3).

◆The Rocks of the Canadian Shield

A very large proportion of the Shield's bedrock consists of granite (with some other igneous intrusives) and gneiss. The Shield is like a sea of granite and gneiss with "islands" of originally sedimentary and volcanic rocks, most of which were more or less metamorphosed. Canadian Shield rocks are shown in Plates 4–8 and 44.

Many early limestones, sandstones, and shales changed to marble, quartzite, slate, schist, and other varieties. Among the metamorphosed sedimentary rocks, or metasediments, is a type called "iron formation," which originated among the Precambrian ocean floor deposits. It consists of chert, a rock composed of silica (quartz), that contains iron minerals (magnetite, hematite, siderite, pyrite, pyrrhotite) (Plate 6). Iron formation is typically banded or finely layered, with darker iron mineral layers and white to gray to red chert layers. The rock occurs in major layers, or beds, that range in thickness from a few inches to more than 100 feet (over 30 m) thick. The deposits in the Lake Superior region, which formed during the Proterozoic Eon, are hundreds of feet thick and extend for hundreds of miles. The predominant iron mineral in these deposits is hematite.

Older, less extensive deposits from the Archean Eon occur farther north. The formations there have been so complexly and tightly folded and so highly eroded that only relatively small areas are available for mining; the deposits outcrop as relatively thin layers and plunge too deeply for efficient operations. The main Archean iron mineral is magnetite.

Some darker volcanic rocks in the Shield islands, such as basalt, have changed with heat and pressure to a compact, dark greenish, finely crystalline rock called greenstone. Its color is due to recombination of the original minerals into chlorite, actinolite, and epidote (Plate 44). Although it is metamorphic, greenstone's volcanic origins are still recognizable because the metamorphosis in many cases did not destroy all the features of the original lava flows. The rocks are still layered, and while the original gas bubble cavities may contain new minerals, the cavities are still in evidence. Where the lava flowed into the ocean, it cooled as a mass of roundish blobs called pillows (Plate 6), and pillowed greenstone outcrops can be found in many places in the Canadian Shield. Continued metamorphism in some greenstones has resulted in a schistose texture, with streaks and layers of flaky mineral grains.

The islands in the Shield also contain many intrusive igneous rocks that have been metamorphosed to gneisses and other rock types.

In some places, series of outcrops show the change from sedimentary into metamorphic rocks or from low-grade to high-grade metamorphics. One example is found along I-35 in Minnesota. In the St. Louis River valley southwest of Duluth, near Thomson and Carlton, the bedrock is metagraywacke ("dirty" quartzite) and slate. Approximately 25 miles (37 km) farther to the southwest, from Moose Lake to Denham, the same formations have been more highly metamorphosed to mica schist.

The Canadian Shield is a different geologic world from that of the neighboring Stable Interior—not organized and sedimentary, but a patchwork of igneous and metamorphic rock. Scattered outcrops and samples from oil and water wells to the west and south of the Shield prove that the Shield material continues under the formations of the Stable Interior, to form the basement complex. Farther west, major exposures of the basement complex occur in the Rocky Mountains.

◆The History of the Canadian Shield

The older the rocks, the more they generally have been changed and eroded, and the record of Precambrian rocks is less complete and detailed than the records of many younger series. Therefore, the larger time categories (eons and eras) are the most useful in describing Precambrian history. Precambrian time is divided into the quite incompletely known Archean Eon and the better understood Proterozoic Eon. Both eons are further divided into Early, Middle, and Late Eras. Geologists can now date many Precambrian rocks within an error range of 1 million to 5 million years and can correlate formations appearing in different areas, but there are still many questions. The fossils that help in the correlation of younger rocks are largely missing in the Canadian Shield, except in the very latest Precambrian formations.

Still, there is excellent evidence for the tectonic cycles in the Sheld, including continental collisions as far back as 3 billion years ago or more. Geologists have postulated that four protocontinents began merging 2.6 billion years ago to form a supercontinent that contained the germ of what is now North America. This continental merging (the Kenoran

Orogeny) and even earlier crustal activity resulted in volcanic, intrusive, and metamorphic rocks from the Archean Era.

The Kenoran Orogeny was preceded by a time of rifting, when faulting created ocean basins as the crust stretched and an even earlier supercontinent split apart. The rifting was accompanied by great volcanic activity, both at sea and on land. For millennia, rivers carried sediments eroded from the land into the ocean basins that became sedimentary rocks interbedded with solidified lava flows and ash falls. During the Kenoran Orogeny, those formations were slowly folded, faulted, metamorphosed to varying degrees, and intruded by granitic material. The result was massive uplift and mountain formation that marked the end of the Archean Eon.

In the Early Proterozoic Era, the Kenoran mountains were eroded. Much of their material was washed into the oceans as sand and silt, and the rest deposited on land. There was rifting and volcanic activity as subcontinents separated again, resulting in another series of marine and terrestrial sedimentary rocks interbedded with volcanic rocks.

Then, between about 2 billion and 1.8 billion years ago, seven microcontinents collided with a larger continent during the Hudsonian Orogeny. Later, between about 1.8 billion and 1.6 billion years ago, a wide strip of crust extending from the present southern California to the present southern Sweden fused with the rest; this strip is now part of the basement complex.

During the following Middle Proterozoic Era, the Hudsonian mountains were eroded, and vast quantities of sand and silt were moved to lowlands and the ocean bottom. Rifting of the Hudsonian supercontinent was accompanied by volcanic activity and the formation of dikes, sills, and other intrusive masses. Middle Proterozoic rocks show evidence of ocean basins and chains of volcanic islands southeast of what is now our continental core, and a large continent may have formed out in the ocean.

The Middle Proterozoic Era ended with the next orogeny, the Grenvillian, when folding, faulting, intrusion, and metamorphism occurred from about 1.3 billion years ago to about 1 billion years ago. At that time, a large continent collided with ours, leaving its mark from Mexico to Sweden, including the Grenvillian rocks of Canada and the Adirondacks. During the orogeny, a midcontinent rift system formed (Figure 44), possibly because of a major upwelling of igneous material that caused cracks in the crust.

The last of the Precambrian eras was the Late Proterozoic

Era, during which there was the usual erosion of older mountains and accompanying deposition of land and marine sediments, along with later rifting and the formation of dikes and small intrusions (stocks). By the end of the era, our continental core was surrounded by subcontinents and island chains with ocean basins in between. Very few Late Proterozoic rocks remain in the Shield, but they are well represented in the Appalachian Province. With the final Precambrian continental rifting, the stage was set for the deposition of sediments in the early Paleozoic seas that covered the Stable Interior and the Appalachian Province.

◆The Provinces of the Canadian Shield

Geologists divide the Shield into seven structural provinces, of which three are within the range of this book (Figure 29). The largest and generally oldest is the Superior, which forms a semicircle around the central basin of the Shield. It occupies most of Ontario and central and northern Québec, with an extension into Minnesota. On the southern and eastern flanks of this province are two others that contain younger Precambrian rocks. The Southern Province is mostly in Minnesota, Wisconsin, and Upper Michigan, along with two contiguous small areas of Ontario. The broad strip of the Grenville Province trends northeast from eastern Ontario through

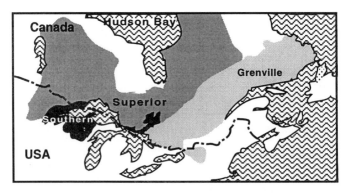

Figure 29. The three Canadian Shield provinces in the area covered by this book: Superior, Southern, and Grenville.

southern Québec to Labrador, with an extension into the Adirondack area of New York and another in Newfoundland.

◆Valuable Minerals of the Canadian Shield

The volcanic and intrusive activity that accompanied the alternating plate collisions and rifting caused innumerable deposits of valuable minerals to form. People have been prospecting here since the early 19th century, and Canadian Shield mines, old and new, working and defunct, are so numerous that only a few can be mentioned here. Many mines are accessible from the highways, and old mine dumps are sources of mineral and bedrock samples.

In the Superior Province, there are three categories of deposits. The first is polymetallic deposits, of copper, zinc, silver, and gold ores mixed together. These occur in volcanic or metamorphosed volcanic rocks and associated intrusives and sedimentaries. There are well-known polymetallic deposits at Chibougamau, Matagami, and Noranda, Québec, and at

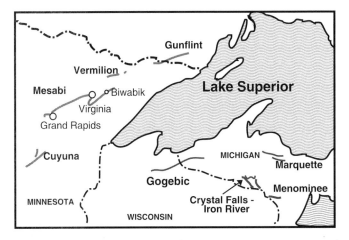

Figure 30. The thick, sinuous lines represent iron ranges of the Lake Superior region, which are sources of iron ore. The ore occurs in folded and then eroded rock layers, which have long narrow areas of outcrop.

Timmins (see Figure 32), Temagami, Kirkland Lake, and Manitowadge, Ontario.

Second are gold vein deposits. Many occur where granitic stocks (small intrusions) contacted and cracked overlying rocks, or where rocks cracked in shear zones. In either case, molten material was forced into cracks; the liquid cooled and solidified typically into quartz veins containing gold. There are gold vein mines at Kirkland Lake, associated with the polymetallic deposits, and at Val-d'Or, Québec.

The third type of valuable Superior Province metallic deposit is iron ore, or iron formation, such as that at Kirkland Lake, Temagami, and Bruce Lake, Ontario, and in the Vermillion Range of Minnesota (Figure 30). Iron formation occurs in areas of originally sedimentary rocks surrounded by granite and gneiss.

The Southern Province has its share of mines, too, including some of the world's most important and famous ones. Northwest of Sudbury, Ontario, is a geological structure that has been a world leader in producing nickel ore, along with lesser amounts of copper, gold, and silver. It is a basin structure of Proterozoic rocks that includes a sill of igneous material. Since the structure is a basin, the sill outcrops all around it. There are mines along the margin of the sill, where its intrusion cracked the rock above it and mineral-rich solutions filled the cracks.

Ontario's Blind River–Elliott Lake area has been a leading producer of uranium ore, which is found in an early Proterozoic quartz-pebble conglomerate. Major silver ore bodies occur in the Cobalt area, also in Ontario, where Proterozoic sills and dikes intrude earlier Proterozoic rocks. As in the Sudbury Basin, mineral-rich solutions migrated into cracks in the bedrock from the intruding molten material.

The Southern Province is a major source of iron ore in the Canadian Shield. Iron formation occurs between other sedimentary rock layers in several structures around Lake Superior (see Figures 30, 33, 43, 46). The iron may have come from the erosion of heavily weathered tropical land masses.

In the Grenville Province, uranium ore occurs in metamorphosed gneiss and intrusives at Bancroft, Ontario, and there are extensive deposits of titanium ore in a small anorthosite stock at St-Urbain, Québec, near the St. Lawrence south of Chicoutimi (not the St-Urbain that is south of Montréal). Metamorphism in the Grenville Province also produced a variety of important industrial materials, such as talc, calcite, graphite, and building stone.

■ THE SUPERIOR PROVINCE

The Superior Province has the oldest bedrock in the Canadian Shield; almost all of it is Archean in age. Islands of what were originally volcanic and sedimentary rocks stretch east–west, surrounded by granite and gneiss (Figure 31). The typical lavas of the islands have metamorphosed into greenstone, so the east-west belts of islands are called "greenstone belts."

Some of the greenstone islands in the Superior Province are erosional remnants of what were once extensive marine sedimentary and volcanic rocks, formed from sediments and lava flows in the Archean ocean. Others originated as small volcanic island groups in the ocean. The Kenoran Orogeny folded, faulted, metamorphosed, and intruded the formations. While the crumpling of the crust caused some areas to rise into mountains, it caused other parts to sink deep below sea level, where great heat and pressure metamorphosed the original rocks. Molten minerals surrounding the sunken parts eventually solidified into the granite and other intrusive igneous rocks.

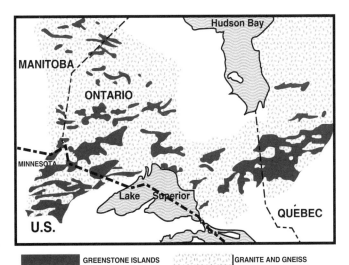

GREENSTONE ISLANDS   GRANITE AND GNEISS

Figure 31. The Superior Province's "greenstone islands" are isolated bodies of metamorphosed sedimentary and volcanic rocks surrounded by an "ocean" of granite and gneiss.

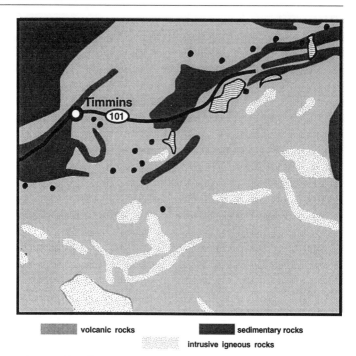

volcanic rocks   sedimentary rocks
intrusive igneous rocks

Figure 32. The Timmins area in Ontario shows the fragmented surface geology of greenstone islands, due to intense folding and faulting of the bedrock followed by deep erosion. The dots are locations of past and present gold mines.

Thus, the surface of the molten material was closer to sea level under the raised mountains than it was under the downfolded zones of crustal rocks. After the end of the Kenoran Orogeny, erosion completely destroyed the mountains. As the crust was uplifted, erosion continued to work its way down into the granite below the mountains. At that level, the rocks of the downfolded zones were exposed as greenstone islands. Figures 32 and 36 show typical surface geology of greenstone islands, and Figures 33 and 37 show the folded nature of the formations.

The islands contain rocks that were originally shale, sandstone, graywacke, conglomerate, basalt, tuff, and volcanic

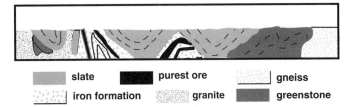

| slate | purest ore | gneiss |
| iron formation | granite | greenstone |

Figure 33. A cross section of the Vermilion Iron Range in Minnesota shows folded Archean metamorphosed sedimentary and volcanic rocks within a greenstone island.

breccia. Some were metamorphosed hardly at all; others were changed to greenstone, schist, and gneiss. The metamorphism in some islands was more intense around the margins than in the interiors, since the marginal rocks were nearer the molten minerals that became granite.

The axes of folding and faulting in the islands trend east and west because the pressure came from the south, presumably caused by crustal plate movement. Now, many rock layers are almost vertical or steeply inclined, dipping southward or northward in the folds.

SUPERIOR PROVINCE HIGHWAY SECTIONS

Ft. Frances to Kakabeka Falls, Ontario: Highway 11 (Figure 34)
The country from Ft. Frances to beyond Atikokan consists mostly of greenstone island material, with more or less metamorphosed sedimentary and volcanic rocks and associated intrusions.

Thin-banded tuff occurs along Highway 11 near Ft. Frances. Syenite was quarried in the past near the west end of the Rainy Lake causeway. At the east end of the causeway and 3.5 miles (5.6 km) farther east, at Commissioner's Bay, mica schist outcrops represent the original shale and sandstone. In some places, the mica schist is interbedded with hornblende schist, which originated as volcanic rock. More hornblende schist outcrops at the mouth of Rocky Inlet of Rainy Lake, west of the bridge. Its rusty appearance is due to weathered iron minerals. Farther along, about 1 mile (1.6

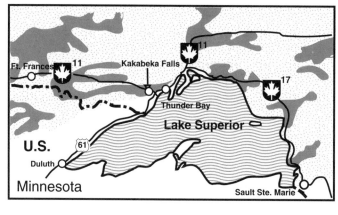

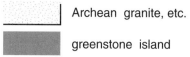

Archean granite, etc.

greenstone island

Figure 34. Highways in the Superior Province's Lake Superior region.

km) east of the side road to Windy Point, you will see more hornblende schist, with veins and lenses of calcite.

Gabbro occurs about a mile west of Bear's Passage. Pink granite shows up at Bear's Passage, followed by mica schist east of the bridge. The granites of the Mine Centre area contain deposits of copper, lead, zinc, and gold. Sediments and volcanics are well represented to the east, in the vicinity of Atikokan. At Little Falls, on the Atikokan River, you can see greenstone with a schistose texture, or greenschist. North of Atikokan, at the Steep Rock iron mines, layers of limestone, dolomite, graywacke, and tuff altered to a schist were folded, along with iron formation and a granite intrusion.

Beyond Atikokan, you enter a granite zone between greenstone islands. Granites and migmatites outcrop along the highway between Atikokan and Kashabowie, and at Huronian Lake. These rocks are part of an extensive batholith formed during the Kenoran Orogeny.

Kashabowie is in another arm of the greenstone island. Greenstone outcrops at Kashabowie Falls, where the river enters Upper Shebandowan Lake, and at the Swamp River, about 18 miles (29 km) east of Kashabowie. Pillow structure is evident here in the old lava flows. Greenstone and green-schist occur at Sunshine, and at Sistonen's Corners, green-stones are vertical. Some of them contain chunks of rhyolite. There is thin-banded tuff near Kashabowie and between Sistonen's Corners and Kakabeka Falls.

Schreiber to Sault Ste.-Marie, Ontario: Highway 17 (Figure 34)

The northeast shore of Lake Superior consists of alternating granite zones and greenstone islands typical of the Superior Province. The island rocks are more or less metamorphosed sedimentaries and volcanics.

Greenstones outcrop in the vicinity of Schreiber to intro-duce an island. Steeply dipping greenschist and slaty rocks show in the Jackfish-Middleton area, the former with pillow structures and gas bubble cavities preserved. The bedrock east of Steel River is slate and graywacke conglomerate.

Conglomerate shows up about 2.5 miles (4 km) east of the Marathon exit. The volcanic and sedimentary rocks con-tinue as far as the intersection with Route 614. About 16 miles (26 km) east of Marathon, between Marathon and the Highway 614 exit, the road passes through the Hemlo gold area. Between 1985 and 1995, over 8.5 million ounces of gold were mined here, a quarter of Canada's gold production. You can see the mining camp from the highway. The volcanic and sedimentary rocks outcrop again where the highway crosses White Lake, and at White Lake Provincial Park, you can see the contact between them and underlying granite.

Beyond the park, the highway traverses a long stretch of between-island granite and migmatite. The roadside parks at South White River, Marion Lake, and Hanmer Lake have good outcrops of migmatite.

The Wawa area is part of another greenstone island, with metavolcanics and iron formation (and old gold mines in quartz veins). At the nearby Michipicoten Falls, metavolca-nic rocks surround a mass of granite; the granite shows at the falls, and the volcanics outcrop downstream.

The Lake Superior Provincial Park provides more outcrops. Near the park's north boundary, iron formation shows con-torted layers of white chert and the black magnetite. Basalt, rhyolite, and tuff with phenocrysts (larger grains) of quartz and feldspar occur along the highway north of the park head-quarters at the south end of Red Rock Lake. There is a layer of iron formation between two of the basalt layers. These Archean rocks were affected by later Proterozoic crustal activity, as evidenced by a diabase dike about 120 feet (37 m) thick, cutting vertically through the tuff and basalt.

About a mile south of the Baldhead River's north branch, broad reddish syenite dikes of Proterozoic age appear, 50 to 100 feet wide (15–30 m). Beyond the dikes are outcrops of an unusual spotted basalt. The spots are large groups of plagioclase feldspar grains in the otherwise fine-grained groundmass.

In the Agawa Bay–Montreal River area, you are once again between greenstone islands. The bedrock here is granite, intruded during the Kenoran Orogeny. At the bridge over the Montreal River, the river flows through a small canyon, following a diabase dike in the surrounding granite, which weathers and erodes faster than the granite. The walls of the canyon mark the edges of the missing dike; the diabase is in the river bed. Uranium ore has been found in the granite and its diabase dikes in this area.

Between Montreal River and Sault Ste.-Marie, Highway 17 passes through an outlier of Upper Michigan's Proterozoic rocks, which sits atop the Archean granite. It is the eastern tip of the Lake Superior Basin, and a bit of the Southern Province. Along the coast, basalt layers outcrop from Mica Bay about 65 miles (105 km) south to near Sault Ste.-Marie. Two miles (3 km) north of Mamainse Harbour, layers of Proterozoic conglomerate and basalt dip southwestward into the lake. The basalt has cavities that were gas bubbles, some of which contain copper; these rocks are an extension of Michigan's Keweenaw Peninsula basalt. Chippewa Falls Park, about 6 miles (10 km) east of Batchawana, has a similar flow at the lower falls.

The older Archean granite shows up again in the vicinity of Sault Ste.-Marie, in the hilly highland between Goulais River and St. Marys River. The rest of the bedrock around Sault Ste.-Marie is Proterozoic or Cambrian. Cambrian sandstone outcrops in disconnected remnants among the basalts along the shore, up to Mica Bay. The sandstone represents the edge of the Michigan Basin in the Stable Interior.

Timmins, Ontario, to Val-d'Or, Québec: Route 101, Trans-Canada Highway 11, Routes 66, 117 (Figures 32, 35, 36) *This route samples the very large eastern greenstone island that extends from Ontario far into Québec, between the Great Lakes and Hudson Bay (see Figure 31). The rocks along the route are more or less metamorphosed Archean sedimentary and volcanic ones, with small igneous intrusions and here and there some Proterozoic diabase dikes.*

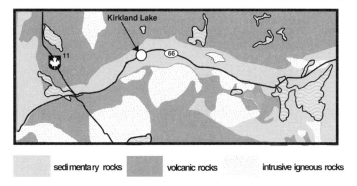

| | sedimentary rocks | | volcanic rocks | | intrusive igneous rocks |

Figure 35. Surface geology and highways in the Kirkland Lake region, Ontario.

Gold is found in dikes and veins in the volcanic rocks around Timmins; there have been many gold mines here.

Figure 32 shows the distribution of rock types in the Timmins area. The volcanics include greenstone and greenschist that were originally andesite, basalt, and a rock whose mineral composition is between the two. Graywacke, argillite (formerly shale), conglomerate, and iron formation make up the sedimentary series.

The majority of rocks along Route 101 from Timmins to South Porcupine are various basaltic ones (greenstones and greenschists). However, between Timmins and South Porcupine is a body of sedimentary rocks, and Timmins itself is built upon a thin strip of sedimentary bedrock. Volcanics continue through South Porcupine to about the eastern edge of Night Hawk Lake. From there to Matheson, the bedrock is all basaltic volcanic.

At Matheson, the route turns southeast on Trans-Canada Highway 11 and continues through metamorphosed volcanics all the way to Kenogami Lake. Then it turns east on Highway 66, where greenschist occurs between the road intersection and Vigrass Lake. Along Vigrass Lake, about one mile west of Swastika, a classic selection of greenstone island rocks shows up—greenstone, graywacke, conglomerate, and argillite, with the usual associated intrusion, in this case syenite. Cross bedded river-type conglomerate occurs on the north side of Chaput Hughes. About 6 miles (10 km) east of Kirkland Lake, or a little over a mile (almost 2 km) west of

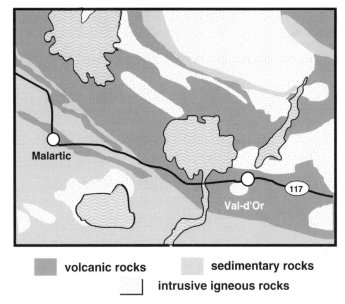

volcanic rocks   **sedimentary rocks**
intrusive igneous rocks

Figure 36. The geology of Route 117 in the Malartic–Val-d'Or region, Québec.

Route 672 to Crystal Lake, the bedrock is reddish gray volcanic porphyry. Farther along, at the bridge, the Misema River flows through greenstone.

The Kirkland Lake region is a region of gold mines, like that of Timmins, and iron formation has been mined south of Highway 66. On the way to the Val-d'Or mining district, you will continue in the more or less metamorphosed sedimentary rocks past Larder Lake to the Québec border. There, Highway 66 turns into Québec Highway 117. For a few miles eastward, the bedrock belongs to an isolated remnant of Proterozoic sedimentary rocks, a reminder that the Proterozoic covering of the Archean rocks was once much more extensive. The town of Arntfield is well inside the next body of Archean volcanics, which extends along the highway past Noranda and Rouyn to McWatters. The Noranda-Rouyn area has polymetallic deposits of copper, zinc, silver, and gold in veins in the metamorphosed volcanic rocks. From McWat-

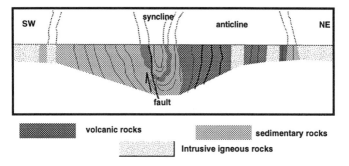

Figure 37. A cross section of the structure between Malartic and Val-d'Or, Québec, shows how rocks have been folded.

ters to Val-d'Or, almost all the bedrock is sedimentary or metamorphosed sedimentary.

Val-d'Or is back in the volcanics (see Figure 36). The gold of Val-d'Or, as its name implies, is found in quartz veins and more or less vertical cracks in volcanic rocks that have been stressed and faulted (Figures 37 and 38). The rocks of the Malartic–Val-d'Or region are basalt, andesite, tuff, graywacke, argillite, and conglomerate, with intrusives of granite, syenite, and granodiorite.

The Northeastern Minnesota Border Country: Route 11, U.S. 53, County Route 12 (Figure 39) *Minnesota has an unusually large number of terminal moraines from Ice Age glaciers. With their associated outwash and till deposits, they cover almost all of the bedrock in the Superior Province part of the state. As a lobe of the last continental glacier retreated northeastward from Iowa, it stopped there periodically and left a series of major curved moraines. The northeastern edge of the state, however, has no moraines and a thinner covering of glacial material. There, you can find some major exposures of more or less metamorphosed Archean sedimentary and volcanic rocks from greenstone islands, and surrounding zones of intrusive igneous material.*

At and around Voyageurs National Park, east of International Falls, the bedrock is typical greenstone island types like those in Canada. Folded metasediments along Rainy Lake near International Falls grade eastward to schist, which outcrops in the village of Ranier and in the northern part of

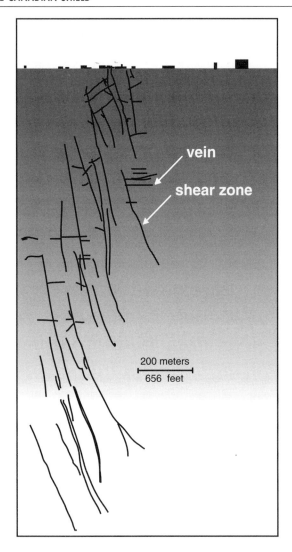

Figure 38. A cross section of a gold mine at Val-d'Or, Québec. Gold occurs in the veins and in the shear zone's cracks. Note the depth of the mine.

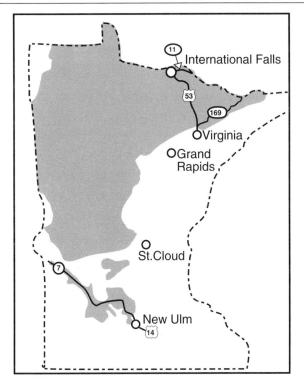

Figure 39. The Superior Province in Minnesota, and some highways that cross it.

the park. Erosion has exposed cross-bedded quartzite and conglomerate west of the park along County Road 109 a little over a mile (almost 2 km) south of Highway 11.

Granite and migmatite (between greenstone islands) form the bedrock in the southern part of the park, also exposed southwest of the park along U.S. 53. Late Archean dikes of dark-colored igneous rocks cut these intrusives. You can see one such dike on County Road 122 south of the park, between U.S. 53 and the town of Kabetogama.

Similar rocks occur in the Boundary Waters Canoe Area, accessible from County Route 12, northwest of Grand Marais. Greenstone with pillow structures is exposed at Knife

Lake and Jasper Lake. Black slate and gray metagraywacke (dirty quartzite) also outcrop at Knife Lake. Conglomerate and quartzite outcrop at Ogishkemuncie Lake.

Virginia to Ely, Minnesota: U.S. 53, Route 169 (Figure 39) *Along this route is the Vermillion Iron Range, a greenstone island with metamorphosed sedimentary and volcanic rocks and some of the surrounding intrusive igneous ones.*

On U.S. 53, 2 to 3 miles (about 3–5 km) north of Virginia, are outcrops of slate followed by granite. The granite at this site must have been close to the edge of a batholith, because it contains pieces of the slate and greenstone bedrock it was intruding. Slate extends along Route 169, from about 5 miles (8 km) north of the state Route 21 intersection to near Tower. West of Tower, quartzite, slate, and metagraywacke outcrop, with a volcanic rock that contains large chunks of older volcanic rock. At the edge of Tower is a conglomerate zone, a part of the slate formation.

From Tower to Soudan, the highway crosses iron formation, slate, and greenstone. All three rock types plus metagraywacke outcrop in the area of the Soudan iron mine. In the Soudan Underground Mine State Park, you can see hematite ore. For about 16 miles (28 km) northeast of Soudan, the highway traverses greenstone and iron formation. From there to Ely are more greenstone and slate. Greenstone with pillow structures lies just west of Ely, and greenstone outcrops in Ely itself.

The Minnesota River Valley, Courtland to Ortonville: U.S. 14, Routes 15, 5, 4, 19, 67, 7 (Figure 39) *This is an outlying piece of the Superior Province. The river has chewed down through Cretaceous sedimentary rocks of the Great Plains and eroded into underlying Archean greenstone island sedimentary and metasedimentary rocks as well as the intrusive igneous and metamorphic rocks that border the island.*

From Courtland, on U.S. 14, to New Ulm, the north wall of the Minnesota River valley consists of Archean sandstone, quartzite, and conglomerate. Near New Ulm, small patches of granite outcrop just upstream from a coarse conglomerate. That marks the end of what may be a small sample of a greenstone island. Upstream from there, the bedrock is granite and gneiss. Some of the gneiss may have originally been sedimentary rock, but the large amount of granite upstream all the way to South Dakota indicates a region between greenstone islands.

The route continues north on state Route 15 to county Route 5 to state Route 4 and Ft. Ridgely State Park. Granite hills extend upstream from the park. Route 4 north and state Route 19 west lead to Morton. Granite and gneiss outcrop between Morton and Redwood Falls. Continuing west on Route 19 and north on state Route 67 will bring you to Granite Falls. There are granite quarries at Sacred Heart, east of Granite Falls. From there, state Route 7 continues up the river to the South Dakota border. Granite has been quarried at Montevideo, north of Bellingham, and in the Ortonville-Odessa area.

■ THE SOUTHERN PROVINCE

The Southern Province has two major series of Proterozoic rocks, which differ greatly in origin and age. One is essen-

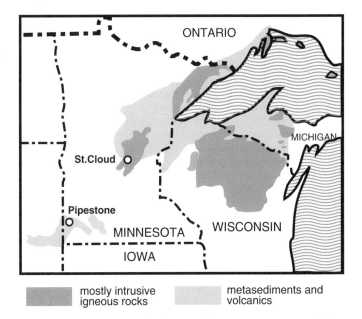

Figure 40. Precambrian bedrock in the western portion of the Southern Province.

Figure 41. The Archean-Proterozoic unconformity, exposed in a glacially polished surface on an island near the north shore of Lake Superior. Proterozoic conglomerate was deposited on an eroded surface of the Archean granite in the lower half of the drawing. The rocks were folded so that their cross section shows now on a horizontal surface. The rock hammer gives the scale.

tially in the Lake Superior Basin (Figure 40). The other occurs in the region northeast of Lake Huron (Figure 48) and south of the Lake Superior Basin in Wisconsin and Michigan (Figure 40). The two portions outside of the Lake Superior Basin are called the Older Southern Province.

◆The Older Southern Province

The rocks of the Older Southern Province were originally Early Proterozoic sediments, lava flows, and volcanic ash. In some places they were deposited on a deeply eroded, hilly surface of Archean rocks (Figures 41 and 42). The resulting sedimentary and volcanic rocks were folded, faulted, and partially metamorphosed during the Hudsonian Orogeny

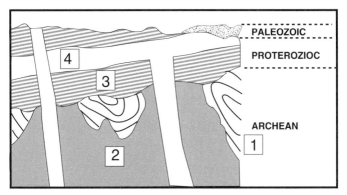

Figure 42. Typical relationships between Proterozoic rocks and those that are older and younger in the Canadian Shield. Folded and faulted Archean sedimentary and volcanic rocks (1) were intruded (2) and metamorphosed, then eroded. Proterozoic sediments and lava flows (3) were deposited on the unconformity. The resulting rocks were intruded by later Proterozoic sills and dikes (4). Eventually, the series was tilted, as in the Lake Superior Basin, and eroded. Finally, Cambrian sand, the first Paleozoic sediment, was deposited.

(Figure 43). Scattered remnants of these Proterozoic rocks in the Superior and Grenville Provinces show that they originally covered more of Canada than they do now; Early Proterozoic seas were much more extensive than they appear to have been at first blush.

The sedimentary and volcanic formations in the Southern Province were intruded and metamorphosed to varying degrees, but on average, they were not as violently folded as those of the Superior Province. There are, however, several major and many minor faults.

The silt and clay sediments of the Older Southern Province became rocks that grade in increasing hardness and brittleness from shale to argillite to slate. Clean sand became sandstone, then quartzite; graywacke and metagraywacke (impure quartzite) evolved from the dirty sands. The formations also include limestone, chert, iron formation, conglomerate, and tillite (solidified glacial till) (Plates 4, 41). Most of these rocks were originally sediments in the Precambrian ocean,

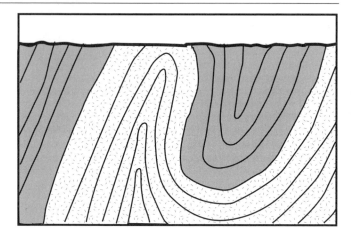

Figure 43. A diagrammatic cross section of the structure in the Cuyuna Iron Range of Minnesota shows how some Proterozoic rocks were tightly folded. The shaded layers indicate chert iron formation. The layers below the iron and chert are iron-containing slate. See Figure 30 for the location of this iron range.

but some of the sandstones and conglomerates were probably terrestrial, and the tillite certainly was. Near Lake Temagami, Ontario, the tillite rests on a polished and striated granodiorite surface, prime evidence of a Proterozoic glacial epoch. Three units of tillite, separated by argillite layers, suggest three advances of the ice sheet. The argillite presumably represents interglacial deposits formed when the ice cap melted back temporarily.

Large masses of intrusive rocks of the Older Southern Province flank the sedimentary and volcanic rocks of the Lake Superior Basin (Figure 40). One of these masses is in central Minnesota, and the largest is in northern Wisconsin. The granites of central Minnesota were intruded during the Hudsonian Orogeny. The mass in Wisconsin indicated in Figure 40 as mostly intrusive igneous is actually a complicated mixture of Archean and Proterozoic granites and metamorphic rocks.

◆The Lake Superior Basin

During the Grenvillian Orogeny there was an episode of local rifting within the old continent called Laurentia, a Precambrian continent illustrated in Figures 105 and 106. It seems illogical that there would be rifting *during* an orogeny, but the evidence indicates that an anomalous upwelling of material from below Earth's crust caused the rifts. The cracking of the crust caused the formation of new highlands and ocean basins, which promoted erosion of the land and deposition in the oceans. The sediments became shale, sandstone, graywacke, and conglomerate, some of which were metamorphosed to slate, quartzite, and metagraywacke. The cracking also caused widespread volcanic activity, and lava

Figure 44. The midcontinent zone of Proterozoic lava flows. To the northeast are those exposed on the flanks of the Lake Superior Basin. From east-central Minnesota to Kansas, the basalts are covered by younger rocks. The bodies of basalt have been offset by strike-slip faults. These lava flows represent a local episode of rifting in the Grenvillian Orogeny.

flowed from major fissures in a zone from what is now Lake Superior to northeastern Kansas (Figure 44). The northeastern end of the series was then gently downfolded into the Lake Superior Basin, a large structural basin that trends northeast and southwest under and around Lake Superior. The formations on the northwest flank, in Minnesota and Ontario, dip gently toward the axis of the lake; those on the southeast flank, in Wisconsin and Upper Michigan, dip gradually in the opposite direction, northwest toward the axis of the lake. The broad Canadian end of the basin is represented by the rocks of the islands along the Ontario shore, from Isle St.-Ignace to Michipicoten Island. To the southwest, the basin pinches out in Minnesota. The rocks that outcrop in the Lake Superior Basin also extend along the direction of the lake's axis into central Minnesota, with a distant outlier in southwestern Minnesota and southeastern South Dakota (Figure 40).

The sedimentary and metasedimentary rocks of the Lake Superior Basin are interbedded with basalt and rhyolite lava flows, many of which extruded into the ocean (Plate 8). All of these Proterozoic rocks were intruded by granites, as well as by gabbro and diabase sills and dikes (see Figure 42).

Volcanic rocks of the Lake Superior Basin outcrop extensively as a massive series of lava flows along the Minnesota shore of Lake Superior and in the Keweenaw Peninsula of Michigan. In Minnesota, the lava series is separated from the underlying Archean rocks by a massive sill of gabbro and other intrusive rock types, which outcrops mostly to the west of the flows. The flows and the sill are dipping toward the center of the lake, so the sill, which is below the lava, outcrops mostly inland to the northwest of the lake, beyond the lava outcrop along the shore.

SOUTHERN PROVINCE HIGHWAY SECTIONS

Pipestone, Minnesota: U.S. 75, Route 30 (Figure 40) *Down in the southwestern corner of the state are outcrops of a Proterozoic formation called the Sioux Quartzite. This formation is the source of rock from which Native Americans of many tribes made their pipes.*

The quarry area is preserved at Pipestone National Monument at the town of Pipestone (U.S. 75 and Minnesota Route

30). A two-mile (3 km) wall of quartzite faces west and trends north and south; it is the eroded edge of a series of rock layers that dip to the east. At either end, the quartzite disappears under the prairie. In a half-mile stretch at the base of the cliff is a shaley layer 15–20 inches thick (38–50 cm) from which pipestone was quarried (Plate 8). The Sioux Quartzite also outcrops in the vicinity of Jasper and Luverne, south of Pipestone, and at Sioux Falls, South Dakota.

The Central Minnesota Granite Zone: Routes 23, 27 (Figure 40)
The bedrock of this portion of the Older Southern Province is a variety of Early Proterozoic intrusive igneous rock, but it is mostly covered with glacial material.

The bedrock outcrops west and southwest of St. Cloud, along Minnesota Route 23 to Cold Spring; east of Little Falls

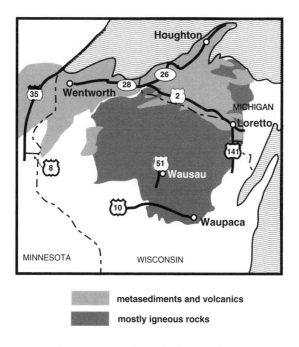

metasediments and volcanics

mostly igneous rocks

Figure 45. Highway routes through the Southern Province in Minnesota, Wisconsin, and Michigan.

(state Route 27); and in the northeast corner of Benton County, northeast of St. Cloud (state Route 23). Granite is quarried in the area south of Mille Lacs Lake and at Warman, southeast of the lake.

Carlton to Taylors Falls, Minnesota: Routes I-35, U.S. 8 (Figure 45) *The highway south from Duluth traverses a good sampling of Proterozoic rocks in the Lake Superior Basin. It is a region of volcanics, sedimentaries, and metasedimentaries.*

Slate, metagraywacke, and quartzite crop out along the St. Louis River at the Jay Cooke State Park, near Carlton. Slate shows up in the town of Carlton, and the metagraywacke at the river's edge at Cloquet, upstream from Carlton. In the area from Moose Lake southwest to Denham, the slate and metagraywacke have turned to schist. Farther south, in the gorge at the town of Sandstone, the Kettle River flows through Proterozoic sandstone.

A side trip to Taylors Falls (U.S. 8) will give you a good view of the volcanics. The St. Croix River there has cut through at least 10 lava flows. The volcanic series is approximately 300 feet (90 m) thick, and the cliffs along the river are about 100 feet (30 m) high (Plate 8). In the Interstate Park Campground, .8 miles south of Taylors Falls on U.S. 8, you can see basalt flows overlain by Cambrian sandstone. At the Interstate Park maintenance area, the basal layer of the Cambrian formation is a conglomerate that contains angular basalt blocks. It represents the surface upon which the sand was deposited in the Cambrian Period. The lava flows are also prominent across the river at the Interstate State Park in Wisconsin, south of St. Croix Falls. In the gorge at St. Croix Falls, Cambrian shale overlies the basalt.

The Mesabi Range—Grand Rapids to Biwabik, Minnesota: U.S. 2, 169, 53, Routes 37, 135 (Figure 30) *The Mesabi Range is the site of some of America's greatest iron mines. In this narrow zone, three originally sedimentary formations outcrop and dip gradually toward the southeast, along U.S. 169. The formations of this Proterozoic rock series are the Pokegama Quartzite, which lies on an eroded surface of older Archean rocks; the Biwabik Iron Formation, above the quartzite; and the Virginia Slate, overlying the iron formation (Figure 46). To learn more about iron mining in this region, visit the Mahoning-Hull-Rust Mine, the Minnesota Museum of Mines, and the Iron Range Interpretive Center in the Chisholm area.*

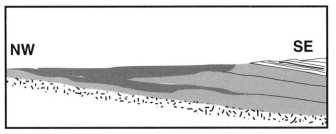

Figure 46. A cross section of the Mesabi Range. The Biwabik Iron Formation is underlain by quartzite and overlain by slate. The dark gray areas show where the richest ore occurred; that ore has been completely removed from open pits. Only lower-grade ore is now mined.

U.S. 2 and the Mississippi River cross the Mesabi Range at right angles, from the west part of Grand Rapids to Cohasset. From Grand Rapids, U.S. 169 reaches the range at Coleraine. From there to just before Nashwauk, the highway follows the Virginia Slate. From Nashwauk to Keewatin the road traverses mostly iron formation. From the vicinity of Keewatin to Mountain Iron, the bedrock switches to Virginia Slate; from there to Virginia it is iron formation. An S-curve in the rock series begins at Virginia. You can follow this curve in the iron formation and quartzite by driving south on U.S. 53 to Eveleth and east on state Route 37 to Gilbert and state Route 135 to Biwabik. Just before the town of Aurora, Route 135 heads north and the iron range continues northeastward.

Duluth to Grand Portage, Minnesota: U.S. 61 (Figure 34) *From this lakeside route, you can survey the rocks that make up the Lake Superior Basin's west flank, a thick series of Proterozoic lava flows and a very large sill, formed in several stages, that forced its way between the volcanics and the eroded Archean rocks below.*

Rock types in the sill range from granite to granodiorite to gabbro to peridotite to anorthosite (almost pure plagioclase feldspar). While the massive sill was still hot, the heavier minerals sank to the bottom to form the anorthosite and peridotite. The lighter minerals floated upward to form the granite and granodiorite. The main mass in the middle became gabbro. The sill dips southeastward under the lake;

volcanic and sedimentary rocks rise out of the lake on the other side of the basin as the Keweenaw Peninsula of Michigan.

In Duluth, the gabbro, capped with lava metamorphosed by the original heat of the sill, forms a bluff. The central and southern parts of the city are built upon the sill. It disappears southward under the sediments of Glacial Lake Duluth, an ancestor of Lake Superior that formed as the last Ice Age glacier retreated. In the northeastern part of the city, the old mostly basalt lava flows lie atop the giant sill, cut by smaller sills and dikes of diabase and granite. There, some dikes and sills outcrop along the lake shore.

Between Duluth and Two Harbors, the shoreline consists of lava flows dipping into the lake. At Two Harbors, on the shore of Burlington Bay just below the city park, five lava flows are visible. Farther north, at Gooseberry Falls State Park, the Gooseberry River flows over three waterfalls formed at the eroded edges of lava flows. From the nearby Split Rock Lighthouse State Park to Beaver Bay and beyond, the bedrock is the diabase of the Duluth sill complex.

Beyond Beaver Bay, rhyolite porphyry flows have eroded into palisades along the coast. You can see the rhyolite in the low cliffs at the mouth of the Baptism River (at the intersection with Minnesota Route 1). Upstream, the river flows through underlying basalt.

Lava flows also outcrop at the Manitou River, in Temperance River State Park, and in Cascade River State Park. About four miles (6.5 km) northeast of the Cascade River, a cliff of red sandstone occurs between lava flows. At the Judge C.R. Magney State Park, the Brule River flows over lava flows in a scenic gorge above the highway bridge, with basalt overlying rhyolite.

Northeast of the Brule River, from Hovland to Grand Portage, the Duluth sill reappears in many hills and ridges of gabbro and diabase. Proterozoic metasediments that predate the lava flows also outcrop in this area. Grand Portage Bay is eroded into slate, which is intruded by dikes and sills. At Pigeon Point, slate and quartzite with a sill in between dip into the lake at the edge of the Lake Superior Basin.

Wentworth, Wisconsin, to Copper Harbor, Michigan: U.S. 2, Routes 28, 64, U.S. 45, Route 26 (Figure 45) *This route traverses the southeastern rim of the Lake Superior Basin, with its Proterozoic sedimentary, metasedimentary, and volcanic rocks.*

At Amnicon Falls State Park, off U.S. 2, 2 miles (3 km) west of Wentworth, a fault has moved basalt onto folded younger sandstone. The basalt causes the waterfall. Geologists believe the sandstone in the gorge below the falls to be latest Proterozoic (or possibly early Cambrian) in age.

East of Wentworth, the bedrock of Late Proterozoic sedimentary rocks is well covered by glacial till, outwash, and lake deposits, the latter from Glacial Lake Duluth. However, a side trip down Wisconsin Route 169 to Gurney and the Potato River Falls will reveal conglomerate, sandstone, and shale of the sedimentary series. Those formations continue northeastward onto the Keweenaw Peninsula.

Farther east, at the U.S. 2–U.S. 51 intersection, you will come upon a different set of rocks, slate and graywacke, that dip northward. These rocks are Early Proterozoic in age, much older than the other rocks of the Lake Superior Basin. They are part of the Penokee-Gogebic Iron Range (Figure 30), which is a narrow strip of folded rock layers trending east–northeast that dip northward under the other basin rocks. Continuing on U.S. 2, hills of volcanic rocks overlook the north edge of Bessemer, Michigan.

Two miles (3 km) east of Bessemer and a half mile south of U.S. 2, miners have left old mine dumps in the local iron formation. More of the slate outcrops at Wakefield, where you turn northeast on Michigan Route 28 and return to the later Proterozoic rocks of the Lake Superior Basin. If you head north on Michigan Route 64 and continue to the lake, you can observe sandstone and conglomerate at the Porcupine Mountains Wilderness State Park. Slightly younger sandstone forms outcrops on Michigan Route 64, along the coast east of Silver City. There is also a covered shale formation. These formations all dip gently northwestward into the basin.

U.S. 45 and Michigan 26 lead to Houghton, on the backbone of the Keweenaw Peninsula. This peninsula is formed by a combination of interbedded basalt, rhyolite, conglomerate, and sandstone. In the late 19th century, an outcrop of this series, a strip less than 30 miles (48 km) long and no more than 3 miles (5 km) wide, was the biggest copper producer in the United States (Plate 45). That strip extends north and south of Houghton, approximately from Painesdale to Mohawk. Most of the copper occurs in the cavities that were gas bubbles in the molten basaltic lava and in cracks that formed after the lava cooled. Miners also removed a large amount of copper from conglomerate with rhyolite pebbles that lies between lava flows. The flows and conglomerate outcrop just north of South Range, south of Houghton.

The lava series is on the south shore of the peninsula's tip; Route 26 to Copper Harbor follows the north coast, where sandstone and conglomerate occur at the shore.

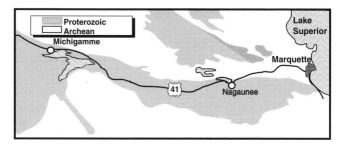

Figure 47. The syncline or basin structure between Michigamme and Marquette, Michigan.

Marquette to Michigamme, Michigan: U.S. 41 (Figure 47) *In the Marquette region, folded Proterozoic metasedimentary and volcanic rocks of the Southern Province overlie older Archean schist, gneiss, and intrusive igneous rocks. The region is an elongate irregular shallow syncline, or structural basin, crossed by U.S. 41, with Archean rocks more or less on the outside and Proterozoic ones more or less on the inside. The basin itself is tilted slightly downward toward the west. The map in Figure 47 shows that the Proterozoic outcrop appears to be pinching out toward the east, suggesting a tilted syncline.*

A series of folded, more or less metamorphosed Proterozoic sedimentary and volcanic rocks occupies the center of the basin. U.S. 41 is in the central portion from Negaunee to Michigamme and beyond. North of the central portion is an upwarping of Archean rocks, and another large area of Proterozoic metasediments that are an extension of those in the central area.

Marquette sits on a formation of Archean lava flows and sandstone that were metamorphosed to greenstone schist, or greenschist, that includes minor beds of quartzite. A little over 3 miles (5 km) south of the city center, where U.S. 41 turns south and leaves the lake shore, exposures of rock (including an old quarry) exhibit the Archean greenschist as well as Proterozoic quartzite, slate, dolomite, and chert.

U.S. 41 westbound crosses the greenschist from Marquette to the vicinity of Negaunee. Pillow structures occur in the greenschist about 5.5 miles (9 km) west of the lake at Marquette, where a lava flow entered the ancient ocean.

Less than a mile north of Negaunee, the highway enters Proterozoic slate and quartzite, and quartzite occurs in the road cut at the northeast corner of Negaunee. If you drive south, you will enter the Marquette Iron Range country (see Figure 30). There, the iron formation overlies the quartzite in the basin and is exposed in the southeastern part of Negaunee (County Road 480) and in the area south and west to Ishpeming (Michigan 28).

Along U.S. 41, west of Ishpeming, you pass through a band of Proterozoic quartzite, then through several miles of volcanics. In the area about 8 to 10 miles (13–16 km) west of Ishpeming are exposures of lava flows, tuffs, and volcanic breccia-conglomerate. From there to Michigamme, the bedrock consists mostly of Proterozoic slate and quartzite derived from the silt and sand of the Precambrian ocean.

Loretto to Wakefield, Michigan: U.S. 2 (Figure 45) *U.S. 2 traverses parts of two Michigan iron-mining regions. The Menominee Iron Range, from Loretto to Iron Mountain, is a series of mostly metasedimentary Proterozoic formations that trends east and west in a set of tight folds broken by faults. The area around Crystal Falls is part of the Crystal Falls–Iron River District, a set of folded metasediments that includes iron formation.*

Loretto sits right at the edge of one of two parallel outcrops of the Vulcan Iron Formation, part of the Menominee Iron Range (Figure 30), and U.S. 2 parallels the rocks to Iron Mountain. Northeast of Loretto, at the Sturgeon Falls Dam, the other strip of the iron formation is visible, as well as quartzite of the metasedimentary series.

West on U.S. 2, from Loretto to beyond Norway, the bedrock is slate. In the mining area just north of Norway, you can see the iron formation and the dolomite that underlies it, along with interbedded quartzite and slate. There is an abandoned mine in the iron formation at Quinnesec. The highway then passes through more slate to Iron Mountain, but the iron-bearing part of the Vulcan Iron Formation is exposed on the north side of U.S. 2, about .7 mile (1 km) west of the intersection with U.S. 141.

There has been much mining at the northeast edge of Iron

Mountain, and you can see the iron formation in the Millie Pit and the former Trader's Mine. At the latter locality, slate occurs above and below the iron formation. Quarrying on the south shore of Lake Antoine, east of Iron Mountain, has exposed the local dolomite, complete with ripple marks, mud cracks, and structures formed by algae. Just north of Iron Mountain, the highway traverses the dolomite. From there to Crystal Falls, the bedrock is greenstone and slate.

The road from Iron Mountain to Crystal Falls leads from the Menominee Iron Range to the Crystal Falls–Iron River District. There, at the dam on the Paint River near Crystal Falls, a new set of metasediments shows up, including another iron formation, the Riverton. In this area, the iron formation exposure indicates a different structure from that of the Menominee Range. The folded iron formation comes to the surface in two arms, one running north–south and the other east–west, the arms joining at Crystal Falls. Slate beds and a graywacke with chert pebbles outcrop at the dam, along with the Riverton Iron Formation.

From Crystal Falls to Watersmeet, the bedrock (under all the glacial till) consists of Proterozoic metasediments. At Watersmeet, you can drive about 9 miles (14.5 km) north to Paulding on U.S. 45, then 2.5 miles (4 km) east to Bond Falls, where there are outcrops of late Proterozoic lava flows.

U.S. 2 between Watersmeet and Wakefield passes Archean greenstone with granite intrusions, as well as Proterozoic slate. Pillow structures show on a glacially polished greenstone surface about one mile east of Wakefield.

Niagara to Crivitz, Wisconsin: U.S. 141 (Figure 45) *U.S. 141 from the Menominee Range into Wisconsin runs through a region of Proterozoic intrusive igneous rocks and volcanics (greenstones). This stretch carries you over the geologic border into the Cambrian Period.*

From Niagara to Pembine, you travel over alternating masses of granitic rock and greenstone; greenstone then granite, then greenstone, then granodiorite. At Pembine, you enter another corner of greenstone outcrop. Five miles (8 km) west on U.S. 8, granodiorite outcrops contain chunks of the volcanics (greenstones), indicating the zone of contact between the intruding granodiorite and the volcanics.

South of Pembine, the greenstone continues, plus a zone of gray and pink granites. Just beyond Amberg, another greenstone area appears, with monzonite dikes. Then, back into the granites until just beyond Wausaukee, where the highway passes onto Cambrian sandstone originally deposited in the ocean that covered the eroded Proterozoic rocks half a billion years ago. Between there and Crivitz is another stretch of granite. At Crivitz, the Cambrian sandstone occurs again.

West of Crivitz about 15 miles (24 km) on County Road W is Thunder Mountain State Park, where you can observe Proterozoic quartzite.

Mosinee to the Wausau area, Wisconsin: U.S. 51 (Figure 45)
This short trip passes a small mass of Proterozoic metasediments in a region of mostly intrusive rocks.

To see intrusives, you can exit U.S. 51 at the Wisconsin 153 exit and cross the Wisconsin River to Mosinee. There, south of the bridge, you will find diorite with included pieces of gabbro.

Quartzite outcrops at Mosinee Hill, west of U.S. 51 at the crossing of the Wisconsin River, and farther north, via County Route N, at Rib Mountain State Park. At the latter site, the layers are almost vertical. Quartzite forms other hills in the area, too. About 5 miles (8 km) north of the Wisconsin Route 29 exit, U.S. 51 passes over County Road WW. Conglomerate shows at the underpass and .1 mile east at the west end of the bridge over the Wisconsin River.

Waupaca to Neillsville, Wisconsin: U.S. 10 (Figure 45)
This is a largely igneous part of Wisconsin, where Proterozoic batholiths and dikes have intruded Archean rocks.

In the Nels Rasmussen Park in Waupaca, greenstone and diorite, possibly of Archean age, are part of the zone that surrounds a batholith. Proterozoic granite dikes cut through some of the rocks in the park. A granite quarry exists north of town off Wisconsin Route 49.

If you cross the Wisconsin River on U.S. 10 and drive south along the river for 3 miles (5 km) you will see granite gneiss in the area below a dam. The gneiss is presumed to have originated as sedimentary and igneous rocks that were metamorphosed during the Kenoran Orogeny at the end of the Archean Eon. The gneiss contains Proterozoic dikes and veins.

About 24 miles (39 km) west of the Wisconsin River on U.S. 10, County Road M leads south to the town of Arpin. The county park at Powers Bluff, which is southwest of Arpin, has a hill of Proterozoic quartzite.

Back on U.S. 10, there is a granite quarry about 2 miles

(3 km) to the northwest of Neillsville. State Route 73 crosses the Precambrian-Cambrian border, where the Cambrian sandstone overlaps the Precambrian rocks just south of town. Where the Black River curves close to the road, it has exposed gneiss that may be Archean in age below the Cambrian sandstone.

The Pigeon River to Schreiber, Ontario: Route 61, Trans-Canada Highway 11/17 (Figure 34) *Proterozoic rocks outcrop in this area, an extension of northeastern Minnesota geology at the edge of the Lake Superior Basin. The area forms an indentation into the southern border of the Superior Province. It is a large region of earlier Proterozoic sedimentary rocks that belong to the Older Southern Province, crossed and separated by later Proterozoic dikes and sills. The dikes and sills are related to the Lake Superior Basin rocks.*

Conglomerate outcrops in the town of Kakabeka Falls, at the intersection of highways 11/17 and 590, and in Kakabeka Falls Provincial Park. At Kakabeka Falls, the Kaministikwia River drops over limestone and dolomite with chert layers and rushes through black shale in the gorge below the falls.

The limestone, dolomite, and chert formation can be seen in Thunder Bay at Hillcrest Park, Trowbridge Falls Park, and Boulevard Lake Park (where iron formation also outcrops). At Mt. McKay, south of the Ft. William part of Thunder Bay, a sill of later Proterozoic diabase caps shale layers.

A bit of the Lake Superior Basin shows up in the Port Arthur part of Thunder Bay, with an old copper mine cut into greenschist (a former lava flow). The greenschist contains copper minerals in the cavities of former gas bubbles and in quartz veins and lenses.

The Proterozoic shale, limestone, dolomite, chert, and iron formation occur from Thunder Bay northeast 25 miles (40 km) to Loon Lake, and southwest into Minnesota. Highway 61 takes you down to the United States border at the Pigeon River. There, at High Falls, the river flows over a diabase dike and cuts a gorge in shale and graywacke. Middle Falls has two diabase sills with shale in between, and three major diabase dikes. One of the dikes causes the falls.

North of Thunder Bay, on the Sibley Peninsula, Proterozoic sedimentary rocks (conglomerate, sandstone, shale, limestone, and dolomite) occur at Sibley Provincial Park. The rock layers dip gradually toward the southeast, on the edge of the Lake Superior Basin. Sedimentary rocks outcrop on Route 387, 2 miles (3 km) east of Highway 11/17, at the north end of Pass Lake, and at Moffat Point and Silver Islet Landing. Nearby Silver Island was the site of rich silver mines. Prospectors found the metal in contact zones between the sedimentary rocks and sills and dikes of the diabase that is so abundant in the area.

The Red Rock Escarpment rises about 7 miles (11 km) south of Nipigon on Highway 11/17. The base of the escarpment shows Archean granite, overlain by Proterozoic shale, dolomite, limestone, and sandstone. The redbeds of this rock series, colored by the iron mineral hematite, give the escarpment its name. The sedimentary rocks are surmounted by a massive diabase sill. The hills around the mouth of the Nipigon River are remnants of the same sill, and in the country east of Nipigon, sandstone, limestone, and dolomite capped by sills of diabase form many hills. Past Schreiber, you enter the Superior Province.

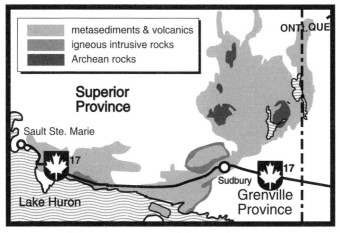

Figure 48. The eastern portion of the Southern Province, Ontario and Québec.

Sault Ste.-Marie to Sudbury, Ontario: Trans-Canada Highway 17 (Figure 48) *Highway 17 passes through the northeastern arm of the Southern Province, which borders and extends north from Lake Huron. Later sills and dikes intruded the earlier Proterozoic sedimentary and metasedimentary rocks of this part of the Older Southern Province.*

After Sault Ste.-Marie, the Southern Province begins between Echo Bay and Desbarats. About 1.5 miles (2.4 km) west of Desbarats, a well-known outcrop of early Proterozoic quartzite dips sharply down to the south. The tops of the layers show clear ripple marks that trend in two directions in different layers, indicating that the shoreline and ocean currents changed directions over time.

Folded limestone and quartzite outcrop at Bowker Point, 1 mile southwest of the town of Bruce Mines, site of the first profitable copper mine in Canada. Copper-rich quartz veins were discovered here in a body of diabase in 1846. About 8 miles (13 km) north of Bruce Mines, just north of Rydal Bank, Route 638 passes a ridge that has outcrops of a conglomerate with quartz and jasper pebbles. Jasper is a brightly colored red, yellow, or brown variety of quartz, highly prized as an ornamental stone.

Cross-bedded quartzite outcrops along Highway 17 about 10 miles (16 km) east of Thessalon. Exposures of the Early Proterozoic Gowganda Formation begin about 25 miles (40 km) east of Thessalon. The formation is most interesting in that much of it is evidently a tillite, that is, a solidified glacial till. The Gowganda Formation also outcrops between Iron Bridge and Blind River. Along with the tillite, the formation contains layers of argillite, quartzite, and conglomeratic graywacke. You can easily recognize the tillite by its miscellaneous assortment of grains, pebbles, cobbles, and boulders of several rock types sparsely arranged in a dark muddy matrix.

At Serpent River, Route 108 leads to the town of Elliot Lake, the site of some well-known uranium mines. The uranium ore is found in an early Proterozoic conglomerate that rests on Archean granite. The series contains arkose and volcanic rocks above the conglomerate. These rocks and their uranium ore occur in a strip from the town of Blind River to Elliot Lake. Back on Highway 17 at Serpent River, quartzite and schist outcrop at the Serpent River bridge.

Between Serpent River and Cutler, the highway cuts through an anticline, passing through schist and other rocks on one limb of the anticline, then granite in the center, then the schist again on the other limb. Road cuts and outcrops at Cutler and a mile west of the town show the schist intruded by granite.

From Cutler to Walford, a diabase sill parallels the highway. At Walford, the sill shows on the highway opposite the west entrance to the town and on the highway at the edge of the town. Interbedded quartzite and schist are exposed at the Walford railroad bridge. Diabase dikes cut all these rocks. In the Massey area, at the bridge over the River aux Sables and in the Chutes Provincial Park, swift rivers have carved gorges into the quartzite and schist.

From Massey to Sudbury, the bedrock and outcrops consist mostly of early Proterozoic sedimentary and metasedimentary rocks similar to those the highway traverses to the west, such as argillite, graywacke, quartzite, and conglomerate, with associated dikes and sills. There is a sill of metadiabase cut by a diabase dike .5 mile west of the Highway 6 intersection, and a dike of olivine diabase 1 mile east of Nairn Centre that contains large feldspar grains. One of the more interesting metasediment exposures occurs about 3.5 miles (5.6 km) west of the center of Sudbury, at Balsam Street, where you can see quartzite with cross bedding and ripple marks, along with argillite.

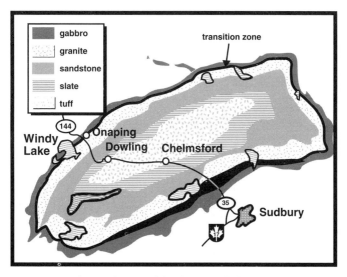

Figure 49. The geology and highways of the Sudbury Basin, Ontario.

Sudbury to Windy Lake, Ontario: Routes 35, 144 (Figure 49)

This route takes you across the Sudbury Basin, renowned because it contains the world's largest deposit of nickel as well as significant amounts of copper, gold, silver, platinum, cobalt, and iron. It is one of the world's greatest mineralogical treasures. Intrusive igneous rock outcrops in an elliptical ring with sedimentary and volcanic rocks inside, including an unusual breccia formed mostly of angular broken rock fragments. These Early Proterozoic rocks are surrounded by Archean ones around the northern margin and by earlier Proterozoic ones along the southern margin.

The latest and most interesting theory of the basin's origin is that it resulted from the impact of a large meteoroid. The oldest rocks of the basin, along the southern margin, are sedimentary formations that were originally deposited on eroded Archean gneiss. After those rocks were formed, the theory goes, a meteroid thundered into the crust and created an enormous crater. The impact promoted the development of a sill-like mass of intrusive igneous rock and broke up the local bedrock into chunks that became the breccia. The igneous rock is dated at about 1.8 billion years ago. Later, rivers and streams carried sediments, which are now slate and sandstone, into the impact crater. The present basin is an erosional remnant of the original one.

The outer ring of the basin is the eroded edge of a sill-like mass of Proterozoic gabbro-type rocks, dark in color, with heavy minerals, grading inward through a transitional zone to granite. It is as if the sill took a long time to cool, and while it was molten, the lighter minerals rose to the top to make the granite, and the heavier minerals sank to become the gabbro. Inside the ring of intrusives is a series of more or less metamorphosed sedimentary and volcanic rocks of Early Proterozoic age, namely slate, sandstone, graywacke, and an atypical tufflike rock.

The city of Sudbury rests mostly on early Proterozoic sandstone and shale. On Highway 17, about 2.5 miles (4 km) west of the city center, rhyolite outcrops on the north side and the breccia outcrops on the south side. The pebbles in the breccia are chunks of rhyolite and graywacke—these chunks may represent original layers shattered by shock waves from the impact of the meteroid. A half mile farther on, at the intersection with Clarabelle Road, which goes to the Copper Cliff Mines, are more metasediments and volcanics. These rocks form the outer edge of the basin.

Clarabelle Road leads to Regional Road 35, where a plaque about 1.5 miles (2.5 km) northwest of Clarabelle Road commemorates the discovery of the Sudbury ores.

The basin's central plain consists of sand, gravel, and clay from a lake caused by an ice dam of the last Ice Age glacier. This lake-bottom material starts just south of Azilda. The bedrock in this area consists of the sedimentary and volcanic rocks that characterize the basin center. At Chelmsford, Regional Road 35 joins Route 144. There are outcrops of graywacke beyond the town, where the highway reaches the top of the hill that overlooks Dowling. Slate occurs partway down the descent to Dowling. About 3 miles (5 km) past Dowling, you come to the Onaping Falls, where a breccia contains fragments that appear to have been thrown into the air by the meteroid impact.

Between Onaping Falls and the town of Onaping, rocks of the intrusive ring's granite zone outcrop. If you turn northeast at Onaping on Route 544, you can see samples of the intrusive transitional zone between Onaping and Levack, grading in color from lighter to darker. Beyond Onaping on Highway 144, past Windy Lake, the bedrock belongs to the surrounding Archean granite and gneiss zone.

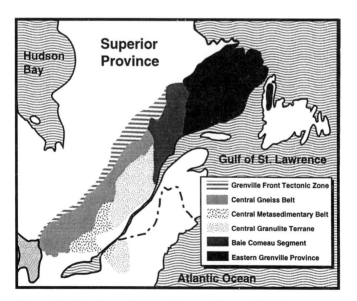

Figure 50. The Grenville Province and its subdivisions.

■ THE GRENVILLE PROVINCE

The formations of the Grenville Province provide striking evidence of the incredibly powerful forces in Earth's crust. Here, rocks of Archean to Middle Proterozoic age have been deformed, highly metamorphosed, and pushed onto the rocks of the other provinces. The Grenville Province occupies southeastern Ontario and eastern Québec, with an extension into the Adirondack region of New York and an outlier in Newfoundland. Movements during the Grenvillian Orogeny shoved a broad band of crust, now extending from Lake Huron to Labrador's east coast, over and across the edge of the continental core. It has been estimated that at one time the crust in the Grenville region was twice as thick as it is elsewhere on the continent, and erosion has since thinned it. As in other parts of the Canadian Shield, the rocks started as marine and terrestrial silt, sand, and other sediments, accompanied by lava flows. The original rocks were metamorphosed to varying degrees and intruded.

Figure 50 shows the zones into which geologists have divided the province.

◆ Baie Comeau Segment and Eastern Grenville Province

In these subprovinces, the bedrock consists of Proterozoic gneiss with large scattered Proterozoic igneous intrusions (batholiths). The Baie Comeau Segment contains schist as well as other rock types. The Eastern Grenville Province outlier that forms the Long Range Mountains of Newfoundland and the backbone of Gros Morne National Park there (Figure 50) is a body of Proterozoic gneiss, schist, and granite.

◆ The Grenville Front Tectonic Zone

This region is characterized by long strips of mylonite, the rock type that marks fault zones, in this case the zone of the great thrust fault action during the Grenvillian Orogeny. To the northwest of the mylonite, some rocks in the Southern and Superior Provinces are only slightly metamorphosed, but southeast of the mylonite, the Grenville gneisses are highly changed from their original types.

◆The Central Gneiss Belt

The gneisses of the tectonic zone continue into the Central Gneiss Belt, where a variety of original rocks were changed by great heat and pressure into several kinds of gneisses with different mineral compositions, along with some intrusive rocks and small amounts of metasedimentary ones. These rocks are Middle Proterozoic in age and were deformed during the Grenvillian Orogeny.

◆The Central Metasedimentary Belt

The Central Metasedimentary Belt is separated in some places from the neighboring gneiss belt by a thin zone that indicates thrust faulting *within* the Grenville Province. Figure 51 shows an exposure with good evidence of thrust faulting. This transitional zone is characterized by gneisses that contain cobbles and boulders from older rock layers that were fragmented in the faulting. The rocks recrystallized during the faulting because of high temperatures and pres-

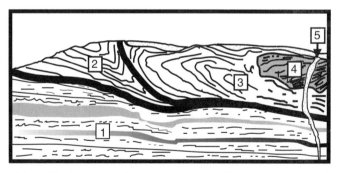

Figure 51. Thrust faulting in the Grenville Province: a rock exposure on Highway 17 between Pembroke and Cobden, Ontario. 1, a mass of gneiss with layers of amphibolite. 2, a detached portion of a syncline in marble, thrust from right to left onto the gneiss. 3, a crushed anticline in the same marble, thrust from right to left across the gneiss and onto the syncline. 4, a block of gneiss that was mixed with the marble as rock formations were crushed and broken. 5, a later dike, cutting through all the other rocks.

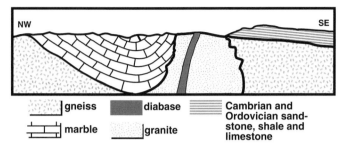

Figure 52. The geologic structure of the Grenville Province, simplified, in the vicinity of Kingston, Ontario. Proterozoic sedimentary rocks were folded, faulted, intruded, metamorphosed, and eroded, then covered by Cambrian and Ordovician sandstone, shale, and limestone.

sures, and therefore do not resemble the mylonites along the Grenville Front. The main body of Proterozoic rocks in the Central Metasedimentary Belt are younger Proterozoic than those of the Central Gneiss Belt, and were apparently thrust over the older ones.

The metasediments began as marine limestone and sandstone, which became the present marble and quartzite that grade with higher metamorphism into gneisses, granulite, and other highly changed types (Figure 52). They are accompanied by volcanic and metavolcanic rocks (greenschist) and intruded by gabbro, monzonite, syenite, and granite (Plate 4).

The Central Metasedimentary Belt rocks were formed before the climax of the Grenvillian Orogeny. There is evidence that during the orogeny the piece of crust that was to become the Grenville Province was significantly shortened by crumpling. Rock layers in Québec dip down toward the southeast, while along the St. Lawrence River they are vertical, and in the Adirondack lowlands they dip toward the northwest. That suggests accordion-like folding from horizontal pressure that squeezed the rock layers. Along the eastern margin of the metasedimentary belt, at the edge of the Adirondack Mountains, another narrow, elongate zone of mylonite represents yet another incidence of thrust faulting. It is further evidence that as the Grenville crustal mass was forced up onto the continental nucleus, it was itself fragmented into blocks that moved onto one another.

◆The Central Granulite Terrane

This subprovince in east-central Québec and the Adirondack Mountains of New York (Figure 53) contains rocks of sedimentary and volcanic origin that were intruded, then metamorphosed. The Middle Proterozoic oceanic sediments and volcanics are now marble, quartzite, migmatite, granulite, and gneiss, along with some granitic gneisses that geologists have interpreted as metamorphosed volcanic rocks.

Some of the intrusive igneous rocks have metamorphosed into gneiss and metagabbro. The Central Granulite Terrane is also known for large areas of anorthosite, a rock composed primarily of plagioclase feldspar, that were metamorphosed to varying degrees. This meta-anorthosite makes up much of the Adirondack Mountains. It also occurs in a similar mass centered about 45 miles (75 km) north of Montréal, in the

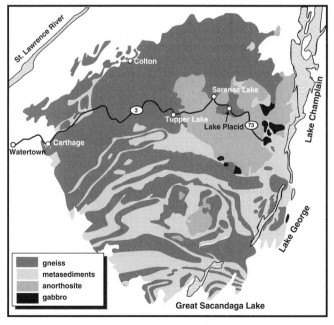

Figure 53. Geology and highways of New York's Adirondack region, the only part of the Grenville Province in the U.S.

Laurentians, and in a third bordering Lac St-Jean, farther north.

The layered rocks of the Adirondack region were folded and faulted during the Grenvillian Orogeny in a very complicated fashion, suggesting that there were discrete fault blocks that moved separately. The evidence suggests that there was a rotational force as well as the main force from the southeast.

GRENVILLE PROVINCE HIGHWAY SECTIONS

Watertown to Underwood, New York: Routes 3, 86, 73 (Figure 53) *This route leads through the Adirondack lowland, with its metamorphosed sedimentary rocks, and the highlands, characterized by the great Adirondack anorthosite body surrounded by metamorphic rocks—gneisses that have a common texture but differ in their mineral compositions.*

Watertown is situated upon Ordovician limestone of the Central Lowland. Just before Carthage on New York Route 3, however, the bedrock switches to Precambrian gneiss that underlies the limestone, at the edge of the Grenville Province. The gneiss is a highly metamorphic rock, formed in the tremendous upheavals that took place during the Grenvillian Orogeny.

At Natural Bridge, the highway approaches the Carthage-Colton mylonite zone, which is exposed between those two towns, extending northeast–southwest. The zone is that of a major thrust fault, on which one large section of the crust overrode another in a northwesterly direction. The zone marks the border between the Central Metasedimentary Belt and the Central Granulite Terrane (see Figure 50).

The mylonite does not outcrop along the highway, but the road passes marble that forms the bedrock to the northwest of the fault zone in the metasedimentary belt and the gneiss that is to the southeast of the fault zone in the granulite terrane.

The small cave for which Natural Bridge is named is an underground channel that was carved in the local marble by the Indian River. The outcrops in the metasedimentary belt seem to have a northeast trend, indicating a force from the southeast that folded the original layers. However, the metasedimentary layers were very complexly folded and faulted during the Grenvillian Orogeny. In the granulite terrane, the formations are not layered, and there is no organized structure.

Between the areas of Natural Bridge and East Pitcairn, the highway more or less parallels the mylonite zone. From Natural Bridge to Harrisville, Route 3 goes through gneiss. Just beyond Harrisville, you come to a valley cut into the marble, but after East Pitcairn, the highway turns and enters the gneiss of the granulite terrane. From there to Star Lake the bedrock is gneiss.

The lake at Star Lake is unusual in that the remains of an Ice Age esker (the gravelly bed of a glacial river) traverse it, forming a long, narrow promontory and an elongate island. From Star Lake to Tupper Lake you will see more gneiss. Just east of Star Lake is an old open-pit iron mine, where prospectors found magnetite and hematite in the gneiss. The bedrock of the towns of Tupper Lake and Saranac Lake is gneiss, but between them, you pass through something different—an arm extending from the main body of Adirondack anorthosite. You will recognize the anorthosite by its igneouslike texture, different from the streaked, pseudolayered appearance of the gneiss (Plate 7).

At Saranac Lake, the route switches to New York Route 86 and heads toward Lake Placid. Although Saranac Lake is built on gneiss, outside of town the bedrock is more anorthosite, which continues almost all the way to Lake Placid. A couple of miles before Lake Placid, the gneiss appears again. The wide valley at Lake Placid was the site of a glacial lake at the end of the Ice Age. In town, pick up New York 73.

Gneiss continues beyond Lake Placid. Cascade Lake, situated in Cascade Pass, is about 8 miles (13 km) east of Lake Placid. The bedrock on the west is gneiss, and that on the east is anorthosite. Weathering and erosion along a northeast-trending fault carved out the pass. From there to Underwood and I-87, the outcrops are all anorthosite. Along the way, on the hill between St. Huberts and Chapel Pond, you get a view of the glacial cirque on Giant Mountain, formed during the Ice Age. The form of the cirque has been emphasized by landslides that have swept away the trees and bushes that formerly grew on that face of the mountain.

Port Severn to the 69/169 junction, Ontario: Route 69 (Figure 54)
The edge of the Central Lowland, with its flat-lying Paleozoic rocks, is a few miles south of Port Severn. Along this

route, erosion has exposed a series of tortured and metamorphosed Precambrian sedimentary rocks. The rocks exposed are a good sample of what you can see in the Grenville Province's Central Gneiss Belt.

As you drive north, you will see gray gneiss that contains garnets and areas of granitic texture, with veins of pink feldspar. In some places in the body of gneiss, there are layers of impure marble. The gneiss is considered to have originally been sandstone and possibly graywacke. Farther along is a pink gneiss with thin layers of hornblende schist. This gneiss probably evolved from a granite with dikes that became the schist.

Norland to Dorset, Ontario: Route 35 (Figure 54) *From Norland to Dorset, Route 35 passes through rocks of the Central Metasedimentary Belt, the Boundary Zone, and the Central Gneiss Belt.*

Between Norland and Moore Lake are about 6 miles (9.5 km) of diorite gneisses, with granite dikes and some migma-

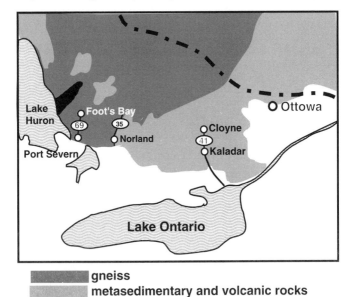

gneiss
metasedimentary and volcanic rocks
intrusive rocks

Figure 54. Highways in Ontario that sample the Central Gneiss Belt and the Central Metasedimentary Belt.

tites (intimate mixtures of granite and gneiss). At the north end of Moore Lake, the highway passes into marble breccia of the Central Metasedimentary Belt. The breccia has a matrix of marble with angular fragments of other rock types. Plate 5 shows an exposure of similar rock. Some pink granite also outcrops in the area.

At the south end of Mountain Lake, about 20 miles (33 km) north of Norland, you reenter the Boundary Zone, where the bedrock is a granite gneiss whose texture indicates thrusting from the southeast. Gneisses extend beyond Carnarvon, with occasional blobs and lenses of black crystalline amphibolite.

Just beyond where the highway crosses Kushog Lake, it enters the Central Gneiss Belt, with pink, gray, and greenish brown gneisses, some of which are metamorphosed granite and syenite. The gneisses occur almost as far as Dorset.

South of Kaladar to Cloyne, Ontario: Route 41 (Figure 54) *The Central Metasedimentary Belt contains rocks that were changed less than those in the gneiss belt. Route 41 traverses exposures of a typical series of such rocks, to illustrate the geology of this belt.*

In an area of metamorphosed sedimentary and volcanic rocks about 5 miles (8 km) south of Kaladar, there is an easily recognizable white crystalline marble. Metaconglomerate outcrops about 2 miles (3 km) north of Highway 7 at Kaladar, with flattened pebbles in a matrix that has a gneissic texture. The long axes of the flattened pebbles are parallel to streaks in the matrix.

North of the metaconglomerate are several outcrops of a granodiorite intrusion that accompanied the folding, faulting, and metamorphism of the original sedimentary rocks. Farther on, about 8 miles (13 km) north of Highway 7, more metaconglomerate appears, with flattened pebbles in a schist matrix. Volcanics occur again at the intersection with Route 506, in the form of greenschist with pillow structures that show its volcanic origin.

The Grenville Front, Ontario: Trans-Canada Highway 17, Routes 539, 539A, 805, 64, 11 (Figure 55) *Several highways cross the boundary fault zone at the northwest edge of the Grenville Province (the Grenville Front Tectonic Zone), but it is difficult to identify exactly where the major faults were. The fault zone is represented by mylonite layers, by rocks of the granite family that consist of broken granitic fragments in a*

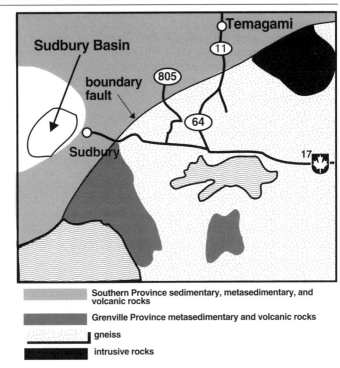

Figure 55. Highways in the Grenville Front boundary fault zone.

finer-grained granitic groundmass, and by gneissic granite. The zone also has highly metamorphosed gneiss and quartzite-type rocks that were originally conglomerate, graywacke, and sandstone, and retain indications of layering. The layering shows them to have been intensely folded, with the folds overturned to the northwest.

Highway 17 from Sudbury crosses into the fault zone at Wahnapitae. Mylonite texture characterizes the fault zone there in several varieties of rock that occur in the area, on both sides of the boundary between provinces. On the Southern Province side of the boundary fault, granite, diabase, and a quartzite that contains feldspar outcrop. On the Grenville Province side of the fault the bedrock is gneiss.

Farther east, at Warren, you can turn north on Route 539 and drive to River Valley. Along the way you will pass graywacke and argillite that have been greatly metamorphosed, becoming gneissic and gaining new minerals. Route 539A continues northwest to become 805; in this area, the road crosses a body of meta-anorthosite. The boundary fault passes under the road about 4.5 miles (7 km) beyond the beginning of Route 805, just north of the railway crossing, marked by the change in bedrock from meta-anorthosite to schists derived from Southern Province sedimentary formations and dikes.

Highway 11 crosses the boundary fault zone south of Temagami, from about 8 miles (13 km) south of town and for about 50 miles (80 km) southward. If you drive north from Highway 17 on Highway 64 and continue north on Highway 11, you will pass the boundary fault about a third of the way from Marten River to Temagami. North of Marten River, the road traverses a zone of granite gneiss. The point at which the bedrock changes from gneiss to granodiorite marks the boundary fault. To the north of the boundary fault is a broad zone of Archean granodiorite and greenstone, with some Proterozoic diabase. The fault crosses the countryside about 20 miles (32 km) south of Temagami.

Val-d'Or to Montréal, Québec: Route 117 (Figure 56) *Route 117 offers a cross section of the Grenville Province between the Superior Province at Val-d'Or and the Stable Interior at Montreal. It crosses the Central Gneiss Belt and Central Granulite Terrane intrusive igneous and metamorphosed sedimentary rocks, evidence of the Grenvillian Orogeny.*

On Highway 117, about 20 miles (32 km) south of Louvicourt, the greenstone, schist, and syenite of the Superior Province become gneiss of the Grenville Province. Gneiss continues from there, with a bit of syenite, for approximately 103 miles (165 km), to Dorval Lodge and beyond.

The gneiss ends about 3 miles (5 km) south of the Parc de La Verendrye's south gate and gives way to a zone of Proterozoic marble that continues to Mont-Laurier, with some variations. In places, pink granite replaces the marble. At Grand Remous, west of the bridge over the rapids in the Gatineau River, you can see quartzite and gneiss of this metasedimentary series. In the Lac Gatineau area, gabbro and granite intrude the marble. A bit less than 4 miles (6 km) south of the bridge at Mont-Laurier is a coarse gneiss with dikes of pink granite. Pink granite outcrops about 11 miles (18 km) beyond the gneiss, presumably intruded during the Grenvillian Orogeny.

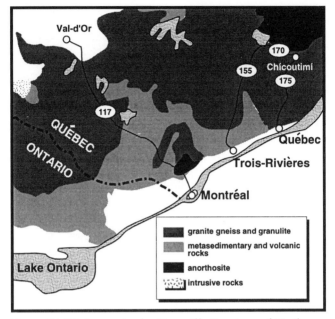

Figure 56. Highways in the Grenville Province of Québec.

Farther along, at either end of the village of Lac-Saguay, gneiss and quartzite outcrop, and more Proterozoic gneiss extends beyond the village. About half a mile south of the junction of Route 321, the older Archean basement shows through the younger rocks, in the form of gray gneiss. You then pass cliffs of granulite with granite dikes along the Rouge River, and about 13 miles (21 km) south of Route 321, you reenter the Proterozoic metasediments.

From that point north of Labelle to the area of St-Jovite, the bedrock is mostly marble, in some places in large cliffs. Gneiss, granulite (between L'Annonciation and St-Jérôme), and quartzite accompany the marble. A pink granite at the north edge of Labelle has a streaky texture resembling that of mylonite, with mineral grains elongated parallel to the streaks. It appears that the granite was stretched, no doubt during the Grenvillian Orogeny. This mylonitic texture occurs here and there in the marble and gneiss of this zone, indicating thrust faulting within the Grenville Province.

Beyond St-Jovite the highway passes exposures of the gabbro that borders the anorthosite mass of southern Québec. You can see the anorthosite at St-Faustin. The anorthosite extends along the highway to St-Jérôme, where the flat-lying Paleozoic rocks of the Stable Interior's Central Lowland overlie the rocks of the Canadian Shield.

Trois-Rivières to Chicoutimi to Québec City, Québec: Routes 55, 155, 169, 170, 175 (Figure 56) *This route passes typical rocks of the Central Granulite Terrane; it is a region of gneisses, intrusive igneous rocks, and varieties of anorthosite.*

The country north of Trois-Rivières contains gneiss whose greenish color is due to green varieties of hornblende and biotite. It is a very highly metamorphosed rock that was originally an intrusive igneous type. The gneiss outcrops on Highway 155 3 miles (5 km) north of where the road that goes to the town of St-Joseph-de-Mékinac joins the highway. Some cracks and fissures in it are filled with contorted crystalline marble.

About 2.5 miles (4 km) north of the road that goes to the Lac Edouard station, outcrops show the contact between the green gneiss and a small body of white, highly metamorphosed anorthosite, whose minerals have been crushed and recrystallized. The gneiss continues from there to Highway 169 near Chambord, with flanking masses of anorthosite and granite.

If you follow Routes 169 and 170 eastward around Lac Saint-Jean, you can take a side road to St-Gédéon, 4.5 miles (7 km) north of which is a quarry in black anorthosite. The rock there shows only minor effects of deformation and metamorphism and contains large grains of plagioclase feldspar. It is typical of the large mass of anorthosite that extends from the north shore of Lac Saint-Jean northward for about 80 miles (130 km) and eastward for about 50 miles (80 km).

Highways 170 and 372 east lead to Chicoutimi; near the south end of the Saguenay River bridge, anorthosite cliffs overlook the road. Most of the anorthosite is relatively fresh and dark, but texture varies to partly crushed and to completely crushed and recrystallized into the white variety.

Highway 175 south traverses a region of gneiss, granite, and granulite varieties that extends to the St. Lawrence River. About a half mile south of the north gate to Laurentides Provincial Reserve you pass an outcrop of grayish pink porphyritic monzonite with a black gabbroic dike. The same monzonite also outcrops about 27 miles (43 km) farther along the highway. Granite occurs about 13 miles (21 km) south of the junction with Highway 169.

About 14 miles (22.5 km) south of the L'Etape lodge, monzonite occurs again, this time greenish, with phenocrysts (extra large grains) of both plagioclase and orthoclase feldspars. And about 3 miles (5 km) south of the Stoneham gate to the park, more monzonite occurs, similar to that above except that its phenocrysts consist only of orthoclase.

Just north of Québec City, the Paleozoic sedimentary rocks of the Central Lowland overlie and cover those of the Grenville Province at the southeastern border of the Canadian Shield.

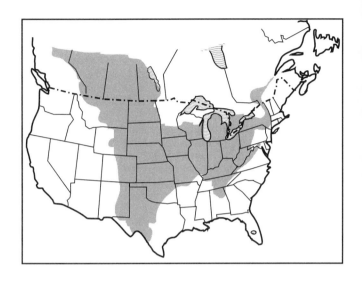

4

THE STABLE INTERIOR

Between the Appalachian Mountains and the Rockies lies an immense interior region of plains, prairies, and low forested hills, a country floored with uncrushed, nearly horizontal layers of sedimentary bedrock. It is called "stable" because its bedrock hasn't been greatly distorted, changed, and uplifted like the rocks in the eastern and western mountainous regions. The Stable Interior extends approximately from Québec City to Alabama to west Texas to the Northwest Territories. It contains, from west to east, the Great Plains, a central lowland, and the plateau country that borders the Appalachians (Figure 57). The Great Plains are outside the scope of this book, so this chapter is limited to the geology of the Stable Interior east of the plains, in the Central Lowland and the Appalachian Plateaus.

The Stable Interior hasn't been entirely stable; actually, pressures in Earth's crust that are not understood in detail but may be associated with plate motion have slightly distorted the sedimentary formations east of the Great Plains, creating broad arches, domes, and basins whose configurations are barely visible in the surface topography. Figure 58 indicates the major structures of the Stable Interior.

Evidently, the crustal forces acted periodically throughout most of the Paleozoic Era. The arches were formed when Paleozoic sedimentary rocks were forced into long, broad, very gradual anticlines by horizontal pressure. Basins developed where rock layers subsided in their central areas; domes resulted from localized uplifts. The Peace River and Transcontinental arches were uplifted, eroded, and then covered by later Paleozoic and Mesozoic rocks, so that the arches don't show on the surface. The Williston Basin is covered by horizontal Mesozoic and Cenozoic rocks. Recent erosion, however, has exposed the other structures shown on the map.

A minor downwarping, the Dunkard Basin, occupies an area between the Appalachian uplift zone and the Cincinnati

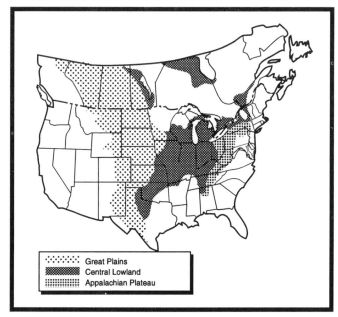

Figure 57. The Stable Interior and its subdivisions. The Appalachian Plateau is further divided into the Allegheny Plateau from New York through West Virginia and the Cumberland Plateau south of West Virginia.

Arch. Its center is approximately where Ohio, Pennsylvania, and West Virginia join. In the area covered by this book, the basins don't form valleys, and the arches and dome don't make uplands; erosion has smoothed off the land surface. In the Illinois Basin and in the northern Ohio part of the Cincinnati Arch, the land surface is generally flat, even though rock layers are slightly tilted (see Figure 59). The basin and the arch are *structural*, not topographic.

■ THE LAY OF THE LAND

The Stable Interior consists of layered sedimentary rocks lying almost flat. The rock ages span the entire Paleozoic Era,

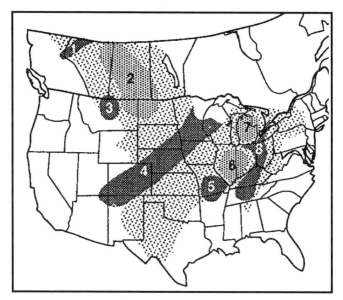

Figure 58. The rock layers in the Stable Interior are gently warped into major structural (not topographic) arches, domes, and basins. 1, the Peace River Arch; 2, the Williston Basin; 3, the Montana Dome; 4, the Transcontinental Arch; 5, the Ozark Dome; 6, the Illinois Basin; 7, the Michigan Basin; 8, the Cincinnati Arch.

from the Cambrian Period to the Permian Period. In the northern part of the province, deposits from Ice Age glaciers cover most of the bedrock and smooth the topography. Pressures in Earth's crust have warped the rock layers into shallow arches, domes, and basins, so the layers are not all truly horizontal.

The Stable Interior is a region of plains, prairies, and plateaus. Over the prairie and plateau portions discussed here, the land surface descends gradually from a maximum elevation of about 1,500 to 1,800 feet (460–550 m) at the western edge of the Central Lowland to about 300 to 400 feet (90–120 m) in southern Illinois and Indiana. It then rises gradually to over 4,000 feet (1,220 m) in the Catskill Mountains and the mountains of West Virginia.

Mostly flat to rolling prairie country characterizes the

Central Lowland from Texas to Manitoba (Plates 10, 11). In the Appalachian Plateaus, erosion has carved the almost horizontal formations into hills and valleys south of the region covered by Ice Age glacial till and outwash (Plate 9). North of the glacial boundary (Figures 64, 66), material deposited by the melting ice caps filled in most of the valleys, rendering the Stable Interior topography relatively flat except for steep-walled valleys cut by the larger rivers and streams (Plates 9 and 38).

In central Ohio, the ice caps left an average of 50 feet (15 m) of deposits on the uplands and an average of 200 feet (60 m) in the valleys. Most of the glaciated topography is gently undulating to lumpy, relieved here and there by long blunt ridges of terminal moraines, smaller sinuous ridges of eskers, and in a few places, elongated hills that are drumlins. Lakes, large and small, also characterize the glaciated region.

The Stable Interior's subsurface structure shows best in the many escarpments, which form long cliffs or elongated rows of hills. Their steep slopes are broken, eroded cross sections of the slightly tilted layers, while the gentle slopes behind them represent the tilted bedding planes or upper surfaces of the layers (see Figure 73 and Plate 13). Gibraltar Island in Lake Erie (Figure 60) has an escarpment at its high end, to the left in the figure.

◆The Rocks of the Stable Interior

Some sandstone, shale, and conglomerate of the Stable Interior started out as river deposits on ancient continents and subcontinents. Other sandstone and shale originated as sand and silt that had been carried out into the oceans surrounding the land masses. Limestone formed with the help of organisms in the warm, shallow oceans. Chemical changes turned limestone into dolomite. Gravel that accumulated along some ocean shorelines solidified into conglomerate. Plates 12–15, 45, and 46 show Stable Interior rocks.

As the Cambrian Period began, massive amounts of sand were being deposited, so thick marine sandstone formations of that age accumulated in the Stable Interior. Younger marine formations have lesser amounts of sandstone. Many of these ocean-bottom formations contain fossils that show what the environment was like (Plate 47). The presence of shellfish and others indicates relatively shallow water; the presence of coral skeletons is evidence that the water was warm. The shallow body of saltwater in which the Stable Interior rocks formed was a continental, or inland sea, not deep open oceans like the present Atlantic and Pacific.

The sediments of other formations accumulated on land or in the ocean close to land. The periodic rising of arches and domes and the rising of highlands to the east during the Paleozoic orogenies gave birth to the land and land-based marine sediments. The sediments turned into conglomerate, sandstone, and shale, many of them redbeds whose color from weathered iron suggests warm, seasonally moist climates. Some sandstones show ripple marks and river channel and delta-type cross bedding (Figure 7), all evidence of onshore and nearshore deposition. Some conglomerates originated on beaches, where heavy wave action removed smaller grains, broke up soft pebbles, and rounded the harder pebbles. The terrestrial deposits include many coal seams that developed in freshwater swamps.

◆Finding Hidden Structures

Aside from the presence of escarpments, evidence for arches, domes, and basins lies in the ages of the rocks. In an anticline or dome, the oldest rocks are in the middle, and in a syncline or basin, the oldest rocks are on the outside, or flanks. Figure 59 illustrates how you can apply this general rule. Suppose you travel from Cumberland, Maryland, across Ohio and Indiana into Illinois. Certain fossils are characteristic of certain geologic periods, so if you find the right fossils, you can determine where a rock layer fits in the geologic time scale. The fossils in outcrops along the route show that from Cumberland to the Ohio River, the rocks get younger. At Cumberland, the fossils are Silurian and Devonian in age. East of Grafton, West Virginia, you pass through a thin strip of Mississippian rocks into a broad area of Pennsylvanian ones. At the Ohio River, you are in the middle of a large area of Permian shales and sandstones. From Cumberland to the Ohio River, you have passed from Silurian to Devonian to Carboniferous to Permian rocks: from older to younger.

As you go westward across Ohio, you see that the rocks become older; the fossils change back to Pennsylvanian, then to Mississippian, and then back to Devonian and Silurian. In passing through Silurian rocks to Permian and back to Sil-

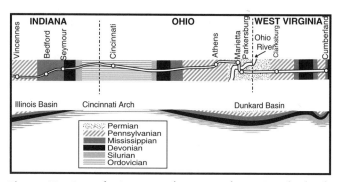

Figure 59. A geologic map of a route from Cumberland, Maryland, to Vincennes, Indiana, with the corresponding geologic cross section. The route traverses a structural basin and a structural arch. The west slope of the Cincinnati Arch is the east portion of the neighboring Illinois Basin.

urian, you have traveled across a structural basin, the Dunkard Basin. The rock layers are warped downward, with the youngest in the middle and the oldest on the outside.

At Cincinnati, the rocks are Ordovician, still older. Then the trend changes again. In western Ohio, Silurian fossils return, and in Indiana, Devonian fossils show up again, in a thin strip of rocks, followed by a strip of Mississippian ones. About halfway between Bedford and Washington, you go into Pennsylvanian sediments, which continue into Illinois. From the Pennsylvanian to Ordovician in Ohio (younger to older), and back to Pennsylvanian in Indiana (older to younger) you have traversed a broad structural arch, with the oldest rocks in the middle. You have then entered another basin.

As you have driven from Maryland and West Virginia to Illinois, you have driven over the Dunkard Basin, the Cincinnati Arch, and into the Illinois Basin. But there are no large-scale ups and downs, just the normal hills and valleys. The basins and the arch are geologic structures that have been leveled by erosion, not topographical ones.

The Cincinnati Arch is a long and broad structure whose axis trends northeastward from west-central Tennessee into southern Ontario (Figure 58). You can see the tilting rock layers of the arch's southeast flank on the islands in Lake Erie near Sandusky, Ohio. One of them is illustrated in Fig-

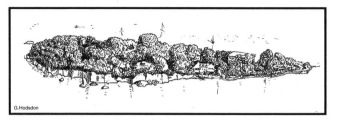

G.Hodsdon

Figure 60. Gibraltar Island, in Lake Erie north of Sandusky Bay, Ohio, shows the degree to which the rock layers on the Cincinnati Arch are tilted. The view is toward the southeast. Bass Island Dolomite of Late Silurian age dips into the lake from left to right, on the southeast flank of the arch.

ure 60. The rock outcrops as cliffs on the northwest coasts of the islands, but on the opposite sides, the limestone and dolomite layers dip gradually into the water. The rocks tilt downward toward the southeast; their southeastern extension is far below the surface in the Dunkard Basin, covered by Mississippian, Pennsylvanian, and Permian layers.

■HISTORY OF THE STABLE INTERIOR

Most Stable Interior rocks east of the Great Plains range in age from Cambrian to Permian. During the Paleozoic Era, the arches and domes were at various times either islands in the Paleozoic ocean or areas of shallower water. Generally, the marine sedimentary rock layers tend to thin on their flanks and crests, and thicken in the basins between them. Where the sedimentary rock layers pinch out entirely, the ocean bottom must have risen above sea level, forming islands.

Geologists have found several lesser arches, domes, and basins tucked in between the major ones. The large-scale rippling of the Paleozoic rock layers suggests that the Stable Interior has not been truly stable. Crustal forces not only caused the arches, domes, and basins, but also periodically uplifted some regions from below sea level to above sea level and lowered sea level to expose other areas: land sank under the sea, or the sea level rose. The stresses of warping and uplifting caused numerous faults.

The folding and uplifting of the Stable Interior is indicated by many unconformities, or indications of missing rock layers. The unconformities show that the land rose or the sea level sank, so ocean bottom became dry land. If any deposits accumulated on the land and were eroded away, there was no record of that time period. The lack of evidence is the unconformity. Nowhere in the interior can you find a complete, unbroken set of Paleozoic sediments, from Cambrian to Permian; no part was under water or above sea level continuously throughout the Paleozoic Era.

Earth's crust is still active in the continental interior. For the most exciting evidence of this fact, you need only go back to the winter of 1811 and 1812, in the vicinity of New Madrid, along the Mississippi River in southeastern Missouri. (New Madrid is actually on the Coastal Plain but is very close to the Stable Interior, whose rocks are underneath those of the Coastal Plain.) From December 1811 to March 1812, a series of earthquakes shook a fault zone that is presumably associated with the breakup of Pangaea. As America, Africa, and Europe have been separating, crustal stresses have affected our continent far inland from the coastline. The New Madrid quakes were serious ones. It has been estimated that the famous San Francisco earthquake of 1906 would have registered 8.25 on the Richter Scale, the modern scale of earthquake magnitude. Using descriptions written at the time and modern evidence, geologists have estimated three of the New Madrid quakes at 8.4, 8.5, and 8.8. That's about as bad as quakes get.

The New Madrid quakes were the most powerful in United States history. The depression filled by Reelfoot Lake, in the northwestern corner of Tennessee, appeared out of nowhere as a fault block dropped, to the astonishment of the local inhabitants. Some effects of the quakes were noted as far away as Washington, D.C.

At the beginning of the Cambrian Period, the ancestral core of North America was a land mass with its highest portion in what is now the Canadian Shield (see the endpaper map). The bedrock of the land mass was a variety of igneous, metamorphic, and sedimentary rocks that had weathered into a thick soil of sand, silt, and clay.

During the Cambrian Period, a shallow sea encroached upon the land from the east, south, and west. The new ocean was less than 100 feet (30.5 m) deep in most places, with some islands in it. By the end of the period, a quarter-moon shaped arc of ocean-bottom sand deposits had developed, an arc that extends roughly from the present Montréal to Kansas to the Northwest Territories. To the south and east, the sandstone that formed from the sand grades to limestone; to the west, it grades to limestone along with shale that originated as silt and clay on the weathering land surface.

During the rest of the Paleozoic Era, the ocean advanced and retreated over the Stable Interior, responding to crustal warping and changes in sea level. At different times, different parts of what is now North America were under the shallow seas. The greatest development of the Paleozoic continental sea occurred during the last third of the Ordovician Period, when saltwater covered most of the crust from what is now the Arctic to the Gulf of Mexico and from the Atlantic to the Pacific. The Devonian inland ocean was also extensive, as was that of the Mississippian Epoch of the Carboniferous Period.

Three times during the Paleozoic Era, islands or larger land masses slowly crashed against the eastern portion of the continental core. These episodes of crustal crumpling were the Taconic, Acadian, and Alleghenian orogenies, which took place during Late Ordovician, Late Devonian and Early Carboniferous, and Late Carboniferous and Permian times. The Alleghenian Orogeny finished the creation of the supercontinent Pangaea. During orogenies, the shallow continental ocean decreased in size as more of the Stable Interior's surface rose above sea level. The surface probably rose partly because the ocean basins were being filled with sediments from land along the margins of the growing continent and partly because of crustal warping.

The westward retreat of the continental sea during the Alleghenian Orogeny is shown by variations in rock types of the same age from east to west. Pennsylvanian age rocks occur in cycles of marine and terrestrial (freshwater) deposition, as shown in Figure 61. The cycle goes from underclay and coal to marine limestone to terrestrial sediments back to underclay and coal, and so forth. Figure 61 shows an example from Ohio, in the eastern part of the Stable Interior. The relative amounts of marine and terrestrial rocks change from east to west. In the east, marine beds are thin and occasional. Most of the rocks consist of terrestrial sediments from the growing Alleghenian Orogeny land mass to the east. To the west, in Kansas and Nebraska, the Pennsylvanian sediments are mostly marine, that area having been too far from new eastern (and southern) uplifts to have risen very often above sea level.

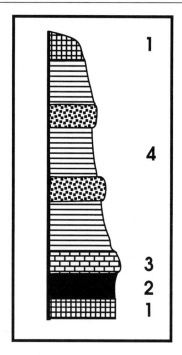

Figure 61. A typical Pennsylvanian outcrop in the eastern coal-mining region. The diagram illustrates one rock cycle and the beginning of another, probably similar to the first. The rock layers are: 1, clay-shale, the "underclay" beneath a coal seam, containing fossil plant roots and much organic matter presumably representing the soil of the coal swamp. 2, a coal seam; 3, limestone, in most cases marine; 4, terrestrial shale and sandstone.

The Alleghenian Orogeny climaxed during the Permian Period, with widespread uplift in the Appalachian Province in the east and south. In the west, uplifts in the Rocky Mountain area started during Pennsylvanian time. The Permian ocean lay mostly in Kansas, Oklahoma, and west Texas. The rest of the interior apparently had been raised above sea level at that time.

In the Great Plains, Mesozoic and Cenozoic rocks cover the Paleozoic continental, or inland shallow sea deposits.

The younger rocks consist of sediments and volcanic rocks derived from uplifted areas in the Rocky Mountain region. The Triassic inland sea was restricted, but during the Jurassic Period, the ocean grew again, to cover much of the Great Plains south of the Canadian border.

During the Cretaceous Period, the inland ocean continued to expand, advancing as far east as Minnesota and Iowa and as far north as Lac des Bois, north of Great Bear Lake in the Northwest Territories. Toward the period's end, the shoreline east of the rising Rocky Mountains moved eastward. River deposits that accumulated inland from the shore contain the well-known late Cretaceous dinosaur remains of Wyoming, Montana, and Alberta, evidence of the large dinosaur population that lived on the plain between the mountains and the seashore.

As the Tertiary Period began, the venerable ocean that had dominated the Stable Interior for nearly half a billion years disappeared. Cenozoic deposits in the Stable Interior, which are mostly in the Great Plains, are all terrestrial.

The Stable Interior can be conveniently divided into three subprovinces: the Appalachian Plateaus, the Central Lowland, and the Great Plains (Figure 57). In the east is the hilly terrain of the two adjoining Appalachian Plateaus, the Allegheny in the north and the Cumberland in the south. The plateaus contain unfolded western extensions of the crumpled Paleozoic rock layers that characterize the neighboring Appalachian Mountains.

What is in many places a distinct escarpment forms the northern and western border of the Appalachian Plateaus. It runs from the west side of the Hudson Valley, curving westward to south of Utica, New York, to south of Buffalo, to the east edge of Cleveland, Ohio, southward to just east of Columbus, to near Portsmouth on the Ohio River, through Kentucky east of Somerset, through Tennessee, and down into northern Alabama.

The western part of the Stable Interior is the Great Plains, bordered on the west by the Rocky Mountains, and extending from Texas to the Northwest Territories. A varyingly distinct escarpment forms the eastern margin of the Great Plains, running (very roughly) from San Angelo, Texas, north to Amarillo, east and north through the Oklahoma panhandle, through Wichita, Kansas, and north to west of Lincoln, Nebraska. The border then curves into Iowa and Minnesota, continues north near the Minnesota–North Dakota line, then northwest through Manitoba close to the west coasts of lakes Winnepegosis and Manitoba and northwest through central Saskatchewan and northern Alberta.

The topography of the Central Lowland, between the Great Plains and the Appalachian Plateaus, is intermediate between the other two subprovinces, with prairies and low hills.

■ THE APPALACHIAN PLATEAUS

Except for New York and northern Pennsylvania, the Appalachian Plateaus bedrock is mostly a vast expanse of Pennsylvanian sediments, with a surrounding strip of Mississippian rocks cropping out around the edge and an island of Permian rocks in the center of the Dunkard Basin.

In the region as a whole, either terrestrial or shallow marine deposits make up the bedrock. Most of the formations

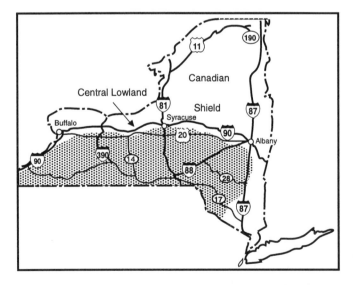

Figure 62. The Allegheny Plateau in New York. The plateau bedrock is almost entirely Devonian in age.

contain sandstone and shale created by erosion of new land that was raised as the eastern subcontinent Gondwana collided with the North American subcontinent Laurentia. That event closed up the earlier Atlantic Ocean and helped form the supercontinent Pangaea during the Alleghenian Orogeny.

◆ The Allegheny Plateau

The Allegheny Plateau comprises the flat-topped highlands of southern New York, northern and western Pennsylvania, eastern Ohio, and West Virginia.

● New York

In New York, the Allegheny Plateau is a table land that has been cut by weathering and erosion into significant hills and valleys and greatly modified by the glaciers of the Ice Age (Figure 62). The plateau occupies most of New York south of I-90 and west of I-87. U.S. 209 from Port Jervis to Kingston and I-87 from Kingston to Albany run right along the plateau's eastern margin. I-90 from Albany to Buffalo parallels the northern margin mostly a few miles to the north of it, and I-90 from Buffalo to the Pennsylvania border is even closer to the margin.

Devonian limestone, shale, and sandstone form almost all of the plateau bedrock in New York. In many places, fossils of marine shellfish and corals proclaim the oceanic origin of the rocks. Farmers in the western part of the state once constructed rock walls from limestone slabs that are studded with the remains of ocean animals. These Devonian formations dip gradually southward from the north edge of the plateau, so they get younger (from Middle Devonian to Upper Devonian) from the northern escarpment to the Pennsylvania border. The Upper Devonian rocks of southeastern New York are the Catskill redbeds, largely red conglomerate, sandstone, and shale that originated as sediments in the great delta system that developed from the highlands caused by the Acadian Orogeny to the east.

● Pennsylvania

Northern and western Pennsylvania are continuations of the Allegheny Plateau in New York (Figure 63, Plate 9). The margin of the plateau in central Pennsylvania is the spectacular Allegheny Front, the escarpment formed by resistant Pennsylvanian Pottsville Sandstone, whose edge towers as far as 1,000 feet (305 m) above the valley floor that borders the Appalachian Mountains. The sandstone accumulated from weathering and erosion of highlands to the southeast. The plateau behind the escarpment has been dissected into hills and valleys by the Allegheny, Monongahela, and Youghiogheny River systems.

New York's south- and southwest-dipping Devonian rocks continue into the northern part of Pennsylvania and disappear under the overlying Mississippian formations. The Mississippian formations outcrop in a jagged and narrow strip, and they soon pass below the Pennsylvanian-age rocks.

North-central and northeastern Pennsylvania have Devonian rocks similar to those of contiguous New York in that

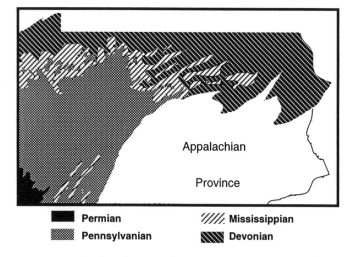

Figure 63. The distribution of Devonian to Permian rocks in the Pennsylvania Allegheny Plateau region.

their sediments came from erosion of the land uplifted during the Acadian Orogeny. They consist of red to greenish gray terrestrial sandstone and shale, with occasional conglomerate lenses and shoreline delta-type sediments. Farther west, the sediments accumulated offshore; the Erie area in northwestern Pennsylvania shows black shale, with some gray shale and sandstone, composed of sand, silt, and clay deposited in the late Devonian sea.

The Mississippian formations of Pennsylvania are mostly sandstones, conglomerates, and shales, some of them redbeds, deposited in the terrestrial environment caused by continuing uplift of the Acadian Orogeny. The rocks grade to a higher percentage of marine origin toward the west.

Most of the state's western half has Pennsylvanian-age bedrock. Its nearly horizontal layers consist of terrestrial sandstone and shale, with coal (Plate 16) and some marine limestone that indicates minor temporary incursions of shallow seas. The coal and the many fossils of land plants in the shales prove that during most of the Pennsylvanian Epoch, what is now the western Pennsylvania plateau was above sea level. By the end of the Pennsylvanian Epoch, the eroded remnants of Acadian mountains were being replaced by new ones caused by the continental collision of the Alleghenian Orogeny.

Down in the southwest corner of the state, two Lower Permian formations cover the Pennsylvanian rocks in the eastern part of the Dunkard Basin. Sandstone and shale, with some redbeds and a couple of coal seams, make up the Permian series. Coal and fossils of amphibians, reptiles, and land plants mark the formations as deposits formed on land created by the Alleghenian Orogeny.

● Ohio

In Ohio, the rock layers dip south, southeast, and east into the Dunkard Basin (Figure 64). The eroded edge of the Mississippian formations is the boundary of Ohio's Allegheny Plateau. It more or less parallels Lake Erie's shore from the Pennsylvania border to about the Lorain-Erie county line, then sweeps in a curve southward past Norwalk, Richland, and Columbus to the Ohio River west of Portsmouth.

Northeastern Ohio is a relatively flat highland, with a few deeply incised river valleys and broad flat higher areas in between (Plate 9). Till and outwash from the glaciers that cov-

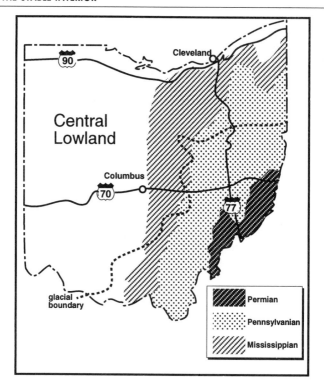

Figure 64. The glacial boundary and distribution of Mississippian to Permian rocks in the Ohio Allegheny Plateau.

ered the region filled in the lower parts of the preglacial topography and rendered it almost level. Glaciers did not reach the rest of eastern Ohio, which therefore has a hilly countryside like that of West Virginia and Pennsylvania, unmodified by glacial debris.

In the Chagrin River valley, east of Cleveland, erosion has revealed a large segment of geologic history in a limited area. The river exposed a section of local bedrock as it cut down through the edge of the Allegheny Plateau (Plate 12). The formations there show how an ocean basin became dry land, as continents collided to the east to form new land. The two oldest formations in the valley are shale, deposited toward the end of the Devonian Period on the bottom of a shallow ocean, far from the shoreline (Plate 12). In contrast, the Late Devonian rocks in Pennsylvania and eastern New York are redbeds and other terrestrial and near-shore deposits associated with the building eastern land mass. What is now Ohio was out in the ocean.

The third formation consists of a Lower Mississippian shale interbedded with layers of fine-grained marine sandstone. The sandstone shows that sand grains were beginning to reach the area of today's Chagrin Valley, instead of only the smaller silt grains and clay. The smaller grains were carried farther west. During the Mississippian Epoch, the shoreline to the east must have been moving westward as mountains on the eastern land mass grew, and sediment from the eroding land filled the neighboring ocean basin.

Going upward, Mississippian sandstone constitutes the next formation, with delta-type cross bedding; deltas indicate the shoreline was nearby (Plate 12). Fossils of leaves and herbaceous stems in the sandstone represent plant fragments that rivers must have carried from the land mass into the sea. The rocks show that as the Acadian Orogeny land mass grew, sediments in Ohio changed from shale to shale/sandstone to sandstone.

You can see these four formations in the area where U.S. 422 crosses the Chagrin Valley. Farther upstream, later Mississippian marine shale outcrops, indicating a temporary episode in which the ocean bottom and nearby land must have sunk, or the sea level rose, and sand no longer reached eastern Ohio.

Then the general progression of sea to land revived. The youngest rock in the area, capping the hills east of the Chagrin Valley, is Early Pennsylvanian conglomerate and sandstone. The formation's lower portion consists of massive layers of conglomerate with rounded pebbles and sand cemented together, which possibly originated as a seashore gravel. Mostly coarse terrestrial sandstone constitutes the formation's upper portion, with what appear to be elongated bodies of solidified gravel from old river channels. The formation contains fossil impressions and natural casts of land plants. According to the testimony of the rocks, the shoreline had resumed its westward movement and had passed by what is now the Chagrin Valley, leaving that area part of the new continent. The edge of the ocean was receding to the west during the renewed continental uplift caused by the Alleghenian Orogeny.

● West Virginia

The Allegheny Plateau subprovince includes most of West Virginia (the eastern edge and northeastern "spout" of the state are in the Valley and Ridge Zone of the Appalachian Province). The topography looks much too hilly for a plateau, until you notice that the hilltops are all at about the same elevation (Plate 9). Streams and rivers have cut deeply into the almost horizontal sedimentary rock layers, and erosion of valley walls has created a closely packed mass of hills. In some areas, it seems as if there is no flat land. This is country where, they used to say, the cows have shorter legs on one side than on the other, so they can graze the hills.

The greatest erosional phenomenon in West Virginia is the New River Gorge, an outstanding sight in the West Virginia mountain country north of Fayetteville. The gorge is 876 feet (267 m) deep at the bridge on U.S. 19, and deepens to a maximum of about 1,400 feet (427 m). You can admire the gorge from the visitor center and museum at the northeast end of the U.S. 19 bridge. Another good view of the gorge is at Hawks Nest State Park, just west of Ansted on U.S. 60.

Northern West Virginia lies in the Dunkard Basin, and Permian shale, sandstone, and conglomerate outcrop there, continuing the formations in Pennsylvania and Ohio. As in those states, the rocks contain fossils of terrestrial animals and plants. The bedrock of the rest of the state is Pennsylvanian sandstone, shale, coal, conglomerate, and limestone whose outcrops surround the younger Permian rocks as you go outward from the center of the basin. The four Pennsylvanian formations dip slightly downward toward the north and northwest, tilted by the orogenic forces that so sharply folded the neighboring Appalachian Mountains.

Most of the Pennsylvanian rocks solidified from terrestrial, or freshwater, sediments deposited in rivers and flood plains, lakes, and swamps of a land area close to sea level. Each of the West Virginia formations contains coal seams interbedded with sandstone, shale, and limestone. The coal and some of the shale evolved from the obviously extensive lowland swamps, with their lush tropical vegetation. Some limestone layers apparently formed in lakes, but occasional layers of marine limestone record minor periodic movements of the sea onto the lowland.

● Evidence of Glaciers

The northern part of the Allegheny Plateau was glaciated in the Ice Age. The glaciers left abundant evidence of their passing, including deposits of till and outwash everywhere.

Except for a small triangular area with Salamanca at its apex, till, outwash, and moraines cover all of New York. The most famous glacial effects in New York are the Finger Lakes and the great drumlin field. The latter covers an area roughly from Syracuse to Batavia and from the Lake Ontario shore down into the Finger Lakes region (Figure 65).

The 11 Finger Lakes occupy unusually long and narrow, almost parallel valleys with high steep sides, a combination of characteristics unique in North America. The lakes are unusually deep, too; the floors of Lake Seneca and Lake Cayuga rest below sea level, yet the elevation of Seneca's surface is 444 feet (135 m) and that of Cayuga is 384 feet (117 m). They probably formed when Ice Age glaciers flowed down valleys and scoured them out by abrading and eroding the weaker rocks in the valley walls and floors.

In Pennsylvania, ice caps covered the northwest corner and the northeast quarter of the state with till and outwash (Figure 66). The glacial boundary, of course, is a moraine. You can see the West Liberty esker and other glacial features at Moraine State Park, on U.S. 422 just east of I-79.

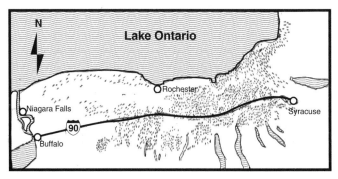

Figure 65. The drumlins in the drumlin field of New York are elongated in the direction of ice flow. They show ice flowing to the southeast, south, and southwest from the Lake Ontario Basin during the last glacial stage.

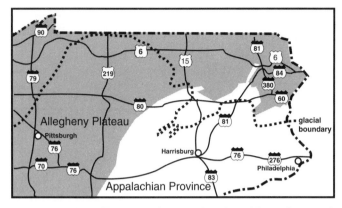

Figure 66. The Allegheny Plateau, the glacial boundary, and highways in Pennsylvania.

North of the Ohio glacial boundary (Figure 64), a clear system of terminal moraines generally parallels the Lake Erie shore, from Ashtabula to south of Cleveland and north of Akron, and extending westward between Ashland and Norwalk. Here, the retreating glacier stabilized periodically to build a series of moraines in one elongate zone, before it melted back farther to expose the Lake Erie Basin (Figure 82).

● Caves

The Allegheny Plateau is not noted for caves, but wherever you see limestone, you are likely to find them. In New York, limestone outcrops around the margin of the Appalachian Plateau, and I-88 between Cobleskill and Central Bridge traverses a strip of the Lower Devonian limestone group that outcrops to the south at Kingston. Howe Caverns and Secret Caverns are a few miles north of the highway from the Howes Cave exit.

West Virginia has some caves, too, in Mississippian limestones along the edge of the plateau. Organ Cave, just off U.S. 219 near Ronceverte, south of I-64, was visited by Thomas Jefferson in 1778. Tradition has it that Jefferson found a large skeleton there, "possibly a dinosaur." More likely he found bones of a mammoth or mastodon, Ice Age relatives of modern elephants. There are no dinosaur fossils in the area.

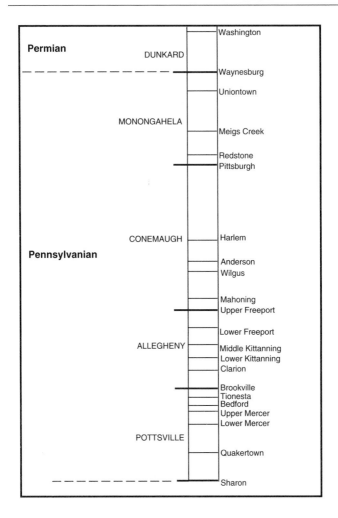

Figure 67. This chart of Ohio coal seams suggests how many plant-filled coal swamps there were during the latter half of the Carboniferous Period and beyond. The coal seams are interbedded with shale, sandstone, and limestone in the Early Pennsylvanian Pottsville Group of formations through Middle and Late Pennsylvanian groups to the Early Permian Dunkard Group.

● Coal

Since the 18th century, people have mined coal on the Allegheny Plateau. In Pennsylvania, coal has been dug from both underground mines and surface strip mines in Pennsylvanian-age rocks, particularly in four certain coal layers. The most famous is the Pittsburgh Seam, a bed consistently 4 to 6 feet (1.2 to 1.8 m) thick. The seam is now mostly worked out in many areas, but it has been a major source of coal from some large underground mines.

Strip pits and underground mines in Ohio have produced Pennsylvanian and Permian coal. Figure 67 shows the major seams. The coal, with its accompanying sandstone and shale, is prime evidence for the Alleghenian Orogeny landmass Pangaea.

The 62 mineable coal seams in West Virginia are another remnant of the growing Pangaea. They indicate the abundance and westward spread of plant-filled swamps that developed as the uplift continued to the east and the ocean retreated to the west. Most of the Allegheny Plateau in West Virginia is coal mining country. The thickness of Pennsylvanian rocks increases as you go southeast in the state, and so does the number of coal seams. The seams reach a thickness of 10–12 feet (3–3.6 m).

ALLEGHENY PLATEAU HIGHWAY SECTIONS

New York; I-90 to New York 17: I-390 (Figure 62) *I-390 from I-90 to State Route 17 affords a north-south geologic cross section of the northern Allegheny Plateau. It starts in Middle Devonian rocks. Since the layers dip southward, you encounter successively younger ones as you travel, ending in Upper Devonian sediments.*

Going south, I-390 cuts through the escarpment formed by lower Middle Devonian Onondaga Limestone, at the Allegheny Plateau's edge. Later Middle Devonian rocks, mostly shale, overlie the Onondaga, and thus outcrop beyond it.

Letchworth State Park, west and southwest of Mt. Morris, includes a gorge of the Genesee River that is 22 miles long and up to 550 feet deep (35 km, 168 m). There are three major waterfalls, where the river cascades over resistant sandstone layers in a mostly shale formation. Late Devonian rivers washed silt, clay, and sand into the ocean from rising eastern land during the Acadian Orogeny. The resulting shale contains fossils of land plants that floated out into the ocean.

Between Mt. Morris and Dansville, I-390 enters the same Late Devonian rock series that occurs at Letchworth Gorge, so between the Onondaga Limestone escarpment and the Dansville area, the route passes from Middle Devonian into Upper Devonian rocks. Along this stretch, I-390 follows a broad old valley cut by a former Genesee River branch at the end of the Ice Age, when a tremendous amount of meltwater flowed from the last glacier. The drainage changed after the ice disappeared, and the once large, turbulent river turned into a tiny creek in the large valley.

Between Dansville and Wayland, the road climbs 700 feet (213 m) over a moraine of the last glacier. More Upper Devonian sandstones and shales outcrop beyond Wayland.

Oneonta to Kingston, New York: Route 28 (Figure 62) *New York Route 28 traverses the northeastern corner of the Allegheny Plateau, between Oneonta and Kingston, and furnishes a northwest-southeast section of the Catskill Mountains. Erosion carved the Catskills into hills and valleys from a broad uplifted syncline in Devonian conglomerate, sandstone, and shale that formed on the flank of the mountains raised by the Acadian Orogeny. Some of the peaks rise over 4,000 feet (1220 m). The eroded edges of resistant rock layers form scarps of varying heights. The highest of these is the eastern margin of the mountains, which overlooks the Hudson Valley.*

A spur of conglomerate and shale that juts out at the base of the Catskills' eastern scarp from between Rosendale and New Paltz forms the Shawangunk Mountains. The Shawangunks are more properly described with the Central Lowland.

Oneonta rests on shale and sandstone of earliest Upper Devonian age. The road climbs into a higher and younger group of shales and sandstones at Meridale, and enters a still younger group beyond Delhi. Redbed sandstones and shales between Delhi and Arkville provide evidence of the land mass formed during the Acadian Orogeny, since they were deposited on land. Road cuts between Highmount and Big Indian show nonred conglomerates and cross-bedded sandstones that are slightly older than the redbeds. The cross bedding in the sandstone suggests it formed in a seashore delta.

Between Big Indian and Phoenicia, the road follows a curving valley in a valley system that forms an almost complete circle with a diameter of about 6 miles (10 km). The valleys may outline a meteoroid crater of Devonian age that has been uncovered by weathering and erosion. According to the theory, the circular valleys originated in the cracked and broken rock of the crater rim, where weathering and erosion worked faster than it did in the less fractured surrounding rocks.

From Phoenicia to Boiceville, the highway returns to older Devonian rocks on the eastern side of the Catskill syncline. As it descends to Kingston, the route passes through Middle Devonian marine shale and sandstone and into Lower Devonian limestone.

New York–Pennsylvania border to Scranton, Pennsylvania: I-81 (Figure 66)

In Pennsylvania, I-81 between the New York border and Scranton has many outcrops of the Upper Devonian Catskill Formation, an extension of the New York Catskill beds. The reddish to greenish gray sandstones and shales are terrestrial river and marine delta deposits from the growing landmass of the Acadian Orogeny.

New Stanton to Somerset, Pennsylvania: I-76 (Figure 66)

As in the rest of the Stable Interior, horizontal pressure warped the Pennsylvania formations into broad gentle folds. The folded structure shows on the surface in narrow elongate strips that extend roughly from Somerset, Nanty Glo, and Homer City to the southwest. The first and third extend beyond the state border, and the middle one stops just before the border. In each of the strips, the formations have been arched so that the oldest Mississippian rocks outcrop along the axis, with successively younger Pennsylvanian ones on the flanks. I-76, the Pennsylvania Turnpike, crosses the central and western anticlines between New Stanton and Somerset. Laurel Hill, through which the turnpike tunnels, is part of the central one.

Rummerfield to Ansonia, Pennsylvania: U.S. 6 (Figure 66)

U.S. 6 between Rummerfield and Towanda crosses a typical Stable Interior gentle anticline. In the center of the anticline the road passes a strip of olive-gray shale, sandstone, and conglomerate that underlies the Catskill Formation. Typical red and green younger terrestrial Catskill beds extend eastward and westward on the nearly level flanks of the fold.

A few miles south of U.S. 6 at Ansonia, Pine Creek has carved a gorge, known as the Grand Canyon of Pennsylvania, or Pine Creek Gorge. The gorge, roughly 600 to 800 feet (180–245 m) deep, cuts through Upper Devonian rocks of the Catskill Formation.

Matamoras to I-380, Pennsylvania: I-84 (Figure 66)

I-84, from Matamoras to I-380, traverses the Pocono area. At Matamoras, the Allegheny Plateau's eroded edge shows as the escarpment on the northwest side of the Delaware River valley. An outcrop of the Silurian Shawangunk conglomerate as in the Shawangunk Mountains to the north forms Kittatinny Mountain, the ridge just over the river in New Jersey. The highway crosses the Front between Milford and Lord's Valley, and the Catskill terrestrial reds and greens appear above the less colorful Middle Devonian marine rocks.

Erie, Pennsylvania to Morgantown, West Virginia: I-79 (Figure 66)

I-79 offers a full north-south cross section of the Allegheny Plateau from Erie south to Morgantown. The beds tilt slightly southward, so from north to south, the road passes over older to younger rocks, from Devonian to Permian.

The route starts in Upper Devonian marine shale and sandstone (mostly covered by glacial deposits). At Meadville, the highway enters the zone of Mississippian marine shale and sandstone. Beyond State Route 358, mostly terrestrial Pennsylvanian sandstone and shale constitute the bedrock. At U.S. 422, near the Moraine State Park, the glacial boundary crosses I-79. From there past Pittsburgh to Washington the rocks are Pennsylvanian. Permian terrestrial sandstone and shale extend from Washington into West Virginia.

Cleveland to Marietta, Ohio: I-77 (Figure 64)

I-77 follows typical plateau structure in Ohio from Cleveland south to Marietta. The rock layers along this highway gradually tilt downward generally to the southeast, so the route crosses from older (Devonian) to younger (Permian) formations.

The bedrock at Cleveland is a blue-gray Upper Devonian shale overlain by a black shale that outcrops along the border between the Allegheny Plateau and the Central Lowland rock series. The black shale, the Cleveland Shale, is most unusual in that it contains the fossilized bones of very primitive fishes not related to any fish living today. The formation also preserves clear impressions of early sharks (Figure 68).

The Cleveland Museum of Natural History has on exhibit examples of these rare fossils.

After leaving downtown Cleveland, the highway climbs into Mississippian marine sandstone and shale bedrock that extends from Cleveland to Akron. Pennsylvanian rocks occur beyond Akron, mostly sandstone and shale, with coal and a little limestone. The terrestrial rocks originated in river channels, flood plains, and low-lying swamps. The sandstone and shale contain fossils of land plants and animals. The road crosses the glacial boundary at Canton.

Southward, the road passes over younger and younger Pennsylvanian rocks. About 10 miles (16 km) north of Mari-

Figure 68. Fishes of the Late Devonian Cleveland Shale; the primitive bony fish *Dunkleosteus* and two individuals of the ancestral shark *Cladoselache.*

etta, the rock becomes Permian shale and sandstone, including redbeds. Here, a few rare fossils of lungfish (identified by their characteristic tooth plates), amphibians, and reptiles prove that the sediments are terrestrial. Lungfish must have used their lungs to survive in the mud when ponds or streams they lived in temporarily dried up. Modern ones burrow in the mud and breathe through air holes.

Columbus, Ohio to Elm Grove, West Virginia: I-70 (Figure 64)

I-70 provides an east-west section of the Allegheny Plateau rocks that extends from the southeast flank of the Cincinnati Arch to the Dunkard Basin. The rocks dip toward the southeast, so from west to east, the highway crosses from older (Mississippian) to younger (Permian) formations.

The Allegheny Plateau margin on this highway appears a few miles east of Columbus; it runs north and south in the area of the Reynoldsburg exit (Ohio Route 256). Mississippian marine formations, mostly shale and sandstone with limestone in the latest beds, form the bedrock eastward from the plateau margin to the vicinity of Gratiot and Brownsville, west of Zanesville. There, the rocks become mostly terrestrial Pennsylvanian sediments.

The highway crosses the glacial boundary between Gratiot and Hopewell. Pennsylvanian rocks along I-70 belong to the same series that I-77 crosses in this area. The overlying Permian rocks begin to outcrop in the middle of Belmont County, in the vicinity of Belmont Lake. However, erosion by the Ohio River and its tributaries has stripped the Permian rocks away along some sections of the highway. So from Belmont Lake to the Ohio River, you can see both Permian and Pennsylvanian formations.

Across the river in West Virginia, road cuts in the eastern outskirts of Wheeling expose late Pennsylvanian shale, sandstone, and coal, surmounted by similar rocks of early Permian age. East of Elm Grove, the highway rises into the Permian series, which contains fossils of land-living reptiles and plants, along with freshwater clams, snails, and fish.

Morgantown to Huntington, West Virginia: I-79, I-64 *In West Virginia, I-79 and I-64 from Morgantown to Charleston to Huntington skirt the Permian formations and pass through Upper Pennsylvanian terrestrial sedimentary rocks.*

Along I-64 in the western part of the state, shales with a high percentage of organic matter and plant fossils are continuations of layers that turn into coal farther north. The area appears to have been a large coastal swamp, periodically flooded by rivers from the east and by occasional storm tides from shallow nearby ocean embayments.

Two marine limestones with some marine shale outcrop in the Huntington area. Freshwater limestones occur above them. Sandstones outcrop both east and west of Huntington, some in massive layers, others showing pebbles and cross bedding from ancient river channels and deltas.

Elkins to Ripley, West Virginia: U.S. 33

U.S. 33 from Elkins west to Ripley crosses the Allegheny Plateau portion of West Virginia, from the east into the Dunkard Basin. You travel from Devonian sandstone and shale at Elkins to Mississippian sandstone and shale in a very thin strip west of town, through the extensive Pennsylvanian series of rocks to the Permian east of Spencer.

◆ The Cumberland Plateau

The Cumberland Plateau, the southern part of the Appalachian Plateaus, extends the West Virginia hills through parts of Kentucky, Tennessee, Virginia, and Alabama (Figure 57). It lies between the southern Appalachian Mountains and the Central Lowland prairies, eroded into hills and valleys by the Cumberland and Kentucky river systems.

The Cumberland Plateau's eastern boundary starts about at Bluefield, Virginia, and goes in a line approximately from Bluefield to Pennington Gap in Virginia, then continues southwestward through Harriman, Tennessee, to a bit beyond Birmingham, Alabama. The western boundary coincides roughly with the western edge of the Daniel Boone National Forest in Kentucky, then with a line from Jamestown to Winchester in Tennessee, and approximately from Huntsville to Russellville to Tuscaloosa to Birmingham in Alabama.

The Cumberland Plateau formations consist mostly of more or less horizontal Pennsylvanian sandstone, shale, and coal, with a fringe of Mississippian limestone, sandstone, and shale along the plateau's western border. In Tennessee, erosion has broken up the Pennsylvanian layers, and Mississipian rocks show up below them in many valleys. The Alabama outcrop area of Carboniferous rocks becomes much wider and more purely Pennsylvanian. Like those of the Allegheny Plateau, the Cumberland Plateau rocks show the transition from Mississippian shallow sea to developing land during the Pennsylvanian Epoch as the Alleghenian Orogeny commenced.

The grand structural feature of the plateau is the Cumberland Overthrust Block, in the area where Tennessee, Virginia, and Kentucky meet (Figure 69). It is a wide syncline, bounded on all sides by faults. The elongated ridge of Pine Mountain forms the northwestern margin, and the ridge of Cumberland Mountain forms the southeastern edge. During the continental collision of the Alleghenian Orogeny, this block of the crust was pushed from the southeast toward the northwest for about 2 to 10 miles (about 3–16 km). At the leading Pine Mountain edge, Mississippian rocks have been thrust over younger Pennsylvanian layers.

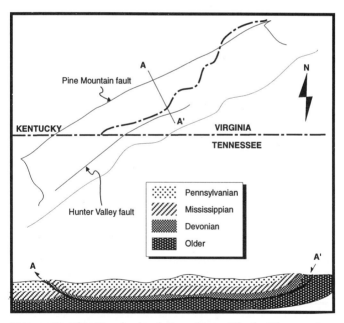

Figure 69. The Cumberland Overthrust Block. The section shows how continued pressure from the southeast arched the fault upward at A' after the block had moved.

The Cincinnati Arch cuts across the Stable Interior west of the Pennsylvanian rocks from Ohio to Alabama. The Mississippian fringe of sandstones, shales, and limestones appears from underneath the Pennsylvanian series all along the arch. In the Tennessee-Alabama area, the Cincinnati Arch and the folded rocks of the Appalachian Mountains approach each other closely, with a basin of Pennsylvanian rocks in between.

Historic Cumberland Gap cuts through Cumberland Mountain between Middlesboro, Kentucky and the town of Cumberland Gap, Tennessee, traversed by U.S. 25E. Geologists believe the gap was carved by an old southward-flowing river that was later captured by a tributary of the Cumberland River. Daniel Boone blazed the Wilderness Road through the gap and opened the country behind the mountains to settlement.

People have been mining Kentucky coal in the eastern Cumberland Plateau, including the Cumberland Thrust Block, since at least 1790. The eastern Kentucky coal field produces from both underground and strip mines. Another mining area occurs in the western part of the state, in the Central Lowland, making Kentucky America's number one producer of coal as recently as 1978.

The Cumberland Plateau portion of Virginia is the only part of that state with active mines, but large amounts of

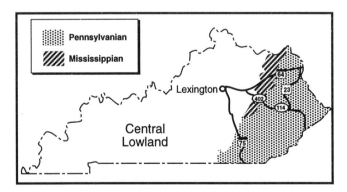

Figure 70. The Cumberland Plateau and highways in Kentucky.

coal come from both underground and strip mines. The thickest seams are about 11 feet (3.4 m).

The Pennsylvanian formations in Tennessee and Alabama also contain coal seams that are mined both underground and in surface strip mines.

CUMBERLAND PLATEAU HIGHWAY SECTIONS

Lexington, Kentucky to Appalachia, and the U.S. 58–U.S. 23 intersection, Virginia: Routes 402 and 114, U.S. 23 (Figure 70)

This is one of several good routes for seeing the Carboniferous rocks deposited on the landmass that was Pangaea, a result of the Alleghenian Orogeny. The road leads diagonally across the Cumberland Overthrust Block.

State Route 402 leads from Lexington east and south to State Route 114 to U.S. 23 and down into Virginia. Route 402 enters the Mississippian rocks in the vicinity of Stanton; Campton is in the Pennsylvanian outcrop area. U.S. 23 crosses the leading Pine Mountain Fault of the Cumberland Overthrust Block (Figure 69) just beyond the point at which U.S. 119 joins U.S. 23 in Virginia crosses the basin on the overthrust block. From Pound Gap in Pine Mountain to Norton are flat-topped hills of massive Pennsylvanian sandstone with underlying coal seams and shale, and scattered thin limestone layers.

Between Norton and Appalachia (Route BR 23), some road cuts have tilted layers characteristic of the neighboring Appalachian Province's Valley and Ridge Zone, and others reveal the almost horizontal layers of the plateau. U.S. 23 continues southward along the border between provinces to the U.S. 58 intersection, which is near the overthrust block's Hunter Valley fault zone.

Morehead, Kentucky to the West Virginia border: I-64 (Figure 70)

Between Lexington, Kentucky, and Huntington, West Virginia, I-64 crosses Mississippian and Pennsylvanian sandstones and shales of the plateau roughly from Morehead, Kentucky, to the state line. After a mixed zone where erosion in valleys has cut through Pennsylvanian layers into older Mississippian layers, the zone of entirely Pennsylvanian rocks starts near the U.S. 60 intersection east of Olive Hill.

Mount Vernon, Kentucky to Jacksboro, Tennessee: I-75 (Figure 70)

South from Lexington, I-75 encounters the Pennsylvanian sandstones and shales near Mount Vernon and traverses the coal country into Tennessee as far as Jacksboro, where it enters the Valley and Ridge Zone of the Appalachian Province.

Manchester, Tennessee to the Alabama border: I-24 *This journey over the Cumberland Plateau shows typical Pennsylvanian and underlying Mississippian rocks exposed here and there by erosion. The route also passes an interesting outlier of the neighboring folded Appalachian Province rocks.*

On I-24, southeast of Nashville, Mississippian rocks overlie Ordovician ones at the edge of the Cumberland Plateau, about 5 miles (8 km) northwest of Manchester. Pennsylvanian layers occur at Monteagle. Between Monteagle and Kimball, Pennsylvanian sandstone and shale show up in the hills, with Mississippian shale and limestone in the valleys along I-24. Between Kimball and the Tennessee River, the road crosses an outlier of the Appalachian Valley and Ridge Zone—a long, narrow, eroded anticline that extends northeastward as far as I-40 between Crossville and Rockwood. The anticline forms the leading edge of a fault block like the Cumberland Overthrust Block, with Ordovician and Silurian limestone, shale, and sandstone thrust up over Mississippian rocks. At the river, I-24 returns to the Mississippian and Pennsylvanian sediments. The road crosses into the Valley and Ridge Zone at I-59, just beyond the state line.

If you wish to follow the anticline, you can drive northeast along it from Jasper, on state Route 28, to U.S. 127 to Homestead, where the highway leaves the structure. U.S. 72 south from Kimball follows the anticline down into Alabama as far as Scottsboro, where the highway turns west. The anticline continues southwestward as far as I-65 north of Birmingham. U.S. 431 crosses the structure where the Tennessee River does, from Guntersville north to the point at which the road turns northwestward.

◼ THE CENTRAL LOWLAND

The Central Lowland occupies the center of the continent, surrounded by the Great Plains, the Canadian Shield, the Appalachian Plateaus, and the Coastal Plain, with a narrow ex-

Figure 71. The regions of the Central Lowland.

tension to the Atlantic Ocean in southeastern Canada (Figure 71). The region can be divided neatly into six regions, as shown on the map.

The Central Lowland has almost nothing but Paleozoic rocks. Erosion has removed all but a few of the rocks younger than Paleozoic, and has exposed only a few rocks older than Paleozoic. As the map in Figure 71 indicates, the sandstones, shales, and limestones have been slightly distorted into broad structures, the most outstanding of which are a major arch, two large basins, and a dome.

◆ The Northeast Corridor

Central Lowland Paleozoic rocks extend northeastward from the Great Lakes region up the St. Lawrence Valley all the way to Newfoundland (Figures 71 and 72). They are tucked in between the Canadian Shield, the Appalachian Province, and the Allegheny Plateau, but their composition and slightly folded structures are those of the Central Lowland, and they are Paleozoic in age, ranging from Cambrian to Devonian. Canadians call their part of the region the St. Lawrence Platform.

The corridor has four major divisions, the Algonquin Arch, the Québec Basin, the Anticosti Basin, and the Ontario-Hudson region. The Algonquin Arch, in the Great Lakes region (Figure 78) extends the Cincinnati Arch northeastward through peninsular Ontario (the triangular region between Lakes Huron and Erie/Ontario). The Québec Basin (Figure 78) lies to the north of the Adirondacks and along the St. Lawrence River as far as Québec City. The rocks on its northern border bent upward as Canadian Shield rocks rose; to the south, the layers were tilted upward by rising rocks of the Adirondacks. Along the southeast margin, crustal forces

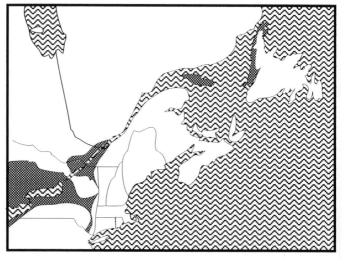

Figure 72. The Northeast Corridor is an extension of the Central Lowland that runs from the Lake Ontario region up the St. Lawrence River into the Gulf of St. Lawrence. The area shown in Newfoundland can be considered a part of either the Northeast Corridor or the Appalachian Province.

distorted the layers by pushing Appalachian Province rocks against them.

The Anticosti Basin (Figure 79) lies mostly underwater in the mouth of the St. Lawrence, between Québec's north and south shores and Newfoundland. Rocks of this basin outcrop on Anticosti Island, in a few small places on the north shore, and along western Newfoundland.

Evidence of faults along the St. Lawrence and lower Ottawa Rivers indicates an episode of rifting in what is now the Québec and Anticosti Basins about 600 million years ago, during the splitting of an earlier continent (see Figure 107). Much later, a second rift valley formed in the area during the Cretaceous Period, when Pangaea was breaking up.

The fourth division comprises portions of Ohio, Pennsylvania, and New York (Figure 72); a fringe area where the Central Lowland rocks dip under those of the Allegheny Plateau and lap up against the Adirondacks.

● The Algonquin Arch

Most of peninsular Ontario belongs to the Michigan Basin and is described with it in this chapter's section on the Interior Basins. The Northeast Corridor starts at the boundary between Ordovician and Silurian rocks near the western end of Lake Ontario and includes the part of Ontario between Lakes Ontario and Erie. The region north of Lake Ontario consists entirely of Cambrian and Ordovician shale and limestone, including the terrestrial Queenston redbeds, dipping gently downward toward the south and southwest.

As the Devonian and Carboniferous rocks of Ohio show the changes from marine to terrestrial that accompanied the Alleghenian Orogeny, so the Ordovician rocks of Ontario show similar changes that accompanied the Taconic Orogeny. From Middle to Late Ordovician, the formations change from marine limestone and shale to terrestrial shale of the Queenston beds, reflecting the changes in environment as highlands rose during the orogeny.

East of Kingston, a neck of Canadian Shield igneous and metamorphic rocks connects with the Adirondacks in New York, which also belong to the Canadian Shield. Along with the Precambrian igneous and metamorphic rocks, it contains some isolated patches of Late Cambrian Potsdam sandstone and Ordovician limestone, erosion remnants of the Paleozoic sediments that once covered this small uplift of Precambrian

Figure 73. The Niagara Escarpment on the skyline, as seen from the Queen Elizabeth Way, along the south shore of Lake Ontario in the vicinity of Grimsby, Ontario. Shown here without power lines and buildings.

rocks. The neck separates the Algonquin Arch region from the Québec Basin region.

The Niagara Peninsula, below the west end of Lake Ontario, has east–west strips of Ordovician, Silurian, and Devonian rocks, dipping southward, that connect with the Michigan Basin formations. Late Ordovician Queenston terrestrial red shales parallel the lake. Younger marine formations south of the red shale show that the ocean later returned to the area, after the Taconic Orogeny; Devonian limestone outcrops at the shore of Lake Erie to the south, and between it and the redbeds are Silurian limestone, dolomite, and shale.

The intermediate rock series includes the massive Niagara Escarpment (Figure 73), surmounted by Silurian Lockport Dolomite, which is an outstanding feature of the area, and illustrates the shallowly dipping structure of the rocks. The escarpment follows the Lake Ontario shore to the west end of the lake, where it turns sharply northward, and Autoroute 403 passes through a spectacular cut in its crest that exposed dolomite with chert nodules. The base of the escarpment marks the beginning of the Northeast Corridor.

Erosion by the Niagara River has exposed a cross section of

several formations between the lakes, in an extension of the Niagara Escarpment. The river, in flowing toward Lake Ontario, encounters the resistant Silurian Lockport Dolomite, which supports Niagara Falls (Plate 10). The formations below the Lockport are mostly shale, with some limestone and sandstone. The shale erodes faster than the dolomite above it. As time goes by, shale breaks and washes away from under the dolomite, until unsupported blocks of dolomite fall into the riverbed below. As the dolomite falls off, Niagara Falls retreats upstream.

The river flows over Ordovician Queenston redbeds from the base of the falls to Lake Ontario, and the layers of shale, limestone, and dolomite overlying the Queenston rocks show clearly on the walls of the gorge. Upstream from the falls, the river crosses the Silurian-Devonian boundary at the rapids near the Peace Bridge at Buffalo, where Devonian limestone crops out.

● The Québec Basin

The Québec Basin's downwarped rock layers consist almost entirely of Ordovician sandstone, shale, limestone, and dolomite, formed in the extensive ocean that covered North America before the Taconic Orogeny. The basin forms an eastward-narrowing triangle from the Precambrian neck to just beyond Québec City.

The "Mont" of Montréal is an unusual feature of the sedimentary Québec Basin, an igneous intrusion. Its rock has resisted weathering and erosion better than the surrounding sedimentary rocks (Figure 74). It is the westernmost in a string of similar hills that extends to the east from Montréal to Lac Brome. The hills contain a great variety of intrusive rock types, including granite, monzonite, and syenite, and the limestone and shale surrounding them contain numerous dikes. The intrusions occurred during the Cretaceous Period, when Earth's crust was cracking as Pangaea split into smaller sections.

The Québec Basin pinches out in the Québec City–Ile d'Orléans area, between the Canadian Shield and the Appalachian Province. Several faults in the basin testify to the Cretaceous crustal movements in the St. Lawrence River Valley. At Montmorency Falls, downstream from the city, faulting action dropped younger Upper Ordovician shale down to the level of older Middle Ordovician limestone that was origi-

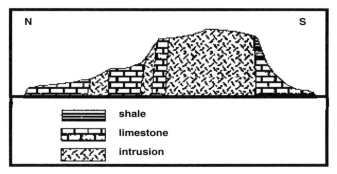

Figure 74. A resistant igneous intrusion into the local limestone and shale forms most of the mountain in Montréal. Movement along a fault on the south side has brought shale from above down to the level of the limestone on the north. The section is about 3 miles (5 km) long.

nally below it. The Montmorency River falls over Precambrian granite upon which the limestone was deposited (Figure 75).

● The Anticosti Basin

In this easternmost extension of the Central Lowland, sedimentary layers dip gradually southward toward the Gaspé Peninsula from the north coast of the Gulf of St. Lawrence and from Anticosti Island. At the east end of the basin, the beds dip westward from western Newfoundland. On Anticosti Island, Silurian limestone and shale overlie Ordovician limestone and shale.

The situation in western Newfoundland is much more complicated. On the Great Northern Peninsula, the Long Range Mountains consist of Precambrian igneous and metamorphic rocks, a disconnected part of the Canadian Shield's Grenville Province. A mixture of Cambrian and Ordovician rocks surrounds them on the west and north coast, jumbled during the Taconic Orogeny by faulting that broke up the usual orderly structure found elsewhere in the Stable Inte-

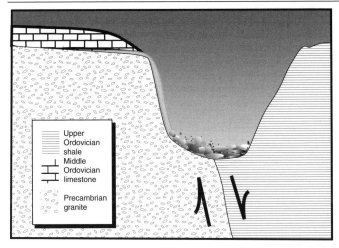

Figure 75. A cross section of the bedrock at Montmorency Falls, east of Québec City. Late Ordovician shale has been lowered to the level of Middle Ordovician limestone (and below) along a normal fault. When the shale was formed, it was far above the granite "basement" rock.

rior. The rocks originated as sediments in the early Paleozoic seas that invaded large parts of our ancestral continent. These Cambrian and Ordovician rocks can be considered either the eastern end of the Northeast Corridor or the local western edge of the Appalachian Province.

Cambrian red to gray arkose, sandstone, conglomerate, and shale outcrop at the eastern margin of the peninsula's northern tip, with associated basalt lava flows around St. Anthony. The west coast has Cambrian limestone, dolomite, sandstone, and shale, with some volcanics between Bonne Bay and Port au Port Bay.

The Ordovician layers grade from earlier dolomite to limestone and dolomite with some beds of shale, to interbedded limestone and shale, to interbedded sandstone and shale. Some of the sandstone shows delta-type cross bedding. On the Port au Port Peninsula, accessible by Route 460 from Stephenville, a conglomerate containing fragments of older limestone overlies the sandstone and shale. Graywacke and shale overlie the conglomerate, in turn. The change from dolomite to sandstone and shale, the cross bedding, the conglomerate of limestone fragments, and the graywacke all indicate activity in the crust and the rising and erosion of nearby land. This is evidence of the Taconic Orogeny toward the end of the Ordovician Period.

The Port au Port Peninsula also includes a small outcrop area of Devonian rock—mainly terrestrial cross-bedded, coarse-grained, red sandstone, with shale and a bit of limestone. This formation presumably represents an early stage of the late Devonian–early Mississippian Acadian Orogeny. To complete the picture, some Carboniferous and even Cretaceous rocks occur on the peninsula.

● The Ontario-Hudson Region

In this southern portion of the Northeast Corridor, early Paleozoic rocks south of Lakes Erie and Ontario extend eastward to lap against the Adirondacks and southward along the Hudson River (Figure 72). The rocks along the south shore of Lake Erie from the Cleveland, Ohio, region to approximately the longitude of Fredonia, New York, are remnants of the inland sea. They continue the Devonian formations that dip off the southeast side of the Algonquin Arch in peninsular Ontario. The oldest bedrock in Cleveland consists of Upper Devonian gray and black shale, all marine. The shale outcrops extend eastward through the coastal strip of Pennsylvania. The Devonian sediments disappear to the south under the Mississippian formations of the Allegheny Plateau.

The most unusual geological feature along the Pennsylvania coast is Presque Isle, the sand spit at Erie (Figure 76). Through the years, currents in Lake Erie have moved sand along its lakeward margins to build it farther and farther along the coast. The local bedrock is upper Devonian shale and sandstone.

In New York, early Paleozoic marine sedimentary rocks of the Northeast Corridor partly surround the Adirondacks. They also border the Allegheny Plateau on its north and east margins, dipping southward under the plateau across the state and westward under the plateau in the Hudson Valley. These bedrocks consist of Cambrian sandstone and dolomite; Ordovician dolomite, shale, and sandstone, including redbeds; Silurian sandstone, shale, conglomerate, and dolomite; and Devonian limestone. The rock types show a gradation from a Cambrian and Ordovician shallow sea to Late

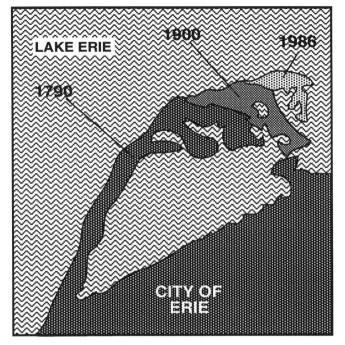

Figure 76. A simplified version of the growth of the sand spit on the Lake Erie shore at Erie, Pennsylvania.

Ordovician land, and back to Silurian and Devonian shallow sea. The terrestrial and near-shore redbeds record the uplift of the Taconic Orogeny.

◆Ice Age Effects

Since the entire Northeast Corridor was glaciated during the Ice Age, till and outwash can be found throughout the region. In Ontario, terminal moraines parallel the shores of Lake Ontario's western tip, a few miles inland. A major moraine north of Lake Ontario extends roughly from west of Newmarket to south of Peterborough and beyond.

Glacial moraines parallel the St. Lawrence River in the

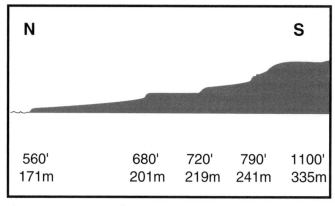

Figure 77. A topographic cross section of Mentor Township, Ohio, east of Cleveland, between Lake Erie and the hills to the south. It shows three major wave-cut terrace levels of the earlier Lake Erie, formed as the Ice Age glacier melted back and the water level dropped. The numbers are elevations above sea level.

Québec Basin. One, about 40 miles (64 km) east of the river at Montréal, extends to about 5 miles (8 km) south of Québec City. A less complete moraine, mostly somewhat closer to the river, starts about 40 miles (64 km) west of Montréal and ends about 50 miles (80 km) southwest of Québec City.

As the last Ice Age glacier melted back into Canada, it left a series of lakes in the Great Lakes Region. When the retreating glacier began to uncover the Lake Erie basin, ice covered the present northern outlet. The water level was higher, the lake was larger, and the early outlet to the southwest was higher than the present one. Later, the melting margin of the glacier uncovered a lower outlet to the east, and the water level dropped. As the process continued, the lake stabilized temporarily at several levels before the last outlet through the Niagara River was uncovered.

While the lake was stabilized, waves cut a terrace along its shores. Each time a new outlet was uncovered, a new terrace formed at a lower level. The old terraces are still visible in some places, between the present lake shores and higher surrounding land. For example, in Mentor township, Ohio, east of Cleveland, there are three major terraces (Figure 77). Ohio Route 306 north from Kirtland to its end descends from the hills around Kirtland onto the flat former lake bottom and over the terraces in the area of the cross section.

Most of New York's great drumlin field (Figure 65) lies within the Northeast Corridor. A multitude of these elongate hills exists between Batavia and Syracuse and from Lake Ontario down onto the Allegheny Plateau. A couple of major moraines parallel the shore of Lake Ontario.

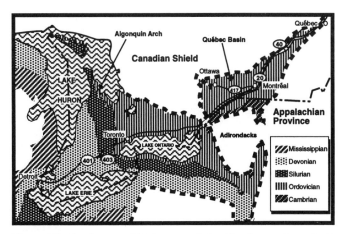

Figure 78. Geology and highways of the Northeast Corridor's western portion.

NORTHEAST CORRIDOR HIGHWAY SECTIONS

Milton to Kingston, Ontario: Route 401 (Figure 78) *In Ontario, Provincial Route 401 from west to east crosses the whole series of tilted Ordovician rocks north of Lake Ontario, from the youngest to the oldest. The west to east shift from redbeds to marine shales to marine limestone illustrates in reverse order, from top to bottom, the advent of the Taconic Orogeny. The limestone is the oldest, from the sea floor; shale indicates erosion of a rising landmass; and redbeds, the youngest rocks in the series, include dry-land deposits.*

The highway enters the youngest Ordovician formations

in the vicinity of Milton, north of Hamilton, which are the Queenston redbeds, remnants of the Taconic Orogeny. The rocks consist mostly of shales, some terrestrial, formed on the newly growing supercontinent that was to become Pangaea. The earlier Queenston beds include some limestone.

The outskirts of Toronto have a bedrock formation of gray shale with minor limestone layers, which underlies the Queenston redbeds. From there on to the east, you will drive over marine beds, part of the great Ordovician ocean that covered so much of what became North America. Shale formations continue eastward, ending with black shale that overlies a group of Middle Ordovician formations that contain limestone.

The limey rocks show up in the vicinity of Oshawa, and the formations contain higher percentages of limestone as you move eastward. First comes shaly limestone, then a mixture of shale and limestone beds, then just limestone.

Hamilton to Fort Erie, Ontario: Queen Elizabeth Way *This route traverses a cross section of southward-dipping sedimentary rocks on the Niagara Peninsula, from oldest to youngest.*

Queen Elizabeth Way, the main route between Hamilton and Niagara Falls, goes along the Late Ordovician Queenston redbeds zone, except where the highway curves south to Niagara Falls. There, the bedrock changes to Silurian.

Between the lakes, the highway south crosses a succession of Silurian sandstone, gray shale, then red and mottled green shale, probably near-shore deposits. Next comes limestone, more shale, and a massive series of four different formations of dolomite, the third of which also contains some limestone and shale. All these rocks show that after the Taconic Orogeny, the sea flooded the land again. Limestone of Early Devonian age shows up at and near the Lake Erie shore.

Kingston, Ontario to Montréal, Québec: Routes 401/20 (Figure 78) *This trip emphasizes the gentle rippled structure of the main Québec Basin, passing over a minor basin and a minor dome in Ordovician marine sedimentary rocks.*

The Proterozoic-Paleozoic unconformity is exposed on Ontario Route 401 just outside of Kingston, by the Route 15 junction. Cambrian sandstone rests on Proterozoic gneiss of the Canadian Shield. Route 401 east crosses the narrow neck of mostly Proterozoic rocks between the Canadian Shield's main body and the Adirondacks and enters Ordovician country and the Québec Basin again in the vicinity of Brockville.

Between Brockville and Vaudreuil, Québec, the route number changes to 20.

Autoroute 20 crosses a minor basin within the Québec Basin, passing over Lower Ordovician dolomite and sandstone, then Middle Ordovician dolomite, limestone, and shale, then the mirror-image repeat of the Lower Ordovician sediments.

The minor basin ends at a small structural dome that surrounds the mouth of the Ottawa River just west of Montréal. The dome has in its central portion patches of Cambrian sandstone and Precambrian gneiss.

Montréal, Québec to Ottawa, Ontario: Routes 40, 417 (Figure 78)

Autoroute 40 toward Ottawa crosses the south edge of the small dome mentioned in the last route, which extends from the St. Lawrence River almost to the Ontario border. The continuation in Ontario, Route 417, passes over Middle, then Upper Ordovician rocks. Along a fault at Ottawa, Lower and Upper Ordovician rocks have come to rest against each other. The fault is a member of the Cretaceous age St. Lawrence fault system. The Canadian Shield begins to the north of Ottawa, directly across the river.

Montréal to Québec City, Québec: Routes 20, 40 (Figure 78)

The larger sub-basin of the Québec Basin extends up the St. Lawrence River to Québec City and is crossed by Autoroutes 20 and 40 on either side of the river. Limestone and dolomite, with shale and some sandstone, make up the rocks of this sub-basin. Middle Ordovician rocks near Montréal lead to a long stretch of Upper Ordovician; Middle Ordovician sediments appear again just before the bend in the river at Portneuf on Route 40. At Québec City, a fault of the system mentioned in the last route dropped Upper Ordovician rocks into a position level with older ones (Figure 75).

Deer Lake to St. Anthony, Newfoundland: Route 430 (Figure 79)

In Newfoundland, the complicated mixture of Cambrian and Ordovician rocks that surround the Precambrian of the Great Northern Peninsula form the bedrock along Autoroute 430. They form the eastern margin of the Anticosti Basin. Routes 440, 450, and 460, west and southwest of Corner Brook, also go into the mixture. Shale, sandstone, limestone, graywacke, slate, argillite, quartzite, schist, marble, andesite, basalt, and greenschist are included in the region's varied

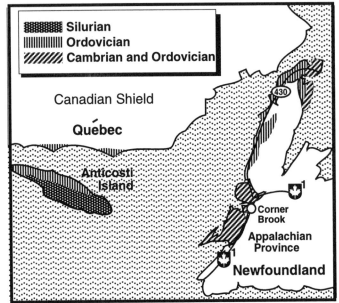

Figure 79. Dry land bedrock of the Anticosti Basin, the eastern extension of the Northeast Corridor.

assemblage of rocks from ocean basins and volcanic islands. The Taconic Orogeny folded, faulted, and in some cases metamorphosed the sedimentary rocks and lava flows as an island chain collided with the continental core.

Cleveland, Ohio to Albany, New York: I-90 (Figure 62) *I-90 crosses a series of marine sedimentary formations that dip slightly downward toward the west and south to disappear under the Allegheny Plateau rocks. Because of the dip, the highway passes from Upper Devonian-age rocks through older Silurian ones and into Ordovician ones. The formations originated on the floor of the sea that existed between the Taconic and Acadian orogenies.*

I-90 from Cleveland, Ohio, to Buffalo, New York, is built on the Devonian rocks, extensions of those in Ontario. From Cleveland into New York, Upper Devonian shale with some sandstone grades eastward into Middle Devonian shale, which in turn overlies the early Middle Devonian Onondaga Limestone at Buffalo. These rocks are exposed in some places along I-90 and north–south roads south of I-90. The shales in New York also outcrop along the Lake Erie shore and in the gorges through which streams approach the lake.

East of Buffalo, Onondaga Limestone forms the bedrock below I-90 to a point north of Batavia, where the highway cuts through the escarpment at the formation's eroded edge. From there to Syracuse, the older Silurian shale and dolomite that underlie the limestone form the bedrock. This part of the highway leads through the famous New York drumlin field. The drumlins begin in the vicinity of the LeRoy–I-490 exit and continue to Syracuse, but they are not readily identifiable from I-90.

I-90 passes into Ordovician shale with a little limestone east of Syracuse, about halfway between Oneida and Utica. These sedimentary rocks extend all the way to the Hudson River, except for a patch of dolomite in the Canajoharie-Randall area.

Rochester, New York

The Genessee River gorge at Rochester offers a cross section of New York geology north of the Allegheny Plateau. At the head of the gorge, the river falls over Silurian Lockport Dolomite, the same formation that causes Niagara Falls. The Lockport sits on top of the Silurian Rochester Shale, below which is a limestone formation. The limestone overlies a sandstone layer that causes the lower falls. Below the lower falls, the river flows over a series of Silurian redbeds. The lowest and oldest rock level outcrops at the mouth of the gorge, where the river has exposed Upper Ordovician Queenston redbeds, including terrestrial sediments from the Taconic Orogeny's mountains.

Herkimer to Middleville, New York: Route 28

Cambrian sedimentary rocks outcrop here and there around the margin of the Adirondacks. North of Herkimer on New York Route 28 is the town of Middleville, in the center of a small area of Cambrian dolomite. The dolomite contains cavities in which have grown small but beautiful, clear, perfectly formed, in many cases double-ended quartz crystals. The crystals are called Herkimer diamonds.

Syracuse to Ellenburg, New York: I-81, U.S. 11, Route 190 (Figure 62)

I-81 between Syracuse and Watertown cuts across Silurian and Ordovician rocks. The Cambrian rocks that underlie the Ordovician ones appear around the north margin of the Adirondacks. There, at the southern edge of the Québec Basin, Cambrian and Ordovician formations lap against the uplifted Precambrian Adirondack rocks. Between the Cambrian sandstone and the Canadian border are Ordovician formations, dipping down to the St. Lawrence River.

North of Syracuse, the Silurian shales give way to older Late Ordovician shales and sandstones in the vicinity of the Central Square–West Monroe exit. Going north, the road passes over successively older Ordovician shales and sandstones until limestone appears south of Adams. Drumlins can be seen along this stretch of road.

On U.S. 11 east from Watertown, the bedrock is Ordovician limestone for about 10 miles (16 km) beyond the city; after that you enter a stretch of Adirondack Canadian Shield gneiss and marble that extends to Potsdam. From Potsdam to Ellenburg the road follows the Cambrian Potsdam Sandstone, except for a few small outliers of Adirondack rocks between Nicholville and Malone. The Potsdam Sandstone is among the earliest Paleozoic sediments in North America; it formed as the rocks of the Proterozoic continental nucleus weathered and eroded during the Cambrian Period, and rivers washed large amounts of sand into the neighboring shallow sea. New York Route 190 southeast from Ellenburg continues through another 15 miles or so (24 km) of Potsdam Sandstone and into the Appalachian Province.

Albany to Keeseville, New York: I-87 (Figure 62)

I-87 north from Albany leads to the imposing gorge of Ausable Chasm, cut through the Potsdam Sandstone. Ordovician marine shales extend from Albany to Glens Falls. Glens Falls is in the Adirondacks; between there and the Keeseville exit are numerous exposures of gneiss and anorthosite, characteristic of the Adirondack portion of the Canadian Shield. From the Keeseville exit, New York Route 9 north leads to one of the isolated patches of Cambrian rocks that ring the Adirondacks, including Ausable Chasm.

Albany to Newburgh, New York: I-87 (Figure 62)

I-87 south from Albany follows the outcrop of marine Ordovician shale and sandstone and Early Devonian limestone and shale that underlie the later Devonian rocks of the Catskill Front. The Catskill Front is the eastern edge of the Stable Interior's Allegheny Plateau, and the Appalachian Province's border is on the other side of the valley, a few miles east of the Hudson River.

West and south of New Paltz, the resistant Silurian Shawangunk Conglomerate forms the Shawangunk Mountains, jutting out between the Devonian rocks of the Catskill Front and the Ordovician rocks of the valley. Beyond Newburgh, the highway enters the Appalachian Province. The Ordovician marine shales and sandstones continue southwest, to pinch out just over the border in New Jersey.

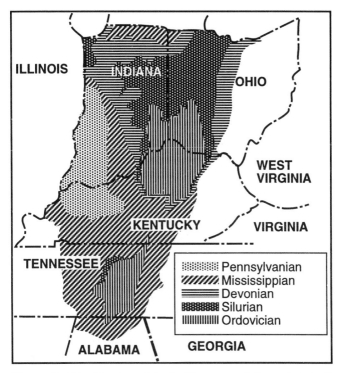

Figure 80. The Cincinnati Arch. The crest of the arch is indicated by the outcrop areas of the oldest (Ordovician) rocks in Tennessee, Kentucky, Indiana, Ohio, and the extension of Silurian rocks into the tip of Lake Erie (see Figure 60).

◆ The Cincinnati Arch

The Cincinnati Arch (Figures 71 and 80) is a large structural arch caused by pressures during the Paleozoic Era that folded rock layers upward along an axis from northern Alabama to Lake Erie. Since the arch trends northeast–southwest, the sedimentary rock layers on its flanks dip slightly to the southeast on one flank and to the west and northwest on the other flank. Figure 59 shows the arch in cross section. The marine rocks that outcrop on the arch range from Ordovician to Mississippian in age and are a veritable historical encyclopedia of the Paleozoic inland seas that from time to time covered so much of what is now North America.

The Cincinnati Arch consists primarily of two oval structural domes, side by side, that have been truncated by erosion so that they don't form highlands. Figure 80 shows where the Ordovician rocks outcrop in the cores of the two domes. The one centered in central Tennessee is the Nashville Dome, and the other is the Lexington Dome. To the southwest, the Cincinnati Arch dips down and disappears in Alabama; to the northeast, the arch pinches out in Ontario. It grades to the west into the Ozark Dome in Arkansas and Missouri, and to the northwest into the Illinois and Michigan basins. To the southeast, its rocks dip underneath those of the Appalachian Plateaus.

In the arch's two domes, younger layers dip away from the central Ordovician rocks (Plate 15) in all directions. The Paleozoic sedimentary rocks of Indiana and Ohio dip away from the axis of the arch, into the Michigan and Illinois basins on one side and under the Allegheny Plateau on the other. The formations in Kentucky dip east, south, and west from the Lexington Dome.

The Nashville Dome's core of Ordovician rocks is surrounded mostly by Mississippian rocks. In most places around the dome, the Mississippian rocks lie directly on Ordovician ones, covering eroded edges of Silurian and Devonian formations. As the Cincinnati Arch was folding, the Silurian and Devonian rocks were eroded from the slowly rising core of the Nashville Dome. Later, the Mississippian ocean advanced to cover their eroded edges, lapping onto the underlying Ordovician rocks. The Mississippian ocean-bottom sediments became the limestone. There are slivers of Silurian outcrops west and northwest of Nashville, along with a couple of even smaller slivers of Devonian rocks, one just west of Nashville along U.S. 70 and another south of Gainesboro that is not along a highway.

The region's early Paleozoic formations, consisting mostly of limestone, are surrounded by an escarpment called the Highland Rim. The Mississippian limestone, sandstone, and shale dip from the Highland Rim and eventually pass under Pennsylvanian rocks of the Cumberland Plateau to the south and east (Plate 15), and under Coastal Plain sediments to the west.

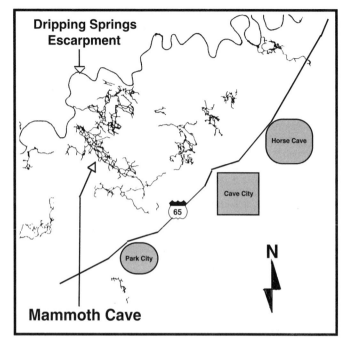

Figure 81. Cave passages in the Mammoth Cave country.

The rocks of the Cincinnati Arch are mostly marine, origi-
nating in the great inland seas of the Paleozoic Era. Limestone formed in quiet waters. As highlands rose during the
collisions of the Taconic and Acadian orogenies, sand and
silt washed into the seas, to become in turn Late Ordovician
and Late Devonian/Early Mississippian sandstone and shale.
Finally, the Alleghenian Orogeny at the end of the Paleozoic
Era caused the region to rise above sea level, resulting in
Pennsylvanian terrestrial sandstone, shale, and coal.

The Silurian and Devonian formations of the Cincinnati
Arch's northwest flank contain a large amount of limestone
and dolomite, and thus, predictably, caves. The cave country
of Kentucky between Elizabethtown and the Tennessee border is an outstanding feature of the Cincinnati Arch. The
area between Munfordville and Bowling Green has the most
and biggest caves. A single system, the Mammoth Cave–
Flint Ridge system, has over 280 miles (450 km) of intercon-
necting passages (Figure 81).

Figure 82. Major moraines in the Great Lakes region. The shaded area was not covered by glaciers during the last (Wisconsin) glacial stage.

The Mammoth Cave area clearly shows one of the most unusual and interesting characteristics of karst topography. Because of the widespread solution of limestone, there are numerous sinkholes in the area, and all the smaller streams flow through these sinkholes into underground tunnels. The only water courses that flow all the way through the limestone zone are the Green River at the northeast end of the zone, and the Barren River and one of its tributaries.

The cave country's Mississippian formations dip gently northwestward and ultimately disappear beneath the Pennsylvanian rocks of the western Kentucky coalfield. The cave zone consists of two thick formations composed almost entirely of limestone. The great mass of pure limestone there has weathered and eroded into karst topography and lends itself to cave formation. To the northwest of the cave zone is the Dripping Springs escarpment, formed by a resistant sandstone layer that was deposited on top of the cave-containing layers. To the southeast of the cave zone are two formations consisting of shaly limestone and shale that underlie the cave zone. The formations above and below the limestone of the cave zone are much less permeable to water and do not dissolve and weather into karst topography.

Figure 83. Kelleys Island in western Lake Erie is famous among geologists for its unusually large pebble- to boulder-size glacial striae, or grooves, in the Middle Devonian Columbus Limestone.

Elsewhere, Olentangy Caverns, in Devonian limestone, are north of Columbus, just off Ohio Route 315, about 5 miles (8 km) north of I-270. Ohio Caverns, in Silurian limestone, are on Ohio Route 245, east of West Liberty. Crystal Cave, on South Bass Island in Lake Erie near Sandusky, Ohio, is unusual in that large crystals of the mineral celestite mostly cover the interior walls. Chemically, celestite is strontium sulfate, and the crystals on South Bass Island are reputed to be the largest known crystals of that mineral. A ferry goes to South Bass Island from Port Clinton, Ohio.

The Tennessee extension of the Kentucky cave zone includes Dunbar Cave, about 4 miles (6.5 km) east of Clarksville, and Jewel Cave, about 12 miles (19 km) northwest of Dickson, on County Route 46, an extension of Tennessee Route 46.

The Illinois Basin melds into the flank of the Cincinnati Arch in the western Kentucky coalfield. There, terrestrial Pennsylvanian rocks contain some well-developed coal seams, such as the No. 9 (Mulford) coal seam, as much as 3.5 to 6 feet (1.1–1.8 m) thick and a continuous seam throughout a large area.

Continental glaciers covered almost all of western Ohio

and most of Indiana during the Ice Age, and the relatively smooth topography there is largely due to erosion and deposition by the glaciers. Till and outwash dominate the glaciated region, with many major terminal moraines that formed as the last glacier retreated (Figure 82). The positions of moraines in Ohio and Indiana indicate that a glacial lobe moved down the axis of Lake Erie into Central Indiana. As it retreated, it left huge ridges of till and outwash where its margins temporarily rested.

Kelleys Island (named for the Kelley brothers, who settled there in 1833), in Lake Erie north of Sandusky, Ohio, offers a world-renowned set of glacial striae—exceptionally large and deep grooves in the Devonian Columbus limestone, made by boulders, smaller stones, and sand frozen in the bottom of the moving Ice Age glacier (Figure 83). Normal-size grooves are illustrated in Plate 46.

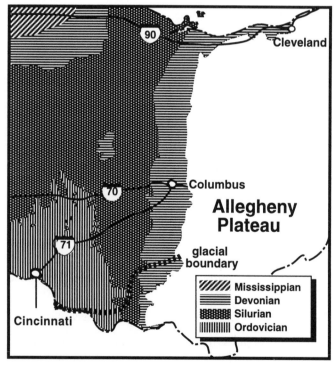

Figure 84. Geology and highways of the Ohio portion of the Cincinnati Arch.

CINCINNATI ARCH HIGHWAY SECTIONS

Brice, Ohio to Cincinnati or the Indiana border: I-70, I-71 (Figure 84) *This route crosses the Cincinnati Arch a bit farther north than the route illustrated in Figure 58. It passes from the Mississippian-Devonian boundary to older rocks at the center of the eroded arch. If you stay on I-70, you enter younger rocks again, toward the Indiana border, on the other flank of the arch.*

East of Columbus on I-70, Devonian marine shales appear from under the overlying marine Mississippian rocks in the Brice-Reynoldsburg area, at the edge of the Allegheny Plateau. Earlier Devonian limestone outcrops in the Columbus area. West of Columbus, there are two choices: I-71 south to Cincinnati shows the maximum number of Ordovician outcrops and road cuts, and I-70 west offers the most complete cross section of the arch.

Columbus is at the boundary between Middle Devonian limestone and Upper Devonian shale. The southbound route starts on the southeast flank of the arch and leads obliquely toward the arch's central axis at Cincinnati. From the Devonian limestone at Columbus, the highway enters the Silurian outcrop area northwest of Derby, about halfway between the U.S. 62 exit and the Ohio Route 56 exit. The Silurian series contains mostly limestone and dolomite, but there is one major shale formation. The Silurian-Ordovician border crosses the highway in the vicinity of the Kingman-Oakland exit; from there to Cincinnati the arch shows only Ordovician shale, with a scattering of thin to medium limestone layers. These are the oldest formations exposed on the arch. The Ohio-Kentucky-Indiana Ordovician region is one of the most prolific fossil-bearing regions in the world, where you can easily see a variety of ancient ocean animal remains.

The direct westward route samples Devonian, then Silurian, then Ordovician, and then Silurian rocks again as it crosses the arch. It leaves Columbus on I-70 and continues over the Devonian limestones to their border north of West Jefferson. Silurian dolomite, limestone, and shale extend to a line approximately between Enon and Snyderville. Beyond that line the geology gets more complicated, because of the Mad, Miami, and Stillwater rivers. They have eroded down through the Silurian rocks to expose Ordovician ones in their valleys.

Ordovician rock begins at the line between Enon and Snyderville and continues to about the Ohio Route 4 exit, in the valley of the Mad River. From there to the Ohio Route 202 exit, the highway briefly returns to the Silurian ocean floor, then to the Ordovician in the Miami and Stillwater River valleys. The country north of the interstate and between the valleys has Silurian bedrock. Layers of Ordovician shale and limestone are visible at the Englewood and Taylorville dams, north of the interstate, from U.S. 40. Silurian bedrock continues along I-70 beyond the Stillwater Valley, from the vicinity of the Ohio Route 49 exit to the Indiana border.

Cleveland, Ohio to the Indiana border: I-90 (Figure 84) *I-90 is the northern route over the Cincinnati Arch, from Cleveland to the Indiana line. It crosses Mississippian through Devonian to Silurian rocks, from younger to older, then back through Devonian into Mississippian, from older to younger, thus outlining the structure of the arch.*

I-90 leaves the Devonian shales of Cleveland's western suburbs and enters the Mississippian shales and sandstones, for a short trip through a corner of the Allegheny Plateau, roughly between Avon and Berlin. The Mississippian rocks are on the border between the Allegheny Plateau and the Central Lowland. The highway then goes back into the Central Lowland and the Devonian shales.

To the west, the shales give way to the underlying Devonian limestone in the Sandusky area. You can turn northwest on U.S. 250, visit Sandusky, and take the ferry to Kelleys Island. There, after you have seen the glacial grooves, you can observe the Columbus Limestone dipping gradually into the lake, and actually see the angle at which the layers are tilted on the Cincinnati Arch (see Figure 60).

I-90 leaves the Devonian limestone about halfway between Parkertown and Erlin to enter the underlying Silurian limestone and dolomite at the core of the arch. About 3 miles (5 km) west of the U.S. 20 exit, Devonian rocks occur again on the other side of the arch. The highway crosses sandstone, dolomite, limestone, and shale zones from the Devonian. From the Wauseon exit to the Indiana line, Mississippian sandstone and shale form the bedrock at the south end of the Michigan Basin.

Cincinnati, Ohio to Lexington, Kentucky: I-75 (Figure 85)
I-75 from Cincinnati to Lexington, Kentucky, traverses only Ordovician shale and limestone. The Ordovician area of Kentucky and Ohio is the center of the Lexington Dome, which is the northern dome structure on the Cincinnati

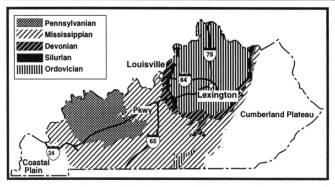

Figure 85. Geology and highways of the Kentucky portion of the Cincinnati Arch.

Arch. In Kentucky, it essentially comprises the Bluegrass Country.

Lexington to Louisville, Kentucky: I-64 (Figure 85) *I-64 from Lexington to Louisville crosses a quick geologic section of the Lexington Dome from the center outward. The rocks change from Ordovician to Mississippian, from older to younger. In between, the highway crosses the thin outcrop areas of Silurian and Devonian rocks that ring the dome structure.*

Lexington lies on Middle Ordovician shale and limestone. Beyond the Kentucky River, the road passes from Middle to Upper Ordovician shale and limestone. The Silurian zone, with dolomite and shale, starts about at I-265, and the Devonian zone, with limestone and shale, goes from Jeffersontown down to the Ohio River bottom. Mississippian sandstone and shale outcrop south of Louisville. Resistant Mississippian sandstone forms a major escarpment called Muldraughs Hill, or the Knobstone Escarpment. It partly encircles the Lexington Dome structure on the east, south, and west and extends northward into Indiana. At the Ohio River, the escarpment is around 600 feet (183 m) high.

Lexington, Kentucky to Kentucky Lake: Bluegrass Parkway, Western Kentucky Parkway, I-24 (Figure 85) *The longer and more complete geologic section of the Lexington Dome from Ordovician in the center to Mississippian on the flank occurs along the Bluegrass and Western Kentucky parkways.*

Lexington's Ordovician shale and limestone extend southwest on the parkway, followed by Silurian limestone, dolomite, and shale in a zone of about 15 miles (24 km) on either side of Bardstown. Devonian limestone and shale occur roughly between the easternmost crossing of the Beech Fork River and the Rolling Fork River, just south and west of their junction. Next is Muldraughs Hill and the Mississippian area, starting with mostly shale and limestone layers, then just limestone, and finally, mostly limestone with some sandstone. These rocks extend to the Kentucky Route 259 (Leitchfield) exit.

At that point the highway leaves the Mississippian sea bottom and enters the dry land of the Pennsylvanian terrestrial sandstone, shale, and coal in the western Kentucky coalfield.

Just beyond the exit to Kentucky Route 109 (Charleston), the rocks are again Mississippian; they continue to I-24 and to Kentucky Lake, at the boundary of the Coastal Plain.

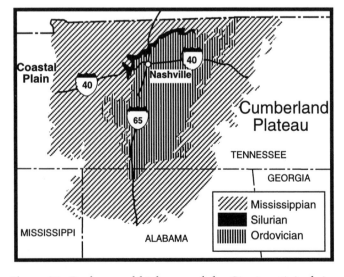

Figure 86. Geology and highways of the Cincinnati Arch in Tennessee, Alabama, and Mississippi.

Tennessee-Kentucky border to Lacon, Alabama: I-65 (Figure 86)

I-65 offers a north-south section of the Nashville Dome and Tennessee geology, from Mississippian (younger) to Ordovician (older) to Mississippian (younger) rocks.

South of the Kentucky border, the bedrock is Mississippian, because the Devonian rocks you might expect to see are covered by Mississippian sediments. Silurian limestone and shale occupy a short stretch approximately between the U.S. 31W and Goodletsville exits, just north of Nashville. From there, Ordovician limestone and shale of the dome's center extend to the Alabama line. The highway reenters Mississippian limestone, shale, and sandstone just into Alabama and finally crosses into the Cumberland Plateau's Pennsylvanian rocks just before the U.S. 31 Lacon exit.

Holladay/Sugar Tree to Monterey, Tennessee: I-40 (Figure 86)

In Tennessee, I-40 cuts across the Nashville Dome east and west. Going east, the route starts at the Coastal Plain margin, near the U.S. 641 exit, about 8 miles (13 km) west of the Tennessee River. Mississippian shale, then limestone, then shale extend to a bit east of the Kingston Springs exit, beyond which are Silurian limestone and shale for about 5 miles (8 km), then Ordovician limestone and shale through Nashville to Mississippian rocks near the Tennessee Route 56 exit. Mississippian shale, then limestone, lead to the Cumberland Plateau boundary near the Monterey exit.

❖THE INTERIOR BASINS

Toward the west, the rock layers of the Cincinnati Arch continue into two large midwestern structural basins, the Michigan Basin and the Illinois Basin (Figure 59). There is no discrete border comparable to the one formed by escarpments of the Allegheny Plateau.

◆ The Michigan Basin

The Michigan Basin (Figures 71 and 87) includes Lower Michigan plus surrounding parts of Illinois, Wisconsin, Upper Michigan, Peninsular Ontario, Ohio, and Indiana. Except in a couple of secondary structures, the basin's bedrock layers all dip toward the center. The map in Figure 87 shows how the surface pattern of the formations consists of concentric rings around a central oval, the oldest rocks outside and the youngest in the middle. With one exception, the rocks range in age from Cambrian to Carboniferous. The Cambrian to Mississippian sediments are marine in most cases, from the great Paleozoic oceans of the Midwest. The Pennsylvanian rocks are mostly terrestrial. One area has a frosting of Jurassic terrestrial sandstone and shale on top of its Paleozoic rocks (Figure 89).

Glaciers covered the Michigan Basin during the Ice Age, leaving till and outwash throughout. The relatively featureless countryside is relieved by major moraines (Figure 82).

The oldest Paleozoic deposits in Michigan are Cambrian sandstone and dolomite that occur in a strip along the Lake

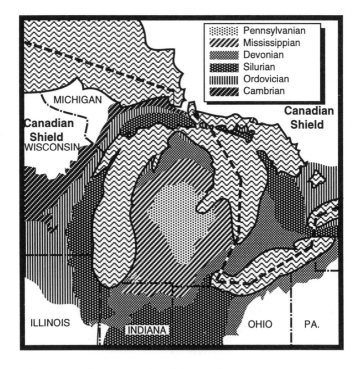

Figure 87. Paleozoic rocks of the Michigan Basin.

Superior shore and inland in the Upper Peninsula. The Cambrian rocks rest on Precambrian ones, which outcrop on the floor of Lake Superior and to the west in the Canadian Shield. In the middle of the Upper Peninsula, an arc of Ordovician limestones rests on the Cambrian rocks. The south shore shows a strip of Silurian dolomite, limestone, and shale.

Two small patches of Silurian rocks occur on the coast of the Lower Peninsula. The Devonian rocks form a ring, partly on the floors of Lakes Michigan and Huron. They consist of Early Devonian dolomite and limestone grading to Middle Devonian shale and limestone, and later shale. Mississippian sandstone overlain by black shale grading to gray shale surround the basin's center. The Cambrian to Mississippian rocks all formed from sediments on the inland sea bottom. Sandstone, shale, and some limestone form the Pennsylvanian sedimentary rocks, with included thin, discontinuous coal layers. There were some mines here in the past, but there are none today. As in other regions with terrestrial Pennsylvanian rocks, the sandstone, shale, and coal originated as sediments and decaying plant matter on the supercontinent Pangaea.

During the Mesozoic Era, sediments washed out of the Rocky Mountains by river systems extended far eastward from the Great Plains. The capping of Late Jurassic red sandstone and shale in the west-central part of the Lower Peninsula is a remnant of that sheet of sediments, most of which has been eroded away. The sediments may have been deposited in valleys that existed at that time. U.S. 10 crosses the area (Figure 89), but unfortunately, there are no outcrops because glacial material covered the Jurassic rocks and they are known only from samples found in the process of drilling wells.

The rocks of Peninsular Ontario originated on the inland sea bottom. They are mostly Devonian, with a wide zone of Silurian at the peninsula's base (Plate 14). The Silurian rocks overlie Ordovician ones that continue far to the east in the Northeast Corridor. Pressure probably associated with the Alleghenian Orogeny forced the peninsular formations into a broad anticline, the Algonquin arch, which trends northeast-southwest, like an extension of the Cincinnati Arch. The now eroded Algonquin Arch is an upfold in the east side of the Michigan Basin, a minor structure superimposed on a larger structure.

Silurian rocks, from the earliest to the latest in Peninsular

Ontario, grade upward from sandstone and shale to limestone and shale to limestone and dolomite. The change from marine sandstone to limestone is evidence of a change in environment. The ocean had earlier retreated from the land during the uplift of the Taconic Orogeny at the end of the Ordovician Period. In Early Silurian times, the region was near the seashore, and sand was being deposited. Then the ocean advanced inland, and the same region ended up even farther from the shore, where the bottom sediment was a calcareous mud that became limestone.

A major feature of Peninsular Ontario is the Niagara Escarpment (Figure 73), which is supported by the resistant Lockport Dolomite of Middle Silurian age. The escarpment extends from Niagara Falls to Hamilton, then north to the south shore of Georgian Bay; it forms the backbone of the Bruce Peninsula and continues through Manitoulin Island into Michigan. Along much of the escarpment, cliffs or hills rise 200 feet (61 m) or so above the country below.

The peninsula's earlier Devonian rocks consist of limestone overlain by shale and limestone, which in turn are overlain by shale. The change from all limestone to all shale is evidence of the Acadian Orogeny, when land to the east was uplifted, and mud washed into the ocean from the new highlands.

Rock formations in northwestern Ohio are continuations of those in Michigan and Ontario and belong to two major structures. They are on the northwest flank of the Cincinnati Arch and also on the southeast portion of the Michigan Basin. Silurian dolomite dips northwest under Devonian sandstone, shale, limestone, and dolomite, which in turn dip under Mississippian sandstone and shale.

The Michigan Basin's Mississippian ring also passes through northern Indiana. Its marine deposits consist of shale, bounded on the south by a strip of Devonian shale, limestone, and dolomite from the end of Lake Michigan eastward. Silurian limestone and dolomite occur farther to the south. The Silurian ring is broken in northwestern Indiana by a broad syncline in which a branch of Devonian rocks remains between Silurian ones. That, like the Algonquin Arch, is a secondary structure on the primary Michigan Basin structure.

In northeastern Illinois, Silurian dolomite forms a broad strip that parallels the Lake Michigan shore. The Silurian area west of Lake Michigan on the map is intertwined in Illinois with the Ordovician area, along a line that goes approximately from between Morris and Minooka (west of Joliet) to Aurora to Harvard. Ordovician shale and dolomite extend westward, beyond the Michigan Basin's margin.

As in the rest of the Michigan Basin, glacial deposits in northeastern Illinois cover most of the bedrock, but to the southwest of Lake Michigan, Silurian rocks outcrop in the valleys of the Des Plaines and Kankakee rivers.

In southern Wisconsin, Ordovician dolomite, sandstone, and shale appear from under the Silurian dolomites and extend off to the west. From Madison to a point on the Menominee River border a few miles north of Marinette, they lap up onto Cambrian sandstone and dolomite.

The Paleozoic inland sea was a warm one, and throughout most of its history, coral animals lived in it. During the Silurian Period, environmental conditions were such that numerous large coral reefs grew from the shallow sea bottom. As time went by, they were surrounded and covered by calcareous mud that ultimately became limestone. Later still, by some unknown process, the limestone and the reefs turned to dolomite. In recent times, the dolomite that came from the limestone surrounding the reefs has weathered and eroded away from the more resistant dolomite of many reefs.

In the region of Silurian rocks to the west and south of Lake Michigan, erosion has left hills that are exposed Silurian reefs, which have been quarried for dolomite. Each of the reefs has a core of amorphous to indistinctly bedded rock that contains many fossils of corals and other animals (Figure 88, Plate 16). Beds of dolomite surround the core, sloping steeply near the core and then flattening out to join the almost horizontal beds of the old sea floor.

Erosion has not entirely exposed any of the reefs, but the

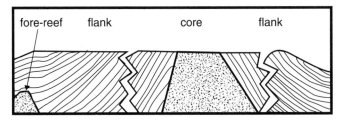

Figure 88. A cross section of the Silurian coral reef at Thornton, Illinois.

visible portions of some of them are large: the core of a reef at Thornton, Illinois, is about 1/3 of a mile (1/2 km) long. This reef, like others, has been quarried for dolomite. It is just southwest of the I-80 intersection with the Calumet Expressway (Illinois Route 394). Where I-80 passes the quarry, dipping flank layers of the reef are visible in a road cut.

In Wisconsin, Silurian dolomites occur in a strip roughly between Lake Michigan and a line from Elkhorn to Green Bay, including the Door Peninsula, and several reefs have been found. As in Illinois, some were quarried until they were more or less destroyed, and most of the quarries no longer operate. At Quarry Lake State Park in Racine, a cross section of a reef core and its flanks can be seen.

MICHIGAN BASIN HIGHWAY SECTIONS

Sault Ste.-Marie, Michigan to the Ohio Border: I-75, U.S. 23 (Figure 89) *I-75 from Sault Ste.-Marie to Flint and U.S. 23 from there to the Ohio border illustrate the downwarped Michigan Basin structure: you pass Cambrian to Pennsylvanian rocks (older to younger) then Pennsylvanian to Silurian rocks (younger back to older).*

The route starts in Cambrian rocks. To the west of Sault Ste.-Marie, along the coast between Munising and Grand Marais, Cambrian sandstone in the Pictured Rocks National Lakeshore has been eroded into fanciful shapes.

On I-75, Ordovician limestone forms the bedrock approximately between Dafter and Cottage Park. From there to the Straits of Mackinac the formations consist of Silurian dolomite, limestone, and shale, with one very interesting exception. The Mackinac Breccia occurs as large inclusions of broken rock embedded in the normal bedrock or as pillars or stacks of resistant breccia where the surrounding rock has been eroded away. Castle Rock, just north of St.-Ignace on I-75, and Sugar-Loaf Rock and Arch Rock on Mackinac Island are examples. The breccia is exposed in road cuts at the north end of the bridge at St.-Ignace, surrounded by contorted Silurian dolomite and shale.

The masses of Mackinac Breccia owe their origin to an ancient salt deposit. In the Silurian Period, evaporating water left beds of salt in areas of very shallow water along the margin of the ocean. Silt washed from the land became interbedded with the salt and eventually turned into shale. Over a

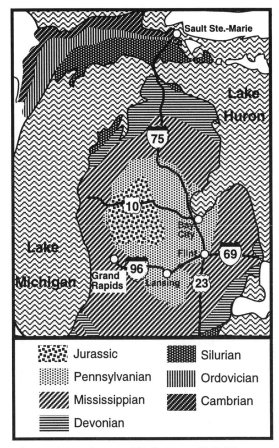

Figure 89. Geology and highways of Michigan.

long period, some salt layers grew to great thickness. Then the ocean bottom sank or sea level rose, and dolomite and limestone covered the salt and shale. Later, in the Devonian Period, an uplift raised the salt and shale above sea level. Ground water dissolved the salt and carried it out of the formation, leaving large caverns. Some cavern roofs collapsed during the Devonian Period, filling in the caverns with fragments of broken rock ranging in size from rock dust to boulders larger than houses. The fragments were later cemented into the solid rock of the Mackinac Breccia.

I-75 crosses the Straits of Mackinac and enters the Devonian dolomite, limestone, and shale of Lower Michigan, which extend to the area of Vanderbilt, where Mississippian sandstone and shale occur. Beyond the Roscommon area, the rocks are Pennsylvanian sandstone and shale all the way to just the other side of Flint. From Sault Ste. Marie to Flint, the highway traverses samples of Cambrian to Mississippian deposits from the great inland seas, plus sediments from Pennsylvanian land.

The formations are successively older on the other side of the basin. Mississippian sandstones and shales form the bedrock a little over halfway from Ann Arbor to Milan. Beyond Milan, Devonian dolomite, limestone, sandstone, and shale extend to a point a couple of miles past the U.S. 223 exit, where the road enters a tiny corner of the Silurian dolomite that extends into Ohio.

Port Huron to Grand Rapids, Michigan: I-69, I-96 (Figure 89) *I-69 from Port Huron to Lansing and I-96 from Lansing to Grand Rapids cross the basin from east to west, from Devonian to Pennsylvanian to Mississippian, from older to younger to older rocks.*

Port Huron lies on Devonian dolomite, limestone, and shale. Mississippian sandstone and shale begin in the vicinity of Goodells, and Flint is just over the boundary into the Pennsylvanian-age center of the Michigan Basin, with its sandstone and shale. Continuing westward, at the Elmdale exit, the road returns to the Mississippian formations, which extend beyond Grand Rapids to the shore of Lake Michigan.

Some of the world's largest sand dunes occur in the Grand Traverse Bay region, at the Sleeping Bear Dunes National Lake Shore. The site is west of Traverse City on Michigan Route 22 between Frankfort and Empire. The dunes, as much as 480 feet (146 m) high, consist of glacial outwash from the Ice Age.

Hamilton to Windsor, Ontario: Routes 2/53, 403, 401 (Figure 78) *Highways from Hamilton to Windsor are built close to the Algonquin Arch's crest. Since the structure plunges toward the southwest, weathering and erosion have exposed the Silurian and Ordovician beds lying beneath the Devonian ones that cover most of the arch.*

Upper Ordovician Queenston terrestrial red shale forms

the bedrock at Hamilton, but Silurian dolomite and lime-stone show up immediately west of the city. The Devonian zone starts about where Route 403 joins Route 401. From there to Windsor, the highway traverses a till-covered Devonian ocean bottom of dolomite, limestone, and shale.

The route provides more than one indication that the rip-pled structures of the Stable Interior are not as simple as they might seem. The Algonquin Arch has a minor down-warping along its crest, exhibited by the Devonian ocean-bottom rocks. From the 403/401 highway junction area to about London, the bedrock is older limestone; from London southwest, the older limestone changes to a younger lime-stone and shale combination, then to a still younger shale. South of Chatham, the rocks change back to the limestone and shale, and farther on, below Lake St. Clair, the older limestone shows up again. The shift from older to younger to older rocks is a sign of a synclinal structure. The syncline is a minor structure superimposed on a larger structure, the Al-gonquin Arch, which is superimposed on the primary struc-ture, the Michigan Basin.

Toledo, Ohio to the Indiana-Illinois border: I-90 (Figures 84, 93)

I-90 leads from the crest of the Cincinnati Arch across the south end of the Michigan Basin, from Silurian to younger Mississippian back to older Silurian bedrock.

In Ohio, I-90 passes from the suburbs of Toledo to the Indiana line through successively younger formations of Silurian to Mississippian age, all remnants of the inland sea bottom. Toledo has Silurian dolomite for bedrock, which gives way to Devonian sandstone about 3 miles (5 km) west of the U.S. 20 exit on I-90. After that, the highway crosses Devonian limestone, dolomite, and shale to about the Wauseon exit, followed by Mississippian sandstone and shale to the Indiana border.

The Indiana rocks along the route are all Mississippian sandstone and shale to approximately 6 miles (10 km) east of Portage, where the bedrock switches to Devonian dolomite, limestone, and shale. Silurian dolomite and limestone occur from about the west edge of Gary to the Illinois line.

South Bend to Rochester, Indiana: U.S. 31 (Figure 95)

From South Bend, U.S. 31 crosses the southern end of the Michigan Basin in northern Indiana from north to south. The rocks all originated as bottom sediments of the inland sea. Mississippian shale extends to the vicinity of Plymouth;

older Devonian shale, then dolomite and limestone continue to just north of Rochester. Still older Silurian limestone and dolomite pass south from there out of the Michigan Basin. The change from younger to older formations outlines the upwarping around the basin structure.

Milwaukee to Madison, Wisconsin: I-94 (Figure 101)

Silurian formations parallel the Illinois shore of Lake Michigan up into Wisconsin. There, I-94 from Milwaukee to Madison crosses the west edge of the Michigan Basin. The westward succession of older marine sedimentary rocks shows the structure. In the Okonomowoc area, the Silurian dolomite gives way to older Ordovician sandstone, shale, and dolomite that extend to the outskirts of Madison. Madison has Cambrian sandstone bedrock.

◆ The Illinois Basin

The map of the Illinois Basin (Figures 71 and 90) doesn't show concentric ovals, as does the Michigan Basin map. What makes the difference is a major unconformity between Pennsylvanian formations and the older ones below, with the Pennsylvanian beds covering most of the basin structure. It appears that the basin structure developed during the latter part of the Mississippian Epoch. The rocks in the basin at that time were a mixture of Ordovician, Silurian, and Devonian sandstone, shale, and limestone. Some of the sediments had accumulated above sea level and others had accumulated below it, as land rose and fell or sea level changed in the Paleozoic inland sea.

Then, a region containing the basin moved above sea level and was eroded. Mississippian sandstones in Illinois show evidence of a river system with rivers flowing to the southwest from highlands in the northeast. The highlands were presumably a result of the Acadian Orogeny, or continental collision. Weathering plus erosion by the river system truncated and exposed the uplifted basin structure, with the oldest rocks then showing on the outside and the youngest ones on the inside. Thus there were Devonian, Silurian, and Ordovician formations in successively older rings around the Mississippian ones. By the end of Mississippian times, a geologic map of the Illinois Basin would have resembled the current map of the Michigan Basin.

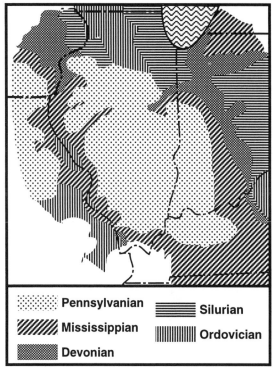

Figure 90. A geologic map of the Illinois Basin.

In the final chapter of the story, Pennsylvanian-age rivers flowed over the region from new highlands in the northeast, as the Alleghenian Orogeny was folding the eastern crust into a mountain range. The rivers deposited silt and sand on the eroded surface of the structural basin. The Illinois Basin was then part of the growing supercontinent Pangaea. The silt and sand became horizontal beds of Pennsylvanian shale and sandstone lying on top of the warped Illinois Basin formations. Since the covering Pennsylvanian layers are not parallel with those of the older basin, they are said to be "not conformable" with the basin layers. The area where the two rock series meet is a major angular unconformity.

Since the Pennsylvanian rocks lap over the older ones,

from Mississippian to Ordovician age, in the north-central part of Illinois, the map of today's Illinois Basin shows an incomplete ring of Mississippian rocks around the Pennsylvanian bedrock area, and partly covered Devonian, Silurian, and Ordovician rings. Shale, sandstone, limestone, and dolomite constitute the Ordovician to Middle Mississippian rocks of the Illinois Basin. The rock types form a typical series of sediments from the wide-reaching Paleozoic inland seas.

In Indiana, the formations of the Cincinnati Arch dip westward into the Illinois Basin and northward into the Michigan Basin. The bedrock there contains evidence for both the Acadian and Alleghenian Orogenies. From Ordovician through Middle Devonian times, muds that became limestone, dolomite, and shale accumulated on the ocean bottom. They give no indication of orogenic activity. However, Upper Devonian and Lower Mississippian marine formations show evidence of the Acadian Orogeny happening to the east, including much sandstone and shale as well as limestone. The sandstone and shale originated as sand and silt washed from the rising mountains. Limestones, showing a return of the ocean, make up the rest of the Mississippian rocks, except for sandstone and shale in the latest layers. Pennsylvanian rocks, as in Illinois, consist mostly of terrestrial sandstone and shale, with coal seams and thin limestone layers. Such rocks represent the supercontinent Pangaea forming during the Alleghenian Orogeny.

In west-central Illinois, there are some sand, gravel, and clay deposits with less variety of rock and mineral fragments than those of the local glacial sediments, that may be Cretaceous and Tertiary in age. Like the Jurassic deposits in Michigan, they may be small erosion remnants of sediments that must have covered much of eastern United States millions of years ago.

● Fluorspar, Karst, and Coal Country

The southeastern tip of Illinois (east of I-57) and the contiguous part of Kentucky are the Illinois-Kentucky Fluorspar District. It extends along the Ohio River from Shawneetown, Illinois, to Paducah, Kentucky, and very roughly as far east as Madisonville, Kentucky. The local structure is a minor arch whose axis trends northwest and southeast, with numerous high-angle reverse faults that trend perpendicular to the arch. The fault zones contain aggregations of minerals that have been mined for industry and gathered by collectors.

Earth's crust here was folded and broken during the Alleghenian Orogeny, and hot mineral solutions from deep below the surface deposited fluorite (fluorspar), barite, and ores of lead and zinc in the fault zones. The Illinois towns of Rosiclare and Cave in Rock are famous among collectors for their fluorite specimens. The area is accessible via Illinois Route 146 east from I-24.

The abundant Mississippian limestones and the karst topography of Kentucky continue into south-central Indiana. The countryside shows numerous sinkholes, disappearing streams (Figure 91), and caves. The Chester Escarpment, which forms the western boundary of the region, is a series of hills upheld by Late Mississippian resistant sandstones that overlie the limestones of the cave country. The escarpment runs roughly from the vicinity of the Crawford-Harrison County line on the Ohio River north and then northwest to the Salt River west of Bedford, then north, where it disappears. It approximates the eastern edge of the Hoosier National Forest. The karst territory's eastern border is approximately a line from a few miles west of Columbus to New Albany on the Ohio River. Karst topography is most highly developed just to the east of the Chester Escarpment.

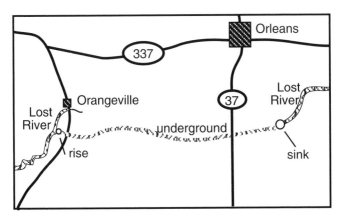

Figure 91. The Lost River in Orange County, Indiana, travels several miles underground through the limestone bedrock.

The Salem Limestone in the Indiana karst region is one of the most widely used building stones in America. It is a classic clastic limestone with an interesting grainy texture because it consists of tiny ocean animal skeletons and sand-size fragments of larger animal remains, cemented by calcite. The limestone is quarried near Bedford and to the south.

Illinois has several major coal seams, furnishing coal from both underground and surface mines. Mines are scattered throughout the Pennsylvanian rock area, but Franklin and Perry counties are the heart of the Illinois coal industry.

Ice Age glaciers covered most of Indiana and Illinois, leaving extensive deposits of till and outwash. Major moraines formed in the northeastern quarter of Illinois (Figure 82). The glaciers never quite reached the southern tip of Illinois, however, and much of the state's northwest portion belongs to the Driftless Area, that part of the Midwest that was never covered by lobes of Ice Age glaciers (Figure 104).

Indiana glacial material consists primarily of till, with silt from glacial Lake Michigan in the north. Major moraines cross the northeastern and northwestern quarters of the state, joining those of Ohio and Illinois.

ILLINOIS BASIN HIGHWAY SECTIONS

Rockford to La Salle, Illinois: I-39 (Figure 92) *Between Rockford and La Salle, I-90 and I-39 cross the oldest rocks in Illinois, which originated as sediments in the inland sea. The route goes southward along a north–south minor arch, with Silurian, then Ordovician, then Silurian formations from east to west, and just a bit of Cambrian in the middle.*

The highway crosses Middle Ordovician limestone and dolomite, then Upper Ordovician shale and dolomite east and west of Rockford. Better exposures occur west of Rockford in the Driftless Area (Figure 102), accessible from U.S. 20. Farther west on U.S. 20 at Galena, named for its lead ore, are remains of old lead mines.

Sandstone and dolomite constitute the Cambrian rock of northern Illinois in a small area that is almost entirely covered by glacial material. The East-West Tollway (I-88) touches a tiny blip of Cambrian in the vicinity of Ashton. A couple of other small exposures occur southwest of the I-39/I-88 intersection.

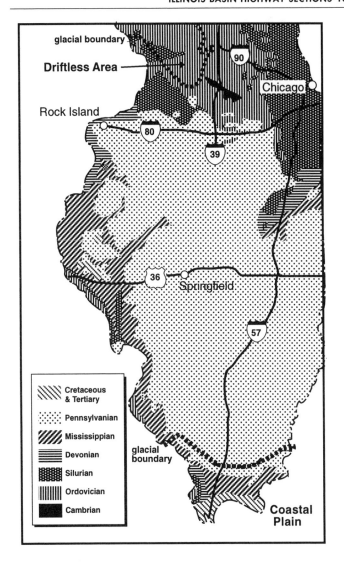

Figure 92. Geology and highways of Illinois.

Chicago to Dongela, Illinois: I-57 (Figure 92) *I-57 from Chicago to the southern tip of the state crosses a complete section of the Illinois Basin. It indicates the broad cover of Pennsylvanian rocks and crosses one area where the Pennsylvanian rocks cover the edge of the original Mississippian basin.*

From Chicago, Silurian dolomite extends about to the Buckley exit, with outcrops in the vicinity of Kankakee. On the way, the highway crosses a small arm in which Pennsylvanian sandstone and shale lie directly on the Silurian rocks, with the unconformity in between. Devonian limestone and shale form the bedrock from the Buckley exit about to the Rantoul exit. A thin strip of Mississippian limestone, sandstone, and shale follows, between Rantoul and Thomasboro. The Devonian and Mississippian formations represent the edge of the original basin. Unfortunately, glacial deposits cover most of the ground from Kankakee to Mattoon.

Beyond Thomasboro the highway crosses a long stretch of Pennsylvanian rocks in the heart of the Illinois Basin, mostly terrestrial sandstone and shale, with coal seams.

I-57 crosses the glacial boundary between Marion and the I-24 exit. Just beyond the latter, Mississippian sandstone and limestone reappear, on the far side of the basin; their outcrop area extends to between the Dongola and Ullin exits. Then, the route passes onto Cretaceous gravel, sand, and clay of the Coastal Plain.

Springfield, Illinois to the Mississippi River: U.S. 36 (Figure 92)
U.S. 36 from Springfield to the Mississippi River furnishes an east-west section of the Illinois Basin's west side. Pennsylvanian rocks as described above occur from Springfield to just before the U.S. 67 intersection. In that area, you enter the rock series for which the Mississippian Epoch was named, that of the central Mississippi Valley. Along U.S. 36, the rocks are limestone (Plate 13). Between Hull and East Hannibal, thin strips of Silurian and Ordovician limestone and shale outcrop along the edge of the valley. To the south of U.S. 36, the Illinois River has eroded through the Mississippian rocks to expose Silurian ones.

Indiana-Illinois border to Rock Island, Illinois: I-80 (Figure 92)
I-80 offers a northern section of the Illinois Basin's margin, between the Indiana line and Rock Island, from older (Silurian) to younger (Pennsylvanian) to older (Devonian) beds.

Classic Silurian dolomites of the area where the Illinois and Michigan basins meet underlie the glacial deposits from the state border to about the I-55 exit. From there to about halfway between the Minooka and Morris exits, the interstate goes close to the boundary between Ordovician shale and limestone of the inland sea and Pennsylvanian terrestrial rocks from Pangaea, where the latter cover the former's eroded surface at the unconformity. Between Morris and the outskirts of Moline the highway crosses the usual Pennsylvanian sandstone and shale. Then it descends over Devonian limestone to the Mississippi River. Near the river in the Rock Island area, outcrops show a complex mixture of Ordovician, Silurian, and Devonian shale, sandstone, and limestone.

South Bend to Indianapolis, Indiana: U.S. 31 (Figure 93) *U.S. 31 traverses the arch between the Michigan and Illinois basins, from South Bend to Indianapolis, from younger to older to younger, from Mississippian to Silurian to Devonian rocks of the inland sea.*

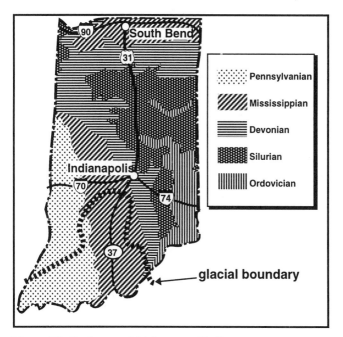

Figure 93. Geology and highways of Indiana.

Mississippian shale extends from South Bend to the vicinity of Plymouth, and Devonian shale grading to dolomite and limestone southward continue to just north of Rochester. There, Silurian dolomite and limestone (with some shale) appear. Silurian coral reefs exist in Indiana as well as in Illinois and Wisconsin, including one exposed by a railroad cut in the city of Wabash (Plate 16). The Devonian formations reappear at about the I-465 exit outside Indianapolis. South of the city, Indiana Highway 37 leads into Mississippian formations about halfway between I-465 and Smith Valley.

To survey the limestone country, continue on Route 37 to I-64 and just beyond it to Indiana 62. The well-known Wyandotte Cave is about 6 miles (9 km) east on Route 62. Route 37 crosses the glacial boundary at Martinsville.

Indiana, from the Ohio border to the Illinois border: I-74, I-70 (Figure 93) *I-74 and I-70 go from the flank of the Cincinnati Arch into the Illinois Basin, from the marine rocks of the inland sea to rocks from the dry land of Pangaea.*

Along I-74 from the Ohio border to New Point, Indiana, Ordovician limestone and shale form the western flank of the Lexington dome on the Cincinnati Arch. Silurian limestone follows from New Point to about 6 miles (9.5 km) beyond the Greensburg exit. From there to Indianapolis, Devonian limestone and dolomite grade to shale in the region of the city, as the formations dip gradually into the Illinois Basin. Mississippian rocks occur at about the I-465/I-70 junction. On I-70, shale grading westward to limestone goes to about halfway between the Indiana Routes 243 and 59 exits. Beyond, the highway crosses Pennsylvanian sandstones, shales, coal, and limestones in the center of the Illinois Basin.

For a traverse of the Illinois Basin's Kentucky portion, see the section on the Cincinnati Arch and read the Bluegrass and Western Kentucky parkways route.

❖THE OZARK DOME AND THE WESTERN TIER

The fourth subdivision of the Central Lowland includes the gently warped Paleozoic sedimentary rock series that passes from the Michigan and Illinois basins westward to disappear underneath the Mesozoic and Cenozoic Great Plains rocks. This rock series includes those of the Ozark Dome in Missouri and northern Arkansas plus the almost horizontal formations of the Western Tier in parts of Texas, Oklahoma, Kansas, Nebraska, Iowa, Wisconsin and Minnesota, with, farther north, an elongate island in Manitoba and Saskatchewan. See Figure 71.

◆ The Ozark Dome

The Ozark Dome is an asymmetrical structural dome situated mostly in Missouri (Figures 71 and 94). The south part of the dome encompasses the Boston Mountains and the plateau country of northern Arkansas.

The hills of the St. Francois Mountains and surrounding

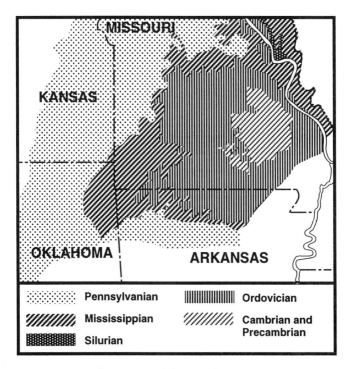

Figure 94. A geologic map of the Ozark Dome.

territory form the dome structure's center, with the oldest rocks exposed southwest of St. Louis. There, Cambrian marine sediments outcrop, remnants of the first Paleozoic inland ocean, with several isolated areas of Precambrian rocks. The Precambrian areas are exposures of the basement complex that underlies the Paleozoic rocks and connects with the Canadian Shield. The undulating upland of the Salem and Springfield plateaus partly surrounds the central region to the north, south, and west. Going outward from the center, the first formations are Ordovician marine sediments, then Mississippian marine sediments. The Mississippian rocks cover the eroded edges of Silurian and Devonian formations that were exposed as the dome grew. Since the Mississippian rocks are not parallel to the bedding planes of the eroded older ones, an unconformity like that in Illinois exists between the Mississippian rocks and the others. More uplands of Pennsylvanian terrestrial and marine sedimentary rocks continue beyond the Mississippian rock zone.

Although the Mississippian rocks cover most of the original exposures of Silurian and Devonian ones, a few thin strips of Devonian rocks do show here and there along the margin of the Salem Plateau, and in the extreme southwestern corner of Missouri, the Elk River and tributaries have cut through the younger rocks to expose Devonian limestone, sandstone, and shale, with a tiny bit of Lower Ordovician. U.S. 71 follows one of the valleys between Anderson, Missouri, and Bentonville, Arkansas. Coastal Plain sediments cover the southeastern part of the dome.

The geologic map of Missouri (Figure 96) shows that I-70 crosses the state close to the margin of deposits left by Ice Age glaciers. Glacial deposits on the margin of the Ozark Dome are mostly till.

● The St. Francois Mountains

The rocks of the St. Francois Mountains are granite, gabbro, and syenite that were intruded below the surface and rhyolite porphyry that was extruded as lava flows 1.48 billion years ago, in the Middle Proterozoic Era. The intrusion and volcanic activity presumably accompanied the breakup of a Proterozoic supercontinent. (See the chapter 3 sections on Canadian Shield history and the rocks of the Southern Province.) As Earth's crust cracked, dikes and sills cut the older rocks about 1.24 billion years ago.

During the Cambrian Period, softer rocks around the resistant intrusive and volcanic ones eroded away, leaving hills. The valleys formed in the softer rocks at the time were invaded by the Cambrian inland sea and now contain essentially horizontal Cambrian sandstone and dolomite. The hills must have been islands in the sea. The Cambrian sediments that accumulated in the sea were later covered by Ordovician sediments that eroded away as the dome's center rose.

The St. Francois Mountains were mined for iron (hematite ore), lead (galena), and barite in the 19th and early 20th centuries. The town names along or near Missouri Route 21 south of St. Louis reflect this: Hematite, Old Mines, Potosi (lead and barite; named for the famous Mexican silver mining town of San Luis Potosi), Irondale, Iron Mountain, and Ironton. Cobalt was mined near the town of Cobalt, and silver west of Fredericktown. The ores condensed from hot mineral solutions, or possibly vapors, deep below the surface. They moved along faults and spread laterally through fracture zones, cavities, and permeable beds. This may have happened during the Alleghenian Orogeny, when intense pressure from the south folded and faulted the rock layers that became the Ouachita Mountains, to the south of the dome.

● The Plateau Country

The plateaus are higher lands on the flanks of the Ozark dome. The Salem Plateau, to the north, west, and south of the St. Francois Mountains, consists mostly of Ordovician dolomite and sandstone. The Eureka Springs Escarpment, atop the Ordovician layers, forms its outer boundary. The escarpment is the eroded edge of a Mississippian cherty limestone. The chert, which is very hard and does not weather readily, makes the limestone resistant enough to form an escarpment. Beyond the Eureka Springs Escarpment is the Springfield Plateau, formed by layers of other Mississippian limestones. Some formations there contain numerous nodules of chert that weather out of the rock and become mixed with the soil.

Caves and springs abound in the limestone and dolomite of both plateaus. The action of the springs, where water issues from underground streams, can be truly impressive. The flow of water from Big Spring, near Van Buren on U.S. 60, from Alley Spring, northwest of U.S. 60 at Winona, and from

Meramec Spring, near St. James on I-44, is measured in hundreds of millions of gallons per day. A scan of a highway map turns up Round Spring, Willow Springs, Mill Spring, Coldspring, Edgar Spring, El Dorado Springs, and more.

Karst features are not highly developed on the plateaus, but at several places in Missouri erosion or highway construction has exposed the remains of ancient caves in the plateau rocks, apparently formed during the Pennsylvanian Epoch. The cave roofs collapsed, filling the caves with rubble that was later covered by Pennsylvanian sediments. In a good example on I-44, at Vichy Road in Rolla, Pennsylvanian sandstone and shale have fallen into a cave in Ordovician Jefferson City Dolomite. There is another example a mile or so to the east.

Iron, zinc, lead, and barite mineral deposits are found in the plateau region (Figure 95). The Tri-State Mineral District, in southwest Missouri and contiguous portions of Kansas and Oklahoma, was once a major lead-zinc producer. Mineral samples can still be found in the tailing piles of old mines in the Joplin–Webb City–Orongo vicinity of Missouri. The plateau mineral deposits originated the same way as those of the St. Francois Mountains area, from hot solutions filling cracks and cavities in the rocks from below.

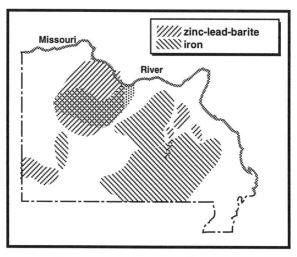

Figure 95. Mining areas of Missouri.

● The Surrounding Country

The Salem and Springfield plateaus extend into northern Arkansas, on the Ozark Dome's south flank, and the geology on the plateau extension resembles that of the Missouri part. South of the Springfield Plateau rise the Boston Mountains, which stretch east-west from eastern Oklahoma to over halfway across Arkansas, just east of Batesville, to where Coastal Plain rocks overlap those of the Ozark Dome. Local rivers have deeply eroded the almost horizontal formations to form valleys as deep as 1,000 feet (305 m). The sediments of most formations accumulated on the inland sea bottom and rising land during the Mississippian and Pennsylvanian epochs; older marine rocks outcrop in the eastern portion.

In Oklahoma, the Springfield Plateau country has been eroded into a landscape of relatively deep river valleys between flat highlands. Here, too, the bedrock originated as sediments in the inland sea. Limestone, sandstone, and shale constitute the Mississippian rocks of the area. One limestone formation contains much chert in layers, lenses, and nodules. As in extreme southwestern Missouri, outliers of Devonian and Ordovician sandstone, shale, and limestone outcrop where rivers have eroded through the covering Mississippian rocks. The only outlier on a major highway is on U.S. 62, halfway between the Arkansas line and Tahlequah.

At the Ozark Dome's southwest margin in Oklahoma, the Pennsylvanian outcrop zone swings north around the older Mississippian zone. The Pennsylvanian limestone, sandstone, and shale dip gently west, southwest, and south along the edge of the dome structure. A few coal seams developed in the rocks, but Oklahoma produces little coal. In the vicinity of Picher, on U.S. 69 north of Miami, there are remains of old lead-zinc mines in the Tri-State Mineral District.

OZARK DOME HIGHWAY SECTIONS

St. Louis to Kansas City, Missouri: I-70 (Figure 96) *I-70 from St. Louis to Kansas City crosses a large area of Carboniferous rocks on the Ozark Dome and more or less parallels the glacial boundary.*

St. Louis rests in the glaciated region on a small outlier of Pennsylvanian sandstone, shale, limestone, and coal that is

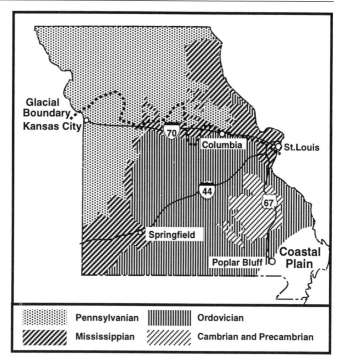

Figure 96. Geology and highways of Missouri.

surrounded by older Mississippian limestone. From St. Louis all the way to Kansas City, the interstate flirts with the glacial boundary and in places with pre-Carboniferous rocks. Beyond Warrenton, it enters a complicated unglaciated area of successively older rocks, with Pennsylvanian sandstone, shale, and limestone, Mississippian limestone, Devonian sandstone, shale, and limestone, and Ordovician limestone, dolomite, and sandstone, along the eroded edge of the Salem Plateau.

A finger of till extends from just past the U.S. 54 exit approximately to the east edge of Columbia. From there, the highway crosses unglaciated Mississippian rocks, mostly limestone, to about the Emma exit. In that part of the journey, I-70 just nicks the end of a small extension of till east of Sweet Springs. From the Emma exit to Kansas City, Pennsylvanian terrestrial sandstone and shale outcrop, south of the glacial boundary, except for a fingertip of till.

St. Louis, Missouri to the Oklahoma border: I-44 (Figure 96) *I-44 offers a section of the Ozark Dome from St. Louis to the Oklahoma line, passing from Carboniferous to older Ordovician and back to Carboniferous over the dome structure, with older rocks in the center and younger ones outside. The route encounters older rocks from the Paleozoic inland sea and younger ones from the rising land of Pangaea.*

At St. Louis, the bedrock is Carboniferous sandstone, shale, and limestone. Beyond St. Louis, the highway is built on Mississippian limestone. Between Eureka and Pacific it crosses older rocks, the St. Peter sandstone of Ordovician age. Ordovician sandstone and dolomite form most of the bedrock from there almost to Springfield, but the highway crosses several isolated erosion remnants of Pennsylvanian sandstone and shale between Leasburg and Rolla. Such remnants indicate that the Pennsylvanian rocks once covered the whole Ozark Dome. Note on the way Meramec Caverns, Onondaga Cave, and Meramec Spring. The Pennsylvanian sinkhole structure visible at Rolla is described above.

The Ordovician dolomite and sandstone extend from Rolla to Lebanon, and the road passes the Onyx Mountain Caverns. The town of Decaturville sits north of Lebanon on Missouri Route 5. A structure just south of the town has been interpreted as an ancient meteoroid or comet impact crater exposed by erosion. Cambrian and Ordovician rocks have been deformed into a circular structural depression surrounded by normal faults, with a central uplifted area. This topography closely resembles many craters on the moon.

Bennett Spring is west of Lebanon on Missouri Route 64, another example of Missouri's well-developed cave and spring systems. On I-44, past Lebanon, the Ordovician rocks continue to about the Northview exit. From there to the Oklahoma border the road crosses younger rocks again: Mississippian sandstone and limestone (with chert nodules) on the west flank of the Ozark Dome. Joplin and Webb City, just east of the border and north of the highway, occupy the heart of the old Tri-State Mineral District.

St. Louis to Poplar Bluff, Missouri: I-55, U.S. 67 (Figure 96) *This route traverses the oldest part of the dome, with its Cambrian and Precambrian rocks.*

I-55 crosses Mississippian limestone to Festus, where it enters the Ordovician outcrop area. Just beyond Festus, the route turns south on U.S. 67, through dolomite and sandstone between Festus and Bonne Terre. Then, the highway enters the Cambrian dolomite and sandstone area in the heart of the Ozark Dome. About 8 miles (13 km) beyond the Missouri Route 72 junction (Fredericktown) the road crosses a small island of Precambrian rock. The Cambrian sediments continue to just before Greenville, where the road goes back into the younger Ordovician rocks. Poplar Bluff is at the edge of the Coastal Plain, where Quaternary river deposits cover the Ordovician rocks of the dome.

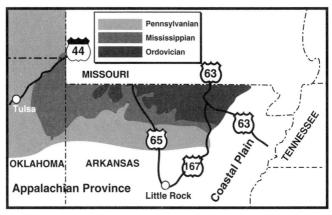

Figure 97. Geology and highways in the Kansas-Oklahoma-Arkansas portion of the Ozark Dome.

Missouri border to Conway, Arkansas: U.S. 65 (Figure 97)

U.S. 65 crosses into the Boston Mountains just south of Leslie, Arkansas. The rocks to the south are Pennsylvanian: first limestone and shale, then sandstone and shale between Clinton and Conway, which is in the Appalachian Province. Along the southern boundary of the Boston mountains, faults and more steeply folded rock layers mark the edge of the Arkansas Valley and Ouachita Mountain zone of the Appalachian Province.

Missouri border to Bald Knob, Arkansas: U.S. 63, U.S. 167 (Figure 97)

U.S. 63 from the state border to Hardy, then U.S. 167 to Bald

Knob cross the Ordovician beds of the ancient inland sea in Arkansas. Ordovician dolomite and limestone occur from the state border to just beyond Cave City. From there, Mississippian limestone and shale extend to Bald Knob, at the edge of the Coastal Plain.

Joplin, Missouri to Tulsa, Oklahoma: I-44 (Figure 97)

I-44 traverses the Oklahoma corner of the Ozark Dome between Joplin and Tulsa. Mississippian marine limestone, shale, and sandstone lie from the state border to about Vinita; from there to Tulsa, mostly terrestrial Pennsylvanian sandstone, shale, and limestone form the bedrock.

Picher, Oklahoma to Pleasanton, Kansas: U.S. 69 (Figure 98)

Southeast Kansas offers a small portion of the Springfield

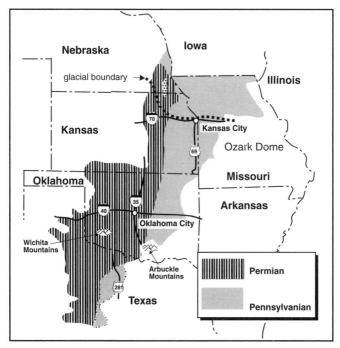

Figure 98. Geology and highways in the southern portion of the Western Tier.

Plateau and a strip of Pennsylvanian, mostly terrestrial rocks from the supercontinent Pangaea that connects the margin of the Ozark Dome in Oklahoma and Missouri. A drive along U.S. 69 north from Picher illustrates the geology of the area where the dome merges with the Western Tier.

The route is entirely on Pennsylvanian rocks up to Pleasanton; to the north and west of that town the Pennsylvanian rocks stretch beyond the fringe of the Ozark Dome.

Picher is in the Tri-State Mining District, where old mine dumps contain mineral samples. Between the Oklahoma border and Ft. Scott, the formations consist of sandstone, shale, and coal. There are a number of strip mines, but Kansas coal production is now slight. A couple of limestone formations remind one that occasionally, shallow Pennsylvanian seas advanced temporarily over the land. Between Ft. Scott and Pleasanton the rocks trend toward shale with a greater thickness of limestone.

◆ The Western Tier

The Western Tier lies between the Michigan and Illinois basins and the Ozark Dome to the east and the Great Plains to the west, from Texas into Canada (Figure 71). The Paleozoic sedimentary rocks extend westward from the great structures without significant warping and disappear under younger Great Plains rocks, to reappear farther west, in and around the Rocky Mountains.

● The Osage Plains and Oklahoma

The Osage Plains stretch beyond the Springfield Plateau, the Boston Mountains, and the Ouachita Mountains in eastern Oklahoma, southeastern Kansas, and west-central Missouri. The region has a relatively flat landscape with occasional escarpments of varying heights. These scarps, or eroded edges of resistant layers, trend more or less northeast and southwest. The rock layers dip gradually west and northwest. The Osage Plains rocks are mostly Pennsylvanian terrestrial sandstone and shale from the growing Pangaea (Figure 100), with some limestone. Middle Pennsylvanian rocks contain a few coal seams, with some surface mines, but Missouri coal production is low. In Oklahoma, the Pennsylvanian lime-

(Text continues on page 199)

PLATES

PLATE 1

THE GROWTH OF NORTH AMERICA

The core of the continent formed over 2.5 billion years ago. Since then, successive orogenies have added land to the margins as subcontinental masses have slowly collided with it. This generalized map indicates the time spans during which the collisions occurred.

North American rocks continue through the Arctic Islands and into Greenland, represented by the white area at the upper right. They are not shown because they are outside the scope of this book.

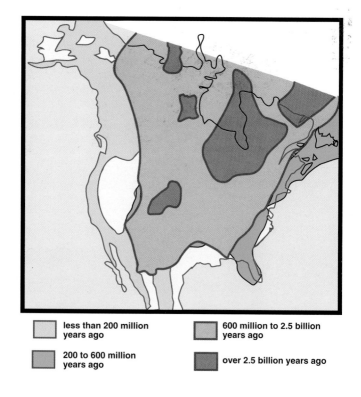

less than 200 million years ago

200 to 600 million years ago

600 million to 2.5 billion years ago

over 2.5 billion years ago

PLATE 2

THE CHANGING EARTH

A hypothetical restoration of land masses 600–580 million years ago. To the west is Laurentia, including ancestral North America and Greenland. To the east is Gondwana, which much later broke into sections, now Africa, Antarctica, Australia, India, and South America. In between are subcontinents that became parts of the supercontinent Pangaea when all these lands fused. To the north is a land mass that eventually became Scandinavia. To the south is an island or a piece of Gondwana that now forms the base of Florida.

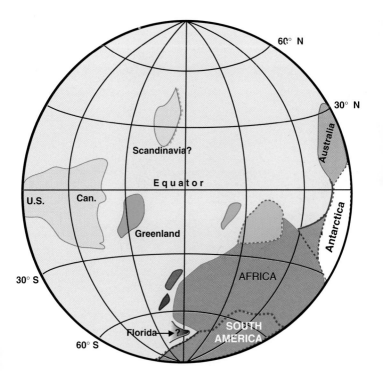

PLATE 3
THE CANADIAN SHIELD

1. Typical Canadian shield country, in the North Bay area, Ontario. Although there are hard rocks in the Shield, weathering and erosion over a very long time plus glacial action have smoothed the landscape. The hill in the foreground shows that the Shield is not entirely flat.

    (Photo: Robert Newby)

2. Eastern and southeastern portions of the Canadian Shield are mountainous; this is a view of Laurentian Mountain scenery in Québec, as seen from Highway 17 in the vicinity of Stonecliffe, Ontario, along the Ottawa River.
    (Photo: R. Michael Easton, Ontario Geological Survey)

3. Adirondack topography, Fulton Chain Lakes. The Adirondacks are another range that relieves the flatness of the Canadian Shield.

2

3

PLATE 4
THE CANADIAN SHIELD

1. Furnace Falls, off Route 503 west of Norland, Ontario. The falls are at the contact between schist and marble that were originally sandstone and limestone. Low falls such as this are common throughout the Canadian Shield in Ontario. They are among the best places to look for exposed bedrock.
   (Photo: R. Michael Easton, Ontario Geological Survey)

2. The Paleozoic-Precambrian unconformity on Trans-Canada Highway 7, west of Madoc, Ontario. Ordovician limestone overlies Proterozoic volcanic rocks that have been somewhat metamorphosed.
   (Photo: R. Michael Easton, Ontario Geological Survey)

3. This granite, near Killarney, Ontario, was intruded about 1.75 billion years ago. The scattered dark blocks in the light pink granite are broken fragments of the bedrock into which the molten minerals that became the granite were forced.
   (Photo: R. Michael Easton, Ontario Geological Survey)

2

3

PLATE 5

THE CANADIAN SHIELD

1. Banded gneiss in the Parry Sound area, Ontario. The light and dark bands are of different mineral composition and have weathered differently. This gneiss may have been formed in a fault zone about 12–19 miles (20–30 km) deep in Earth's crust.
   (Photo: R. Michael Easton, Ontario Geological Survey)

2. Gray gneiss along Route 169 north of Bala, Ontario, that has been injected with pink granite.
   (Photo: R. Michael Easton, Ontario Geological Survey)

3. Marble that contains inclusions of gneiss and schist. During the process of metamorphism, the marble flowed into cracks in adjacent rocks, fragments of which became trapped in the marble. This mixed rock is common in the Kingston-Westport, Bancroft-Minden, and Parry Sound areas of Ontario.
   (Photo: R. Michael Easton, Ontario Geological Survey)

1

2

3

PLATE 6

THE CANADIAN SHIELD

1. A polished slab of iron formation. The darker layers are the iron-bearing ones; the light layers are chert. The rock sample is about 8 inches (20 cm) wide.
   (Photo: R. Michael Easton, Ontario Geological Survey)

2. Evidence for Proterozoic glaciation. This outcrop of the shale variety called mudstone in the Espanola, Ontario, area contains pebbles, cobbles, and boulders of other rock types. The mudstone is associated with tillite, and the original sediment presumably formed when the "foreign" rocks melted out of ice floes and dropped into lake-bottom mud.
   (Photo: R. Michael Easton, Ontario Geological Survey)

3. Archean pillow lava northeast of Wawa, Ontario. The dark strips outline the blobby forms the lava took as it erupted underwater, on the ocean floor. Weathering, erosion, and glacial polishing have produced this cross section of the lava flow.
   (Photo: R. Michael Easton, Ontario Geological Survey)

2

3

PLATE 7

THE CANADIAN SHIELD

1. This cliff on Pokomoonshine Mountain in the northeastern Adirondacks consists mostly of granite gneiss. It is cut by a slightly darker tabular body of diorite that extends diagonally from lower left to upper right.

   (Photo: James F. Olmsted)

2. A road cut in anorthosite on Route 3, about a mile west of Saranac Lake, New York.

   (Photo: James F. Olmsted)

3. A closeup of the anorthosite shown in photo 2.

   (Photo: James F. Olmsted)

2

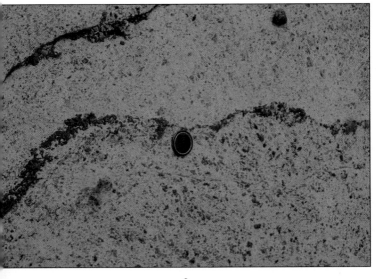

3

PLATE 8

THE CANADIAN SHIELD

1. Intensely folded gray marble and darker layers of a related rock type with a more complicated mineral composition. The rusty-colored layer at the lower right part of the exposure is a granitic material. This outcrop, south of Ausable Forks, New York, is a good example of the small outcrops that can sometimes be found in wooded areas where one would not expect to find bare rocks.

   (Photo: James F. Olmsted)

2. Proterozoic lava flows exposed along the St. Croix River near Taylors Falls, Minnesota.

   (Photo: David Crawford)

3. An outcrop of the Sioux Quartzite formation in Pipestone National Monument, Pipestone, Minnesota. The upper layer, with the more irregular broken surface, is quartzite. The more angular layer below it consists of the shaly rock that Native Americans used for making pipes. The Sioux Quartzite originated as sand, silt, and clay derived from the weathering and erosion of highlands during the Proterozoic Eon.

   (Photo: NMN, Inc.)

2

3

PLATE 9

THE STABLE INTERIOR

1. The Allegheny Plateau in southwestern Pennsylvania.

2. Allegheny Plateau topography in northeastern Ohio. The deep and steep-walled valley of the Chagrin River crosses from right to left in the middle distance but is hardly evident in the photograph. This view shows a "dissected plateau."

3. The Allegheny Plateau in central West Virginia, where there are no glacial deposits to fill in the valleys. The tops of the hills are all about the same elevation and represent the earlier plateau surface, but weathering and erosion have carved steep, narrow valleys everywhere.

2

3

PLATE 10

THE STABLE INTERIOR

1. Niagara Falls and the gorge downstream typify the Stable Interior in that flat-lying rock layers are exposed underlying a relatively flat landscape. Structurally, Niagara Falls is a part of the Central Lowland.

(Photo: Robert Newby)

2. The Central Lowland in northeastern Kansas; classic farmland around Brown County Lake, near Hiawatha.

3. Even in the Central Lowland, extra-resistant rocks form hills in the otherwise flat topography. These hills are in the St. Francois Mountains area, near Centerville, Missouri.

2

3

PLATE 11
THE STABLE INTERIOR

1. A Central Lowland landscape in north-central Texas. The weathered shale in which the grass and mesquite grow holds the petrified bones of Early Permian fish, amphibians, and reptiles, as well as fossils of the kinds of plants associated with Pennsylvanian and Permian coal seams elsewhere.

2. A restoration of the Early Permian landscape in the region pictured in photo 1. The "sails" on three of the reptiles shown may have been temperature-regulating organs.
   (Photo courtesy New York State Museum, Albany, NY)

3. Palo Duro Canyon, southeast of Amarillo, Texas, where Triassic and Tertiary (Pliocene) rocks of the Great Plains overlie Permian rocks of the Central Lowland. The Permian and Triassic rocks are mostly redbeds.

2

3

PLATE 12

THE STABLE INTERIOR

1. A typical outcrop of shale (Devonian) with a couple of fine-grained sandstone layers in the Chagrin River valley in northeastern Ohio. The sandstone layers stand out because they are more resistant to weathering than the shale is.

2. An outcrop of Mississippian sandstone in a tributary ravine to the Chagrin River in Ohio (see photo 1). Note the two layers showing delta-type cross bedding, which indicates that the sand accumulated in the ocean near the shore.

3. Silurian limestone in an Ohio quarry. The quarry face shows how limestone tends to break into rectangular chunks, which distinguishes it from most sandstone and shale. Because of this characteristic, many geological diagrams have bricklike rectangles to indicate limestone.

2

3

PLATE 13

THE STABLE INTERIOR

1. The Onondaga Limestone escarpment, at New York Route 19, south of the New York Thruway.

2. Chert nodules in Devonian limestone, in western New York. The quarter at upper right gives the scale.

3. Mississippian limestone and chert layers, where U.S. 36 descends into the Illinois River valley. The chert is the more resistant rock.

2

3

PLATE 14
THE STABLE INTERIOR

1. It is not necessary to have a large outcrop to interpret geologic history. The students in the photograph dated limestone in this relatively unobtrusive outcrop as Pennsylvanian in age, on the basis of the marine fossils visible in it.
   This outcrop borders U.S. 36 in north-central Missouri.

2. Devonian shale with resistant limestone layers, along the
   Ausable River near Arkona, Ontario. This particular variety
   of shale weathers easily into small pieces and silt that cover
   the outcrop, hiding the bedding planes.

3. Alternating shale and dolomite layers of Ordovician age
   along the New York Thruway. Apparently, some ocean-bottom action rippled the layers slightly before they turned into
   solid rock. Note that the layers at the upper portion of the
   road cut are normal and unwarped.

2

3

PLATE 15

THE STABLE INTERIOR

1. Ordovician limestone and shale at the core of the Cincinnati Arch in Kentucky (I-75).

2. Pennsylvanian sandstone and shale in an unusual road cut in northeastern Kentucky on I-64, which shows the cross section of a river channel. The channel filled with sand and silt and was eventually covered with sand, which became the light-colored sandstone at the upper level of the cut.

3. Permian sandstone and shale along U.S. 36 in northeastern Kansas. Many of the best road cuts are on highways other than interstate highways.

2

3

PLATE 16

THE STABLE INTERIOR

1. Pennsylvanian sandstone and shale, with a coal seam in between, along I-80 in west-central Pennsylvania. Bits of coal from the outcrop have slid down and partially obscured the shale below.

2. Some joints in bedrock are irregular; others are surprisingly straight. This outcrop in an abandoned quarry shows that the quarry workers removed rock up to a joint and stopped. The upper surface is a bedding plane. The rock is a Silurian sandstone in northwestern New York.

3. This railroad cut revealed a cross section of a Silurian coral reef in north-central Indiana. The limestone layers slant downward on the flanks of the reef.

2

3

PLATE 17
THE APPALACHIAN PROVINCE

1. Long parallel ridges that are all about the same height characterize Appalachian Mountain topography. The ridges and intervening valleys are erosion remnants of folds in Paleozoic sedimentary rock. This photograph shows the country around the border of Virginia and West Virginia.

2. In some places, entire folds were tilted. These anticlines and synclines in central Pennsylvania dip down to the right.

(Photo: Richard Blount)

3. The Great Valley in Virginia as seen from the Blue Ridge Parkway, looking west toward Goshen Pass and Little North Mountain.

(Photo: Gerald P. Wilkes,
Virginia Division of Mineral Resources)

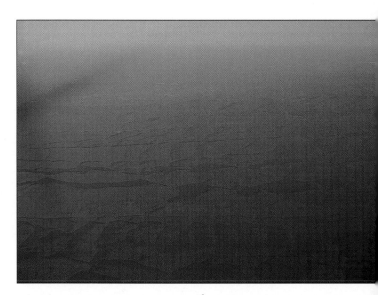

2

3

PLATE 18

THE APPALACHIAN PROVINCE

1. The Blue Ridge as seen from the front yard of James Madison's home near Orange, Virginia, in the Piedmont. Erosion has smoothed off the Piedmont's folded and faulted Paleozoic sedimentary and metamorphic formations.

2. Piedmont topography, as seen from I-77 in North Carolina. Only one hill in sight, off in the distance.

   (Photo: Grant Hodsdon)

3. Ouachita Mountain scenery, from U.S. 259 in southeastern Oklahoma. The hills contain folded and faulted Paleozoic sedimentary rocks in structures similar to those of the Appalachian Mountains.

2

3

PLATE 19
THE APPALACHIAN PROVINCE

1. Rolling topography in the Arbuckle Mountains, Oklahoma, off U.S. 77 near Turner Falls. This is the eroded surface of the Arbuckle Anticline, in Paleozoic sedimentary rocks. The light-colored rocks that resemble gravestones in the grass in the right foreground are the broken edges of almost vertical limestone layers.

2. The hills of the Wichita Mountains in Oklahoma rise sharply out of the flat Central Lowland prairies. They are erosion remnants of Cambrian rocks that were forced up through later Paleozoic formations. Their erosion, in turn, contributed to the Permian sedimentary rocks that now surround them.

3. The Llano Uplift area in central Texas is not mountainous, but its irregularly eroded Proterozoic igneous and metamorphic rocks contrast greatly with the flat-lying younger sedimentary rocks that surround it. This view is of schist and gneiss exposures in the Inks Lake State Park, southwest of Burnet.

2

3

PLATE 20

THE APPALACHIAN PROVINCE

1. A portion of the White Mountains in east-central New Hampshire, as seen from U.S. 302 near Glen. Mt. Washington, the highest peak in the northeastern United States, is on the skyline. The range is largely folded and faulted Paleozoic metasedimentary and metavolcanic rocks, with Paleozoic and later intrusives and volcanics.

2. The biggest ridge in the Massachusetts section of New England's Triassic Basin, north of Holyoke. A resistant basalt layer supports the ridge (Mt. Tom) extending away in the photograph. The gentle slope to the right is underlain by layers on top of the basalt, in a series of formations that dips down to the east. The steeper slope to the left shows the broken, eroded edges of the basalt and sedimentary layers that underlie it.

3. Mountainous country along the Restigouche River, the border between New Brunswick and Québec, at Matapédia, Québec. The bedrock consists of folded and faulted Early Paleozoic sedimentary formations.

2

3

PLATE 21

THE APPALACHIAN PROVINCE

1. The cliffs along the north shore of Chaleur Bay, in the Miguasha area of the Gaspé Peninsula, Québec, show Devonian terrestrial sedimentary rocks. The sediments indicate rising lands during the Acadian Orogeny.

2. A Newfoundland landscape along Trans-Canada Highway 1, on the Avalon Peninsula west of St. John's. The rocks in the area are mostly folded and faulted Proterozoic sedimentary and metasedimentary rocks.

3. Tilted Pennsylvanian sandstone and shale on the shore at Joggins, Nova Scotia. The rocks are part of a terrestrial series that contains coal seams.

2

3

PLATE 22

THE APPALACHIAN PROVINCE

1. This spectacular road cut in Sideling Hill, western Maryland (I-68), shows a syncline in Mississippian conglomerate, sandstone, and shale.

    (Photo: Grant Hodsdon)

2. Folded Ordovician limestone and shale in a road cut on U.S. 322 near Reedsville, in central Pennsylvania. Typically, smaller folds like these occur on the limbs of larger ones, where layered rocks have been folded by horizontal pressure.

    (Photo: Paul Kenepp)

3. The folds of two synclines and an anticline are exposed along U.S. 259 in the Ouachita Mountains, southeastern Oklahoma. The rocks are Late Devonian or Early Mississippian shale and chert of the Arkansas Novaculite.

2

3

PLATE 23

THE APPALACHIAN PROVINCE

1. Tilted Cambrian or Ordovician limestone on the south limb of the Arbuckle Anticline, on U.S. 77 in Oklahoma.

2. Tilted Late Proterozoic quartzite in the Piedmont, at Willis Mountain in Buckingham County, Virginia. Piedmont rocks such as this and the one in photo 3 are related to rocks farther north in the Appalachian Province.

   (Photo: John D. Marr, Jr.,
   Virginia Division of Mineral Resources)

3. Migmatite in the Virginia Piedmont. Migmatite is an intimate mixture of igneous and metamorphic rock materials, in this case granite and gneiss. This particular rock is approximately 1 billion years old; it belongs to the Grenville Proterozoic series that occurs also in New England, Newfoundland, and in a large area of the Canadian Shield. Compare this with the rock in photo 2 of Plate 25.

   (Photo: John D. Marr, Jr.,
   Virginia Division of Mineral Resources)

2

3

PLATE 24

THE APPALACHIAN PROVINCE

1. The lumpy surface of this metabasalt layer indicates that it was extruded onto the ocean bottom. It formed during volcanic action associated with an island chain during Early Paleozoic times. The brook and its basaltic bed are in northwestern Massachusetts.

2. An exposure of typical schist, probably Late Silurian in age, northwest of Sanford, Maine (Routes 109/11). Dark and light contorted layers reflect the pressure it underwent. The dark layers are rich in dark minerals, such as biotite, and the light layers contain light-colored feldspar and quartz.

3. A schist outcrop in the woods of Hoosac Mountain, western Massachusetts. The age is probably Cambrian.

2

3

PLATE 25

THE APPLACHIAN PROVINCE

1. Proterozoic gneiss from Northfield Mountain in northwestern Massachusetts. Gneiss differs from schist in that its dark minerals do not occur as flakes. The texture shows that the minerals have slowly flowed. Aggregations of lighter minerals (feldspar and quartz) shaped like those in the photograph are sometimes called "augen," from the German word for eyes.

2. This contorted mixture of Ordovician gneisses dramatically demonstrates how the minerals flowed under heat and pressure. The exposure is in the bed of the Deerfield River in Shelburne Falls, northwestern Massachusetts.

3. Another variety of gneiss (of Silurian age), this one containing unusually large garnet crystals whose flat faces, or surfaces, have not been preserved. The site is between Springfield and Southbridge, Massachusetts.

2

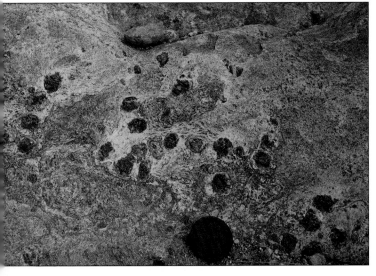

3

PLATE 26
THE APPALACHIAN PROVINCE

1. Late Proterozoic quartzite in the town of Lee, in western Massachussetts. The rock is part of a Berkshire region thrust fault slice; the laminated appearance is due to the pressure exerted during faulting.

2. Folded layers of quartzite at Ogunquit, on the Maine coast. Many features of the formation show it was originally sand washed into the ocean. Geologists are uncertain about the age; it may be Late Proterozoic, Cambrian, or Ordovician.

3. The Stockbridge Marble of western Massachusetts is Cambrian or Ordovician in age. Like the rock in photo 1, it was moved westward in a thrust fault slice during the Taconic Orogeny.

(Photo: Grant Hodsdon)

1

2

3

PLATE 27

THE APPALACHIAN PROVINCE

1. This Late Proterozoic or Cambrian phyllite in eastern New York near the Massachusetts border cleaved as slate does, under horizontal pressure. The flat surface in the photograph is a joint plane. The slaty cleavage planes slant down to the right. A sandstone layer, which did not cleave, in another part of the road cut shows that the bedding planes of the original sediment are at a low angle to the cleavage planes.

(Photo: Grant Hodsdon)

2. Even a small outcrop can give valuable evidence of geological history. This tiny exposure at the edge of a country road in the Virginia Piedmont shows vertical layers of a variety of schist.

3. Where outcrops are rare, as in the Piedmont, one can look for stone walls composed of local bedrock. This wall on a Piedmont farm is made up of phyllite quarried in the area.

2

3

PLATE 28

THE APPALACHIAN PROVINCE

1. Latest Proterozoic basalt or andesite lava flows from the subcontinent Avalonia, at Hingham, on the coast of Massachusetts. The light-colored zones were probably layers of volcanic ash mixed with the lava.

2. Avalonian gravel, now conglomerate, associated with the lava flows in photo 1.

3. Cambrian shale in Newfoundland, from the ocean off Avalonia. This outcrop is along the Manuels River, near the east coast of Conception Bay.

2

3

PLATE 29

THE APPALACHIAN PROVINCE

A restoration of Avalonia during latest Proterozoic times. The middle layer of the painting is a cross section of the ocean, and the lower layer is a section of the oceanic bedrock. Volcanic mountains form the coastline. Rivers deposit gravel and sand in their valleys and carry sand and silt into the ocean to form deltas. The silt is carried farther out to become part of the ocean-bottom mud. The light color in the valleys suggests the possibility of valley glaciers; some Avalonian rock may have originated as glacial till. Molten minerals flow out on the ocean floor through cracks in the bedrock below the ocean. The molten material is also forced into cracks between rock layers. When it cools, it will later appear as marine lava flows, dikes, and sills in the underlying rock.

(Painting: Elizabeth Mehlin)

PLATE 30

THE APPALACHIAN PROVINCE

1. With no trees and very little grass, a Triassic Basin is easily recognizable in east Greenland. The rocks in the foreground are horizontal Jurassic shale and sandstone in the down-dropped block of the basin. The bluish rocks in the background, older and more unevenly eroded, were originally extensions of the rocks below the sandstone and shale. In between is a fjord, no doubt formed from erosion along the fault zone between the different rock units.

2. Jurassic sandstone dips into the Connecticut River north of Holyoke, Massachusetts, showing how the whole series of Triassic and Jurassic formations tilts down to the east in the New England Triassic Basin.

3. Dinosaur tracks occur in some Triassic Basins. These are in Jurassic sandstone at Dinosaur State Park, south of Rocky Hill, Connecticut.

(Photo: Dinosaur State Park)

2

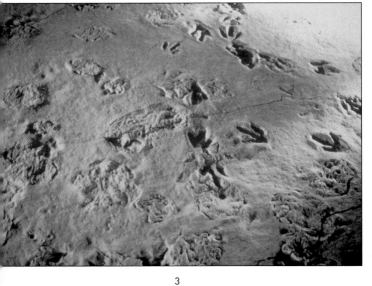

3

PLATE 31

THE APPALACHIAN PROVINCE

1. Jurassic Granite, Conway, New Hampshire. The outcrop is cracked by variously oriented joints.

2. Mt. Katahdin in Maine is a mass of Devonian granite exposed by erosion. It is a good example of how the top portion of a batholith may be uplifted above the surface of Earth's crust.

3. A hill of Cambrian granite in the Wichita Mountains, Oklahoma. Note how the surface has weathered into pillowlike sections between the joints. This condition is common in weathered granite.

2

3

PLATE 32

THE APPALACHIAN PROVINCE

1. A normal fault in Mississippian sandstone on U.S. 70, east of Hot Springs, Arkansas. The fault shows in the center of the photograph, slanting down to the left. The darker rock layers have moved down relative to the lighter layers.

2. A normal fault on Massachusetts Route 2, Athol. The fault slants down to the right. A Silurian biotite schist on the right has moved down relative to the older Ordovician rock on the left, a complicated mixture of gneiss, amphibolite, and granitelike rock. In other places, the schist overlies the gneiss and its companions. The fault, which formed in the Mesozoic Era, is much younger than the rocks. See photo 3 on Plate 45.

3. The edge of a down-dropped fault block in the Arbuckle Mountains of Oklahoma along U.S. 77 (see Figure 127). The two young women on the left stand next to the exposed fault. A strip of fine-grained material fills the fault zone, slanting up to the right. To the left of the fault zone is Pennsylvanian conglomerate, originally gravel that filled the valley formed by the faulting. To the right, the rock is Cambrian limestone.

2

3

PLATE 33

THE COASTAL PLAIN

1. A typical Coastal Plain landscape in St. Mary's County, Maryland, looking east at the face of a small escarpment.
   (Photo: Maryland Geological Survey)

2. A scene along I-95 on the Georgia Coastal Plain.
   (Photo: Grant Hodsdon)

3. Escarpments on the Coastal Plain near Mt. Willing in Lowndes County, Alabama. The one in the foreground is supported by resistant rock of Cretaceous age; those in the background have resistant Early Tertiary (Paleocene) beds.
   (Photo: T. Markham Puckett, Alabama Geological Survey)

2

3

PLATE 34

THE COASTAL PLAIN

1. Cretaceous sandstone near Arlington, Texas.

2. Cross bedding in Cretaceous sandstone near Fort Worth, Texas. The young man is sitting on the thin horizontal bed at the top of a delta. A portion of a later delta shows above the horizontal layer.

3. The cross section of a Cretaceous river channel in the Lake Arlington spillway near Arlington, Texas. The river cut the channel in dark sandstone and shale; the channel fill is light-colored sandstone.

2

3

PLATE 35

THE COASTAL PLAIN

1. The Texas Coastal Plain at Dallas. Note the flat horizon. The road cut exposes light-colored Cretaceous Austin Chalk with two normal faults.

2. The Cretaceous Demopolis Chalk, on the Alabama River at Whites Bluff, Dallas County, Alabama.
   (Photo: Charles W. Copeland, Alabama Geological Survey)

3. A mass of Cretaceous oyster shells, Valliant, Oklahoma.

2

3

PLATE 36

THE COASTAL PLAIN

1. Cretaceous and Tertiary sand, silt, and clay at Gay Head, Martha's Vineyard, Massachusetts.

2. Tertiary sandstone and shale in the cliffs at Calvert Cliffs State Park, Maryland, on the west shore of Chesapeake Bay.
(Photo: Maryland Geological Survey)

3. Tertiary gravel in a quarry west of Charlotte Hall, Maryland.
(Photo: Maryland Geological Survey)

2

3

PLATE 37

GLACIAL EFFECTS

1. Pleistocene glacial till, set down on the surface as the glacial ice melted. In comparison with the gravel in photo 2, it has a larger variety of rock fragment sizes (granules, pebbles, cobbles, and boulders) and a higher percent of fine-grained material. In some tills, the fine matrix is silt and clay; in others, it is sand. Framingham, Massachusetts.

2. Glacial outwash gravel deposited by the fast-running waters of a river. The largest rock fragments are pebbles and small cobbles. The source of the river was the retreating last ice cap of the Ice Age, or Pleistocene Epoch. Wellesley, Massachusetts.

3. Sandy glacial outwash with some pebbles and boulders in Conway, New Hampshire. Note the wind-caused ripple marks in the sand.

2

3

PLATE 38
GLACIAL EFFECTS

1. This road cut in western Ohio (I-70) shows what appears to be a cross section of a preglacial ravine cut into Silurian dolomite that was filled by till during the Ice Age.

2. The lumpy ridge in the background is the Kettle Moraine in Washington County, Wisconsin, a ridge of till formed by the retreating last glacier of the Ice Age. The glacier temporarily stopped melting back and built up an elongate pile of debris before it resumed its retreat.

(Photo: Kent M. Syverson)

3. The smooth, sinuous ridge on this Washington County, Wisconsin, farm is an esker, formed of gravel from the bed of a river that flowed in the last Ice Age glacier.

(Photo: Kent M. Syverson)

2

3

PLATE 39

GLACIAL EFFECTS

Eskers are channel fills of rivers that flowed on or in Ice Age glaciers (see Plate 38 and Figure 19). These three photographs show a segment of a large esker in Weymouth, Massachusetts.

1. The profile of the esker, rising abruptly from a generally flat area.

2. The steeply sloping side of the Weymouth esker, from the narrow summit.

3. An exposure of the river gravel that makes up the esker.

2

3

PLATE 40

GLACIAL EFFECTS

1. A glacial erratic, a boulder that was carried to this spot from elsewhere by the Ice Age ice cap. The granite boulder sits on a diorite outcrop on top of a hill. There is no higher outcrop nearby that it could have fallen from, so it must have been carried here by flowing ice. Sudbury, Massachusetts.

2. Cathedral Ledge, overlooking Echo Lake in Conway, New Hampshire. The Pleistocene ice cap flowed more or less from left to right across the hill. It carried loose granite away from the "downstream" side in a process called glacial plucking, creating a precipitous cliff.

3. Tuckerman Ravine, on the flank of Mt. Washington, New Hampshire, is a cirque, a large concavity carved into the bedrock by ice action at the point at which a valley glacier originated. The glacier carried loose pieces away as it flowed, chewing into the mountainside. It developed before the last ice cap covered the area.

2

3

PLATE 41

MINERALS

1. Quartz has irregular broken surfaces and reflects light like broken glass.

2. Orthoclase feldspar breaks with flat reflective surfaces in two planes at right angles to each other.

3. Crystals of biotite mica in a piece of granite. Each tabular crystal consists of sheets of the mineral that easily separate along the flat reflective surfaces (cleavage planes). Biotite cleaves in only one plane.

2

3

PLATE 42

ROCKS

1. A Jurassic conglomeratic sandstone from the New England Trias-sic/Jurassic Valley at Turners Falls, Massachusetts. Along with normal sand grains, it contains tiny chunks of feldspar and small pebbles. The composition indicates fast deposition of freshly weathered material from nearby. If there had been heavier river action over a long time, the feldspar would have been ground up into invisible mineral grains.

2. Jurassic samples from the New England Triassic/Jurassic Valley that indicate variety in redbed colors. The sandstone on the left, from Holyoke, Massachusetts, shows reflections from tiny white mica flakes (the mineral muscovite). The sample on the right, from Turners Falls, Massachusetts, contains interbedded shale and fine-grained sandstone layers.

3. This Silurian shalelike rock, from Rochester, New York, contains a very high proportion of the iron mineral hematite, which gives redbeds their color. This ironstone outcrops from New York along the Appalachian Mountains to the Birmingham, Alabama, area. People have mined it for iron since Colonial times, but the only significant modern operation has been the one at Birmingham.

2

3

PLATE 43

ROCKS

1. Mylonite forms along thrust fault zones from rock dust and larger broken fragments as major masses of Earth's crust move past each other. It shows a finely laminated texture. The sample is from a fault zone in Weston, Massachusetts.

2. Tillite is the solid rock form of glacial till. It differs from conglomerate in that its pebbles, cobbles, and boulders vary greatly in size, rock types, and degree of roundness, and are relatively widely separated in a fine-grained matrix of silty or sandy material. The sample in the photo is from a Late Proterozoic sediment layer near Boston that some geologists believe is a tillite. Others consider it to have been originally an ocean-bottom mud into which pebbles and larger rocks somehow became included.

3. Breccia is a conglomerate with included pebbles that are not rounded. This sample is a volcanic breccia of Jurassic age from the White Mountains of New Hampshire, formed as lava engulfed pebbles on the land surface.

2

3

PLATE 44

ROCKS

1. Samples from metamorphosed Ordovician lava flows in western Massachusetts. The upper rock is amphibolite, consisting mostly of the mineral amphibole. It was originally a basalt. The other is fine-grained gneiss made up of minerals that identify it as having been a rhyolite lava. The paper clip is a medium-size one.

2. Amphibolite from the Adirondacks, with parts of three broken garnet crystals. Aside from the garnets, the minerals are mostly amphibole.

3. Greenstone, a metamorphosed basalt lava rock from western Massachusetts. Ordovician metamorphism changed the original minerals into green ones.

1

2

3

PLATE 45

ROCKS

1. Native copper, just as it occurred in the rock, from the 19th-century mines on the Keweenaw Peninsula, Michigan. The green and white spots are bits of the enclosing rock still adhering to the copper. The sample is about 5 inches (13 cm) long.

2. Tumble-polished pebbles of flint, a colorful form of chert, from Flint Ridge, Ohio.

3. Samples from both sides of the fault illustrated in photo 2 of Plate 32. The sample on the left is the granitelike rock of feldspar and quartz; the one on the right is the schist. Their differences and relative positions help identify the fault.

2

3

PLATE 46

ROCK STRUCTURES

1. Glacial striae on a glacially polished surface of Devonian limestone from Ohio. The scratches were made by pebbles frozen into the bottom of the glacier.

2. Fossil current ripple marks in Permian freshwater red sandstone from north-central Texas. The shadowed right-hand slopes of the ripples are much shorter than the lighted slopes on the left side, indicating that the river current flowed from left to right.

3. Flute casts on a fine-grained Devonian sandstone from the Gaspé Peninsula, Québec. The photograph shows the lower surface of a layer; the convex flute casts are formed of sand that flowed over and filled depressions in a mud layer. An ocean current scoured out the elongate depressions. The steep, pointed ends of the flute casts show that the water flowed from left to right.

2

3

PLATE 47

FOSSILS

1. The right lower jaw of a primitive Devonian ocean fish related to *Dunkleosteus* (Figure 68), embedded in a piece of shale from northern Ohio. The bone is about 5 inches (13 cm) long. Such rare fossils should be given to museums; this one was donated to the United States National Museum.

2. A bit of Devonian sea floor (limestone) from western New York. It contains fossils of four different kinds of coral.

3. Crinoid fossils on a piece of Mississippian limestone from central Iowa. Crinoids are shallow-ocean animals with cylindrical bodies upright at the ends of long stalks. The body has a ring of feathery arms, and the animal resembles a blossom on a long stem. The stalk is made up of discs and is like a long stack of tiny coins.

2

3

PLATE 48

FOSSILS

1. A Cambrian trilobite from the Avalon Peninsula of Newfoundland. Trilobites were small animals that fed on shallow ocean bottoms. The impression in shale is of the hard covering of the many-legged animal, which was like the covering of a modern crab, lobster, or crayfish. The trilobite was 1.5 inches (4 cm) long. Its closest relatives are found today as fossils in Europe, and it lived in the sea surrounding the ancient subcontinent of Avalonia. It arrived here when Avalonia collided with the core of North America and was left behind when Europe split away during the breakup of the supercontinent Pangaea.

2. A Silurian coral head from the Gaspé Peninsula of Québec. The fossil is a mass of tiny tubes; each tube was occupied by a coral animal.

3. A piece of Pennsylvanian black shale from the coal-bearing beds of Nova Scotia, with the carbonized impression of part of a fern frond. Such land plant fossils are remains of the areas raised above sea level during the final continental collision of the Paleozoic Era.

2

3

stone, sandstone, and shale dip gently westward under Permian sandstone, shale, limestone, and dolomite (Figure 98).

In western Oklahoma and the adjoining part of the Texas Panhandle, massive horizontal pressure during the Alleghenian Orogeny in Late Pennsylvanian times warped Pennsylvanian formations and those below them downward into the deep Anadarko Basin, whose long axis trends northwest and southeast under the surface. The basin doesn't show on the surface now because it was filled in and covered, first by limestone and later by shale, including many redbeds, during the Permian Period. The present relatively featureless plain is therefore included with the Central Lowland and not the Appalachian Province because its surface bedrock layers are close to horizontal, like the layers in the rest of the Stable Interior; Appalachian layers are strongly folded and faulted.

Two structures in southern Oklahoma interrupt the serenity of the Osage Plains; the Arbuckle Mountains and the Wichita Mountains. Both resulted from the Alleghenian Orogeny. The folded and faulted Paleozoic layers of the Arbuckle Mountains form part of the plain's eastern margin. Farther to the west, the Wichita Mountains, composed of Cambrian metamorphic, intrusive igneous, and volcanic rocks, were thrust up through the surrounding Paleozoic sedimentary rocks. Both ranges of hills are described in chapter 5, on the Appalachian Province, to which they properly belong.

The Permian redbeds of Oklahoma include many layers and smaller bodies of gypsum (hydrated calcium sulfate). One of the state's most interesting geological features, Alabaster Cavern, contains passages in white gypsum and pink alabaster (the common name for noncrystalline gypsum), and walls lined with gypsum crystals. Alabaster State Park is northeast of Enid, between U.S. 64 and U.S. 412, east of Oklahoma Route 50.

● Texas

Permian shale and sandstone continue from Oklahoma into north-central Texas, overlapping Pennsylvanian shale, sandstone, and limestone. These predominantly terrestrial rocks were formed from sediments deposited as weathering and erosion attacked the mountains that were slowly rising while the ancestral South America docked against the continental core during the Alleghenian Orogeny.

As in other areas, terrestrial fossils in the rock show that these sediments in Texas accumulated on land. Pennsylvanian rocks contain coal and fossils of land plants. Early Permian sandstone and shale hold fossils of land plants, amphibians, reptiles, and freshwater fish (Plate 11). U.S. 281, U.S. 277, and Texas Route 114 roughly border the outcrop area where the animal fossils weather out of the sediments in Texas. The Permian rocks to the west and northwest consist of limestones, dolomites, sandstones, and shales, some marine and others terrestrial, formed as later Permian seas advanced and retreated with changes in sea level at the end of the Alleghenian Orogeny.

The Coastal Plain borders the Texas Western Tier on the east, and on the west, a striking escarpment marks the edge of the Great Plains. The "Break of the Plains" rises as much as 1,000 feet (305 m) above the Permian country. Upper reaches of the Red, Colorado, and Brazos rivers have cut canyons into the edge of the escarpment. The most famous is Palo Duro Canyon, where Palo Duro Creek, which becomes the Prairie Dog Fork of the Red River, has pierced younger rocks and exposed the underlying Permian sediments.

At the state park at Palo Duro Canyon, 12 miles (19 km) east of Canyon at the end of Texas Route 217, an imposing section of Earth's crust up to 1,120 feet (341 m) deep is revealed. Palo Duro is a colorful place to see the Great Plains rocks overlying the Permian rocks of the Central Lowland (Plate 11). The rounded slopes at and near the bottom are brick red to vermilion Permian shale and sandstone, with layers and isolated strips of white gypsum. The steeper middle slopes consist of maroon, lavender, and yellowish Triassic shale and sandstone. Pliocene sand, clay, and gravel of ordinary hues form the upper 100 feet (30 m) or so.

● Kansas and Nebraska

Along the west flank of the Ozark Dome in Kansas, the Pennsylvanian formations dip gradually westward and are covered by later Permian ones (Plate 15). The Pennsylvanian and Permian rocks continue the Oklahoma surface geology, and they themselves continue northward into the southeast corner of Nebraska (Figure 98). The Pennsylvanian bedrock includes sandstone and shale plus a relatively large amount of limestone from the inland sea, for what is now Kansas was far away from the rising mountains to the east and south, and sand and silt at times did not reach the area. Nonetheless, some terrestrial sandstone and shale, with coal, remind us of the ever-changing sea-land boundary.

The Permian formations of Kansas and Nebraska consist of limestone and shale, mostly marine, including gypsum-bearing redbeds. Some later Permian rocks in Kansas originated in shallow ocean embayments where the water temporarily evaporated. The evaporation left gypsum and great quantities of underground salt.

The first and second of the Ice Age's four major glaciers covered southeastern Nebraska, leaving large expanses of till and outwash to hide bedrock outcrops.

Figure 99. A geologic map of the middle portion of the Western Tier.

● Northern Missouri and Iowa

On the north flank of the Ozark Dome, Pennsylvanian sandstone and shale, with occasional limestone layers and coal seams, extend northward into Iowa (Figure 99). To the northeast, the central Western Tier continues up into Wisconsin

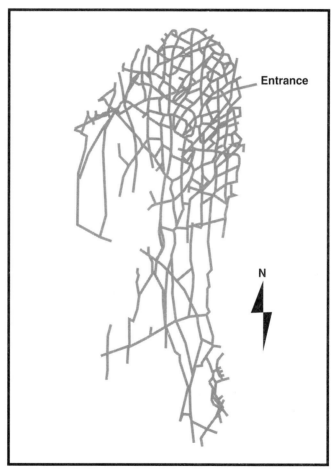

Figure 100. The intricate passages of Cameron Cave, near Hannibal, Missouri. The pattern is controlled by joints in the limestone along which water flowed.

and Minnesota. The geologic map shows Paleozoic rocks from Cambrian to Carboniferous age dipping away from northern Wisconsin along the Transcontinental Arch (Figure 58), which dips southwestward. The eroded northeast end of the arch shows in Wisconsin, where early Paleozoic rocks lap up onto the older Precambrian core that occupies the northern part of the state. To the west of the arch, Great Plains deposits cover the Paleozoic formations; to the east of it, the Paleozoic rocks grade into the Michigan and Illinois basins.

A strip of Mississippian limestone and dolomite with lesser amounts of shale that borders the Pennsylvanian rocks along the Mississippi River contains caves. Two famous ones, Mark Twain Cave and Cameron Cave (Figure 100), are on Missouri Route 79, about 2 miles (3 km) south of Hannibal.

Ice Age glaciers covered all of northern Missouri and all of Iowa and left a mostly complete covering of till and outwash. A series of arc-shaped major moraines crosses north-central Iowa, marking the retreat of a glacial lobe that flowed down from Canada (Figure 82).

The Pennsylvanian formations of northern Missouri and central and south Iowa contain shale and sandstone with minor amounts of coal and varying amounts of limestone. Most of the rocks originated from terrestrial sediments of Pangaea. The strip of Mississippian rocks that borders the Pennsylvanian ones in Iowa consists primarily of limestone, with some shale and dolomite. Some of the limestones have chert nodules. In this zone, quarries as well as streams expose the bedrock.

Devonian limestone, dolomite, and shale outcrop here and there in Iowa. Shale and thin limestone layers in one of the Devonian marine formations are exposed in Iowa's Hackberry Grove State Preserve in the Clay Banks Recreation Area. The area is southeast of Mason City, on the south bank of the Winnebago River, about 7 miles (11 km) east of U.S. 65 and 2 miles (3 km) west of County Road S 70. Devonian dolomite outcrops all around the Coralville Reservoir, south of Cedar Rapids.

Iowa's Silurian rocks consist of sandstone and dolomite. Silurian dolomite outcrops at the Palisades-Kepler State Park, southeast of Cedar Rapids where the layers have been folded into minor anticlines and synclines along the east side of the Cedar River. Silurian dolomite forms an escarpment south and west of Dubuque, through which there are road cuts on U.S. 20 and U.S. 151. One of the Silurian formations contains several coral reefs, similar to those in Wisconsin, Illinois, and Indiana.

Other formations in Iowa originated in the extensive Ordovician sea; they consist of sandstone, shale, limestone, and dolomite. Ordovician shale underlies the Silurian dolomite in the escarpment southwest of Dubuque. At Guttenberg, upriver from Dubuque, impressive road cuts show off Ordovician layers north and south of the town on U.S. 52 and County Road X 56. The widespread midwestern St. Peter Sandstone of Ordovician age lies on Cambrian rock at Pikes Peak State Park, south of McGregor, on the Mississippi River across from Prairie du Chien, Wisconsin. Cambrian sandstone, shale, and dolomite from the earliest Paleozoic inland sea outcrop along the river below the Ordovician ones, although the simplified map in this book does not show their area of outcrop. Erosion has exposed Ordovician shale, limestone, and the St. Peter Sandstone in the northeast corner of the state, along Iowa Route 9 north of Churchtown.

Caves have formed in the Ordovician limestones and dolomites of northeastern Iowa. Spook Cave can be reached from U.S. 52 about 7 miles (11 km) west of McGregor, and Wonder Cave is off U.S. 52 north of Decorah. Sinkholes and caves also occur in one of the Silurian formations.

● Minnesota and Wisconsin

The Devonian rocks in south-central Minnesota (Figure 101) all belong to the Cedar Valley Formation, which consists of dolomite and limestone.

The Ordovician formations are mostly dolomite, limestone, and shale, along with the St. Peter Sandstone. The sandstone occurs in massive layers, as much as 155 feet (47 m) thick in some places in Minnesota.

The Cambrian rocks of the region consist mostly of sandstone, with a small amount of shale, limestone, and dolomite. They lap up onto the Canadian Shield's Precambrian rocks, their sand and silt grains having come from the weathering of the continental core that existed at the beginning of the Cambrian Period. At that time, the continental core was one of several separate subcontinents that started to come together later, during the Taconic Orogeny at the end of the Ordovician Period.

Glacial till and outwash cover most of the bedrock in this

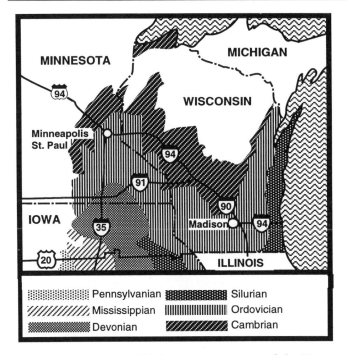

Figure 101. Geology and highways in a portion of the Western Tier in Minnesota, Iowa, and Wisconsin.

region, and major curved moraines outline the retreating lobes of the last glacier in central Minnesota and eastern Wisconsin (Figure 82, Plate 38). The ice left numerous drumlins and eskers (Plate 38) in southeastern Wisconsin. The Ice Age National Scientific Reserve lists nine sites in Wisconsin where the remains and effects of the last great ice cap can be seen. They are Interstate State Park, Chippewa Moraine, Mill Bluff State Park, Devil's Lake State Park, Cross Plains, Kettle Moraine State Forest, Campbellsport Drumlins, Horicon Marsh, and Two Creeks Buried Forest.

As the last glacier melted and retreated, meltwater became trapped between highlands and the ice in the central part of the state, forming the large, temporary Glacial Lake Wisconsin. Silt, clay, and sand from its bottom cover most of Juneau County, with extensions into Monroe and Jackson counties.

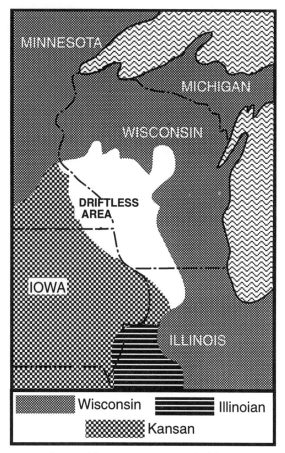

Figure 102. The Driftless Area, surrounded by regions of till and outwash. Of the four glacial stages during the Ice Age, the Kansan Stage was the second, the Illinoian the third, and the Wisconsin the fourth. By chance, none of the ice caps covered the Driftless Area.

One of the most interesting "effects" of the Ice Age glaciers in the northern Midwest is a spot that they missed. A region in Wisconsin, Minnesota, Iowa, and Illinois was bypassed by all four ice caps of the Ice Age, as shown in Figure

102. At any time, the margin of an ice cap is very irregular, with lobes and arms sticking out from the main mass of ice. Purely by chance, no part of any of the four glaciers covered the area, which is called the Driftless Area, "drift" being a general term for glacial deposits. The Driftless Area is the largest area in the glaciated part of the Central Lowland with exposed bedrock outcrops.

In Minnesota, erosion has exposed Cambrian sandstones downstream from Hastings along the Mississippi River south of the Twin Cities, and where tributary valleys cut down to the river. Ordovician dolomite caps all the higher areas.

The Twin Cities lie in a shallow basin containing Ordovician rocks, which show in the bluffs of the Mississippi and Minnesota rivers, where the St. Peter Sandstone and three formations of limestone and shale outcrop.

A cut near the intersection of U.S. 63 and U.S. 61, at the west end of Barn Bluff near Red Wing, shows a fault in Cambrian rocks. It involves the yellowish Jordan Sandstone and the greenish Franconia Sandstone, which was formed earlier and normally lies considerably below the Jordan; in the normal sequence, the top of the Franconia is as much as 65 feet (20 m) below the bottom of the Jordan. At the fault, the older greenish Franconia has moved upward to the same level as the yellowish younger Jordan, and only a 3-foot (1 m), almost vertical zone of small rock fragments and rock dust separates the two kinds of sandstone.

In Goodhue County, south of Red Wing, is the Red Wing–Rochester Anticline, that was eroded toward the end of the Mesozoic Era and capped with Cretaceous sand and clay. Along the arch's core, the erosion cut deeply into it and in some places through the St. Peter Sandstone, exposing the dolomite that underlies the St. Peter. The much younger sediments filled in the low spots during the Cretaceous Period. More recent erosion has removed the sand and silt that covered the higher areas, leaving a small outlier of Cretaceous material surrounded by Paleozoic rocks, like the Mesozoic outliers in Illinois and Michigan; it is another indication that Mesozoic sediments originally covered much of the Central Lowland. Minnesota Route 58 crosses the structure between Red Wing and Zumbrota.

Recent erosion has created karst topography in the Ordovician dolomites and limestones of extreme southeastern Minnesota, from Rochester south and southeast. Mystery Cave is southeast of Spring Valley, south of Minnesota Route 16 and east of U.S. 63. Niagara Cave is about 1.5 miles (2.5 km) north of the Iowa border, south of Harmony, on U.S. 52.

Caves in Ordovician limestones and dolomites of Wisconsin include Lost River Cave and the Cave of the Mounds, off U.S. 151 and 18, between Mt. Horeb and Barneveld, and Crystal Cave, on Wisconsin Route 29, 1 mile southwest of Spring Valley.

Mineral deposits occur in the Ordovician dolomites of southwestern Wisconsin, particularly in Grant, Iowa, and Lafayette Counties. Mineral-bearing solutions deposited minerals of lead and zinc, along with small amounts of iron and copper, in cavities and seams formed by ground water. Miners actively worked the area during the 19th century.

● Manitoba and Saskatchewan

Figure 103 shows the Western Tier's northern portion. This portion was separated from the rest of the tier by the uplifting of the Transcontinental Arch's northeastern end (Figure 58). The Paleozoic rocks that originally may have covered the arch's crest have been eroded away, exposing the underlying Precambrian Canadian Shield rocks.

Paleozoic sedimentary rocks of the tier's northern portion dip gradually toward the west and southwest and pass under the sedimentary formations of the Great Plains, as they do elsewhere in the Western Tier. To the east and north, they can be seen to overlap the Precambrian rocks of the Canadian Shield.

This part of the tier was glaciated, and glacial till and outwash cover much of the area. But in a large part of Manitoba, lake-bottom silt and sand cover the till and outwash. The sand and silt were deposited in Glacial Lake Agassiz, the largest single body of fresh water to be formed in North America, at least since the beginning of the Pleistocene Epoch, or Ice Age. The lake was named for Louis Agassiz, a 19th-century Swiss naturalist who was a pioneer in recognizing the remains of preexisting glaciers. The map in Figure 104 shows the full extent of this majestic lake, which formed the way other late Ice Age lakes did, from meltwater as the last ice cap retreated. It originally drained down the Minnesota River to the Mississippi, but as the ice melted back, it uncovered a lower outlet, which led to Hudson Bay. When the ice cap disappeared from the region, most of the lake drained away, leaving remnants in the deeper portions

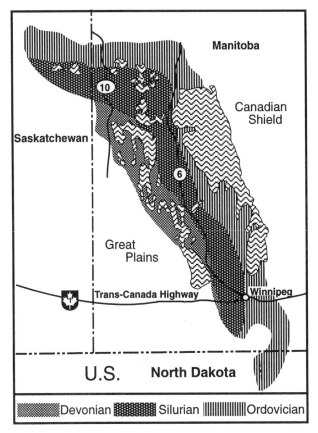

Figure 103. Geology and highways in the northern part of the Western Tier, which is mostly in Manitoba.

of its basin: Lakes Winnipeg, Winnipegosis, Manitoba, and others.

Glacial Lake Agassiz covered almost all of the northern portion of the Western Tier, but erosion since the Ice Age has uncovered the bedrock in many places. The rocks originated in the Paleozoic inland sea. It appears that if there were any

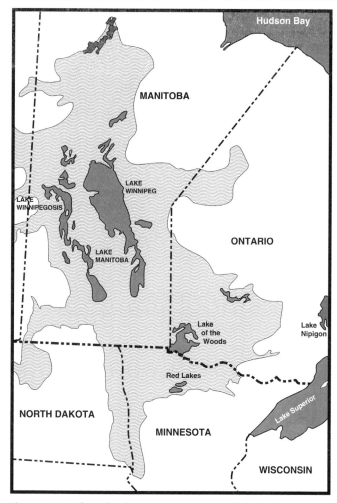

Figure 104. Glacial Lake Agassiz, largest of the lakes formed against retreating Ice Age glaciers, and its modern remnants.

Cambrian rocks in the past, they were eroded away before the Ordovician sediments were deposited, for the earliest Ordovician formation rests directly on Precambrian rocks, most of which are granite. Sandstone and shale form the basal Ordovician sediments, but the rest of the Ordovician series consists almost entirely of dolomite. An escarpment of dolomite up to 100 feet (30 m) high marks the line between Ordovician and Precambrian rocks to the northwest of Lake Winnipeg's north end.

Dolomite forms the Silurian bedrock, with local limestones and some reefs. Gypsum deposits at Gypsumville may be Silurian evaporites, since they are within the Silurian outcrop area, but they have not been precisely dated.

The oldest Devonian rocks are Early Devonian dolomites and limestones. The Winnipegosis Formation, along the shores of Lakes Manitoba and Winnipegosis, contains large reefs. Later Devonian limestone and dolomite, interbedded with shale and anhydrite, overlie the earlier Devonian formations. Anhydrite is an evaporite mineral related to gypsum; the former is pure calcium sulfate, and the latter is the same compound with a water molecule attached to each of its molecules. In other words, gypsum is hydrated anhydrite. The Dawson Bay Formation, exposed at the northwest end of Lake Winnipegosis, has some massive beds of anhydrite, indicating significant evaporation of ocean water in some areas back in the Devonian Period.

WESTERN TIER HIGHWAY SECTIONS

Warner, Oklahoma to Shamrock, Texas: I-40 (Figure 98)
I-40 crosses Middle Pennsylvanian limestone, sandstone, and shale in Oklahoma from Warner to a few miles beyond Okemah; after that the bedrock is Upper Pennsylvanian limestone, sandstone, and shale to about U.S. 377, where the Permian begins. The highway continues west over younger and younger Permian, mostly limestone and shale layers into the Texas Panhandle, where Great Plains sediments cover the Permian about 9 miles (15 km) west of Shamrock.

Oklahoma-Kansas border to Davis, Oklahoma: I-35 (Figure 98)
The main north-south route through Oklahoma is I-35, from southern Kansas to the Arbuckle Mountains. It follows the outcrop of Middle Permian marine and terrestrial sandstone and shale all the way. The sediments accumulated as a shallow sea retreated from the area, while highlands of the Al-

leghenian Orogeny arose to the south and east and other up-lifts occurred to the west.

Morgan Mill to Wichita Falls, Texas: U.S. 281 (Figure 98) *U.S. 281 crosses Pennsylvanian and Permian formations at the Texas end of the Western Tier, from Morgan Mill, north of Stephenville, to Wichita Falls.*

Morgan Mill rests on Cretaceous limestone of the Coastal Plain. About 9 miles (14 km) south of the Brazos River, the Pennsylvanian rocks of the Western Tier appear, dipping gently northward. Near the river, a silty limestone with quartz pebbles outcrops. The silt and pebbles suggest weathered material being deposited in an ocean by nearby rivers from eroding land, which in this case would be the growing Pangaea. The highway traverses limestone and shale of the Pennsylvanian ocean from the river to Jacksboro. Then it enters rocks from Pangaea; about 20 miles (32 km) northwest of Jacksboro a black shale occurs, with plant fossils and some coal that was formerly mined. The town of Newcastle, northwest of Graham, occupies the same zone of late Pennsylvanian coal-bearing rocks. Beyond the black shale you see the later Permian terrestrial redbed shales and sandstones, to Wichita Falls and beyond.

Kansas City to Salina, Kansas: I-70 (Figure 98) *I-70 in Kansas has many road cuts that illustrate the late Paleozoic layers dipping toward the Great Plains.*

From the state border to the Lawrence exits, the bedrock consists mostly of limestone, with a little shale and sandstone and one thin coal seam, a combination of marine and terrestrial beds. From Lawrence to beyond Topeka, marine limestone and shale constitute the bedrock.

About 9 miles (14 km) west of the Topeka exit and 1 mile (1.6 km) east of the Willard exit is a gray limestone that contains rice grain–like skeletons secreted by fusulinids, one-celled organisms from the Pennsylvanian sea. These fossils form an interesting contrast to the usual limestone fossil fauna of shellfish, snails, and corals.

Limestone and shale west of the Maple Hill exit form the first Permian exposure, and just east of the Wabaunsee exit, the highway crosses evidence of the Nemaha Uplift, a subsurface fault zone below the Pennsylvanian rocks that extends north and south through most of Kansas. The limestone there dips slightly downward to the east, instead of to the west as the rest of the area rocks do. The atypical dip is a surface manifestation of the deep structure. Precambrian granite on the west side of the fault moved upward, bending and carrying sedimentary formations above it into a high ridge. The ridge later sank, and sediments of the inland sea were deposited above it. Continued movement along the fault zone domed the later rocks upward, which caused the eastward dip of the limestone along the highway. The significant local earthquakes of 1867, 1906, and 1929 here may have been produced by recent movement at the fault zone.

West of the Wabaunsee exit, limestone with chert nodules outcrops. The hard nodules, weathering out in large numbers, give the Flint Hills their name, since "flint" is a common name for some varieties of chert. Permian limestone and shale continue as the bedrock until about the New Cambria exit, or about 10 miles (16 km) east of Salina. There, Great Plains Cretaceous rocks overlap the sediments from the varying but long-lived Paleozoic ocean.

Missouri River to Beatrice, Nebraska: U.S. 136
The Permian limestones and older Pennsylvanian sedimentary rocks of Kansas continue into southeastern Nebraska. Around Tecumseh, Pennsylvanian sandstone, shale, limestone, and conglomerate surrounded by Permian rocks have been exposed by weathering and erosion at the crest of a minor anticline. Some Pennsylvanian layers contain fossils, and a few show a rare condition: they are so loaded with fusulinid fossils that the fusulinids seem to make up the whole mass of the rock. U.S. 136 crosses the anticline between the Missouri River and Beatrice, where Great Plains Cretaceous beds cover up the Permian ones.

Kansas City, Missouri to the Iowa-Minnesota border: I-35 (Figure 101)
I-35 from Kansas City north crosses the Pennsylvanian, Mississippian, and Devonian zones. Pennsylvanian, mostly terrestrial deposits extend from Kansas City to the vicinity of Ames, Iowa, just south of the Mississippian bedrock zone, which consists mostly of limestone. The highway crosses from Mississippian rocks to older Devonian limestone, dolomite, and shale about at the Franklin–Cerro Gordo County line, just north of the Sheffield exit. From there to the Minnesota line, the bedrock is Devonian.

Mississippi River to Roelyn, Iowa: U.S. 20 (Figure 101)
U.S. 20 furnishes an east-west cross section of the older for-

mations in Iowa. The Dubuque area has Ordovician marine limestone, sandstone, and shale bedrock. The escarpment capped by Silurian dolomite, which is about 2.5 miles (4 km) west of Dubuque, marks the eastern edge of the Silurian zone. Devonian marine limestone, dolomite, and shale occur in the vicinity of Independence and continue westward past Waterloo to Ackley, where Mississippian limestone, dolomite, and shale overlie them. At Ft. Dodge, Pennsylvanian, mostly terrestrial sandstone and shale outcrop along the Des Moines River. A few miles to the west, approximately at Roelyn, Cretaceous Great Plains rocks overlap the Pennsylvanian sediments, but glacial till covers the overlap.

Illinois-Wisconsin border to Imogene, Minnesota: I-90 (Figure 101) *I-90 crosses the eroded end of the Transcontinental Arch, passing first over younger Ordovician, then older Cambrian, then younger Ordovician and Devonian rocks.*

I-90 enters Wisconsin from Illinois on Ordovician dolomite and limestone bedrock, but what is visible is mostly glacial material along the highway northward. A kettle lake shows up west of the highway about 23 miles (37 km) north of the state border and 1 mile south of the Wisconsin Route 59 exit. A moraine causes lumpy topography around the Route 59 exit, and 7 miles (11 km) north of the U.S. 51N exit, the highway passes drumlins to the east.

About 3 miles (5 km) north of the U.S. 12 and 18 exit, a quarry east of the road shows Ordovician bedrock, the St. Peter Sandstone and overlying dolomite. Beyond that, a couple of miles past the junction with I-94, the highway enters a drumlin field.

A Cambrian sandstone outcrop zone starts near the U.S. 151 and U.S. 51 exits, and extends along I-90 until it approaches the Mississippi River. The sandstone outcrops on a few hills and ridges and is exposed at the bridge over the Wisconsin River.

About 13 miles (21 km) west of the Wisconsin Route 33 exit, I-90 enters the last glacier's terminal moraine. Beyond that is the Driftless Area: there is no more glacial till along the highway until it gets to Minnesota. The exit at Wisconsin Route 13 leads to the main road to the Wisconsin Dells, a scenic area where the Wisconsin River, with entering tributaries, flows for 7 miles (11 km) between cliffs of cross-bedded Cambrian sandstone. The cliffs rise as much as 100 feet (30 m) above the water.

From a couple of miles beyond the Route 13 exit all the

way to Tomah, the highway traverses the flat plain that was the bottom of Glacial Lake Wisconsin. At the County Road C exit, there is a roadside park next to Castle Rock, an imposing butte of Cambrian sandstone.

About one mile beyond the junction with I-94 near Tomah, I-90 leaves the glacial lake floor and enters a valley walled by Cambrian rock at the lower levels and later Ordovician at the higher levels. The rocks belong to the edge of a Cambrian-Ordovician escarpment of dolomite and sandstone, and they continue on to the Mississippi River.

In Minnesota, once you get west of the Mississippi River valley, the bedrock is all Ordovician, as described above. Just beyond the U.S. 63 exit, you enter the Devonian dolomite and limestone region and also the Driftless Area's western edge. From there to North Dakota, Ice Age deposits essentially hide the bedrock along I-90. The Devonian region ends somewhere between the Albert Lea–Manchester exit and the Alden exit, and the highway returns to Ordovician bedrock. Somewhere between the Blue Earth and Imogene exits, Great Plains rocks overlap Ordovician ones, and the Great Plains begin.

Tomah, Wisconsin to Monticello, Minnesota: I-94 (Figure 101)
A northern route starts in the Driftless Area and follows I-94 beyond Tomah. This route traverses Cambrian and Ordovician rocks near the edge of the Canadian Shield, including one small outlier of the Shield.

A few miles beyond Tomah, the highway leaves Glacial Lake Wisconsin's bed and enters a more rolling country with Cambrian sandstone bedrock. At Black River Falls, the Black River has cut through Cambrian sandstone to expose Precambrian metamorphosed sedimentary rocks in its valley. A few miles northwest of Osseo, in the vicinity of the Foster exit, I-94 leaves the Driftless Area, although there are Cambrian outcrops beyond that point. Ordovician dolomite and sandstone overlie the Cambrian sandstone just east of the St. Croix–Dunn county line, near the Knapp exit.

Rivers have exposed Ordovician sediments in their valleys as far west as the Twin Cities area; beyond there, you cannot see the rocks along I-94. The edge of Cambrian bedrock, which is the edge of the Western Tier, is somewhere in the vicinity of Monticello. The nearest major bedrock outcrop is between St. Cloud and I-94, about 30 miles (48 km) to the northwest. There, erosion has exposed Precambrian granite of the Canadian Shield.

Hadashville to High Bluff, Manitoba: Trans-Canada Highway 1 (Figure 103)

Trans-Canada Highway 1 in the Winnipeg area traverses the Paleozoic section from east to west. It crosses the Precambrian-Ordovician boundary in the vicinity of Hadashville and Prawda. Ordovician shale, sandstone, and dolomite continue from there through Winnipeg to between the outskirts of the city and St. François Xavier. A short stretch of Silurian dolomite follows, after which is a shorter stretch of Devonian marine limestone and dolomite in the Poplar Point–Fortier–Eustache area. Oakville and High Bluff lie just over the border on the Great Plains rocks that overlie the Devonian ones.

Winnipeg to Ponton, Manitoba: Route 6 (Figure 103)

Going north from Winnipeg, Provincial Route 6 leaves the Ordovician shale, sandstone, and dolomite about at Grosse Isle and crosses the Silurian dolomites to near St. Laurent. The highway then curves through the Devonian limestone, dolomite, shale, and anhydrite to Lundar. North of there, it follows closely the Silurian-Devonian border, just inside the Silurian zone, all the way to Fairford Reserve, where it turns almost directly north and traverses Silurian country up to just beyond the intersection with Provincial Route 60. Beyond Route 60, the road returns to the Ordovician bedrock zone. At Ponton, Route 6 turns northeast and enters the Canadian Shield.

Mafeking to Cranberry Portage, Manitoba: Route 10 (Figure 103)

Provincial Route 10 goes over a more complete section of Devonian dolomite, limestone, shale, and anhydrite than can be seen on the above two highway sections, from the vicinity of Mafeking to about the intersection with Route 60. The route passes Dawson Bay, for which the anhydrite-bearing formation mentioned above was named. Continuing northward, Route 10 crosses the Silurian dolomites from the Route 60 intersection to Wanless, where the Ordovician zone begins. Almost all of the Clearwater Lake Provincial Park is underlain by Silurian rocks; the neighboring Cormorant Provincial Forest is mostly on Ordovician marine bedrock. At about Cranberry Portage on Route 10, and in the Grass River Provincial Park, volcanic, metamorphic, and intrusive igneous rocks of the Canadian Shield emerge from under their Ordovician cover.

Winnipegosis to Camperville, Manitoba: Route 20

Route 20 from Winnipegosis to Camperville traverses Devonian dolomite and shale, with reefs, that outcrop along the shore of Lake Winnipegosis.

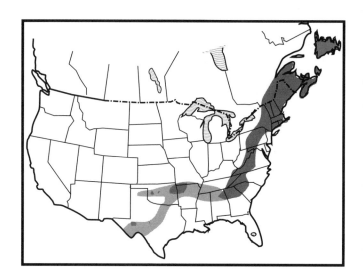

5

THE APPALACHIAN PROVINCE

The Appalachian Province is a related series of Late Proterozoic, Paleozoic, and Mesozoic formations that extends from
Newfoundland to southwest Texas. The province can be
readily divided into five subzones: southeastern Canada and
New England, the Valley and Ridge Zone (the Appalachian,
Ouachita, and Arbuckle Mountains, and the Llano Uplift
and Marathon Region of Texas), the Blue Ridge, the Piedmont, and the Triassic Basins (Figures 110, 119, 126, 131, and
138). The Appalachian Province rocks are continuations of
those in the Stable Interior, but in contrast to the Stable Interior, where the rocks offer only indirect evidence of orogenies, or continental collisions, the Appalachian Province
gives *direct* evidence of orogenies, as well as direct evidence
of rifting, or separation, between orogenies.

Coastal Plain and Stable Interior sediments cover much of
the Appalachian Province's southern portion west and
southwest of Alabama, but rock types, ages, and geologic
structures make it evident that the Ouachita Mountains of
Oklahoma and Arkansas are geologically related to the central Appalachian section that runs from Pennsylvania to Alabama. Farther to the west and south, localized uplifts have
preserved other samples of the province in the Arbuckle and
Wichita Mountains of Oklahoma and the Llano Uplift and
the Marathon Region of Texas.

The testimony of the rocks shows that for billions of years,
supercontinents have been breaking up and re-forming in
multimillion-year cycles (Figure 28). Periods of mountain-
making—folding, thrust and reverse faulting (Figure 14),
metamorphism, volcanic activity, and intrusion—occurred
as continents, subcontinents and island chains collided to
form supercontinents. Minor local orogenies have crumpled
and changed rocks as well, as movements within plates
caused sections of Earth's crust to push and grind against
each other. North American rocks indicate six major oroge-

nies and several minor ones in the 3.9 billion years of our continent's history. The Precambrian orogenies listed on the geologic time chart were all major ones. In more recent times, the Taconic, Acadian, and Alleghenian orogenies combined made up the major Appalachian one. In between major orogenies, crustal sections moved apart during times of rifting, accompanied by normal faulting (Figure 14), volcanic activity, and igneous intrusions. The stretching and faulting promoted volcanic activity and intrusions, just as the compressive stresses and faulting of collisions did; weakening of the crust permitted molten minerals to rise to or near the surface. Canadian Shield rocks preserve evidence of the Precambrian orogenies and rifting periods, but the rocks there are so old that much of the original evidence has been destroyed by crustal action and erosion through the eons of time. Thus, Precambrian history is more of an outline, with many details missing.

The post-Precambrian rock record is much more complete. Paleozoic rocks show many details of the last major orogeny, the Appalachian. It is possible to distinguish the three minor collisions that made up the major orogeny. "Minor collision," of course, is a relative term, for the crustal activity in any orogeny is almost inconceivably massive. Mesozoic and Cenozoic rock records in eastern North America clearly show the faulting, intrusion, and volcanic activity that characterized the rifting that followed the Appalachian Orogeny.

■ THE LAY OF THE LAND

Tilted rock layers and a variety of rock types characterize the Appalachian Province. Throughout the province, rock layers are tilted at all angles. Each outcrop or road cut is a small piece of a large structure in which formations have been folded and faulted. Layers dip north and south or northwest and southeast on the limbs of folds or in tilted fault blocks.

The part of the province closest to the Stable Interior has essentially only sedimentary rocks; in the rest of the province, metamorphic and igneous rocks abound, mixed with the sedimentaries. The Appalachian region has been so affected by folding, faulting, and intrusion that in large areas it seems as if every road cut and outcrop shows a different kind of rock. Ages of the rocks range from Late Proterozoic to Cretaceous, the majority being Paleozoic. From New York City and northern Pennsylvania northward, the province shows all the typical glacial effects; you can even see cirques in the mountains.

Plates 17–21 show landscapes in the Appalachian Province. The Piedmont region, to the east of the Blue Ridge, has a relatively flat, gently rolling landscape, with only a few high hills. It has been highly eroded, and a thick layer of topsoil hides its complex rock structure. The rest of the Appalachian Province is a land of mountains, hills, and valleys, formed by the weathering and erosion of its folds and fault blocks, with more resistant rock types standing out as hills and softer rocks eroding into valleys.

■ THE ROCKS OF THE APPALACHIAN PROVINCE

Most rocks of the Appalachian Province originated as sediments on land and on ocean bottoms, or as lava flows and volcanic ash, in Late Proterozoic, Paleozoic, and Mesozoic times. During the orogenies that crumpled the crust of the province, some of the original rocks were changed by heat and pressure into metamorphic types, particularly in the Piedmont and in the Appalachians north of New York City. The pressure folded and faulted some formations so deeply into the crust that they melted and recrystallized in the heat of Earth's interior. Shale, sandstone, and limestone turned into schist, quartzite, and marble. Volcanic rocks heated again; rhyolite became gneiss and basalt became greenstone or amphibolite. Also during the orogenies, masses of molten minerals were forced upward as folding took place, to leave batholiths, stocks, and sills that cooled into many kinds of intrusive igneous rocks. The crustal rifting that followed the Appalachian Orogeny caused numerous major dikes and other intrusions to form, adding to the igneous rocks. Plates 21–28, 30–32, and 42–44 are photographs of Appalachian Province rocks.

Other evidence for the major Appalachian Orogeny and subsequent rifting is visible in Appalachian Province structures. Folded rock layers (Plates 21–23) are exposed as segments of upwarped and downwarped series of anticlines and synclines, called anticlinoria and synclinoria (see Figure 13). Many enormous segments of crust moved miles along thrust faults and piled against each other, overlapping like shingles on a roof. Reverse faults developed, and strike-slip faults dislocated both kinds. As Earth's crust stretched during the rifting that followed the Appalachian Orogeny, normal and vertical faults formed, and blocks of rocks tilted and dropped downward (Plate 30).

Consider a typical piece of complicated surface geology in the northern Appalachians or the Piedmont. Horizontal pressure of an orogeny has forced the upper crust into an anticlinorium that consists of sedimentary and volcanic rock layers. As the folds occurred, granite intruded them, forming batholiths. The sedimentary rocks near the batholiths and those folded deeply down were metamorphosed; those farther away from the batholiths and those folded upward were metamorphosed only slightly, if at all. As the pressure continued, the crust cracked to form thrust faults, with blocks moving up and over each other. Perhaps a rotational component of the orogenic force added some strike-slip faults. Weathering and erosion have exposed some of the rocks that formed far below the surface.

If you drive across the ancient anticlinorium, you will see a variety of rocks, the layered ones tilted in various ways. In some cases, a later orogeny added a second generation of folding, faulting, intrusion, and metamorphism. Geologic maps of the Stable Interior show rock structure with relatively simple patterns. As a contrast, imagine how complicated a geologic map showing the different rocks of the eroded structure in Figure 134B would look.

■ THE HISTORY OF THE APPALACHIAN PROVINCE

The Appalachian Province was formed by an impressive series of activities in Earth's crust following the development of the Canadian Shield. The final major act in the Canadian Shield's Precambrian geologic history was the Grenvillian Orogeny, which ended about 1 billion years ago and helped to build a supercontinent (Figure 105).

About 800 million years ago, during the Late Proterozoic Era, continental masses and chains of islands began to separate from each other, and the crustal stress promoted volcanic activity. Late Proterozoic rocks in the Appalachian Province show that the ocean bottom, islands, and larger land masses all contained volcanoes and fissures through which lava flowed. The evidence indicates that the Grenvillian supercontinent split into several large land masses (Figure 105):

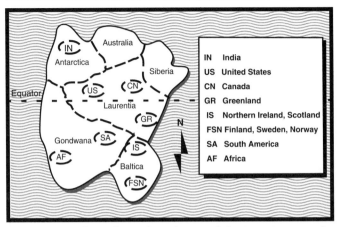

Figure 105. A hypothetical rendering of the Late Proterozoic supercontinent that spawned the subcontinents Laurentia and Gondwana. Dashed lines show where the supercontinent rifted. Note how the area that was to become North America changed orientation.

Laurentia (now the central United States down to Texas with a westward extension to southern California, plus the Canadian Shield and Greenland), Baltica (part of Europe), Gondwana (most of Africa and South America), Australia, Antarctica and India, a mass that became Siberia, and possibly others. There were also smaller subcontinental land masses and island chains.

Geologists have conceived more than one theory to describe eastern North America's Paleozoic history. All theories employ the concept of moving plates, as described in chapter 2. One suggests that the land mass called Laurentia collided with South America during the Ordovician Period, then moved away, and later returned to "sideswipe" South America during the Devonian Period. The continents, on the moving surfaces of plates, separated again, according to the theory, and then moved together for a third time, as Laurentia collided with Gondwana and Baltica.

The theory described in the following paragraphs (Figure 106) is one that, according to experts, best fits what geologists know about eastern North America, Europe, and Africa. As students of Earth history continue their investigations, details and even generalities may change.

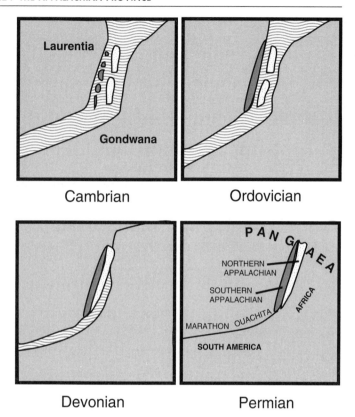

Figure 106. Plate tectonic activity in the Atlantic Ocean. During the latest part of the Proterozoic Eon, the supercontinent in Figure 105 split, and major continents Laurentia and Gondwana separated. At the end of the Cambrian Period, there were smaller land masses between the two large continents. By the end of the Ordovician period, the western island chain had collided with Laurentia. By the end of the Devonian Period, the other minicontinents (now called Avalonia) had almost finished joining Laurentia, and part of Gondwana had collided, too. The process continued into the Permian Period, and the old version of the Atlantic Ocean disappeared in the formation of the supercontinent Pangaea. Later, Pangaea started to break up, and the modern Atlantic Ocean was born. The rifting process continues today.

As of about 750 million years ago, during the Late Proterozoic Era, oceans existed between the large subcontinents. An island chain and a subcontinental mass (possibly a couple of large islands) developed in the ocean between Laurentia and Gondwana. Over millions of years, volcanic activity built up the island chain and the subcontinent, and lava flowed into the ocean. The land areas (Laurentia, the island chain, the subcontinent, and Gondwana) weathered and eroded, and sediments accumulated on land and in the ocean. The result of all this building up and tearing down was a series of Late Proterozoic, Cambrian, and Ordovician sedimentary and volcanic rocks. Geologists call the subcontinent Avalonia, from the Avalon Peninsula of Newfoundland, where part of it can be seen today. (See Plate 29.)

The first orogenic episode, or collision, in Appalachian history occurred between about 480 million and 440 million years ago, during the Ordovician Period. It is called the Taconic Orogeny after the Taconic Mountains of western New England and southeastern New York. In a condensing phase of the plate tectonic cycle, the ocean basin between Laurentia and the island chain became smaller and smaller and finally disappeared as the eastern margin of Laurentia slowly moved under the section of Earth's crust that held the islands. (See Figure 25 for a modern example of such action.) From Middle Ordovician time to the beginning of the Silurian Period, the collision spawned folding, faulting, volcanic activity, metamorphism, and intrusion. Mountains grew, leaving an enlarged Laurentia.

During the Silurian Period, various islands and the small continental blocks of Avalonia occupied the ocean between Laurentia and Gondwana and Baltica. Volcanic and sedimentary rocks continued to form as the inexorable action of plates in the tectonic cycle worked to close the ocean. Eventually, early in the Devonian Period, somewhere around 400 million years ago, Avalonia began to "dock" against the remains of the island chain. The docking destroyed another ocean basin and added another zone to the eastern margin of Laurentia. Some formations folded during the Taconic Orogeny were refolded and faulted anew. Many formations metamorphosed and were intruded by molten minerals. Volcanic activity continued. The major crustal crumpling started in the north and slowly worked its way southward, to end about 350 million years ago, early in the Carboniferous Period. That cataclysm has been named the Acadian Orogeny, for the Acadian region of southeastern Canada, where its effects are pronounced. At that time, the ocean basin between the northern parts of Avalonia and Gondwana closed, leaving open ocean from about the latitude of southern New England on down. That left Laurentia, an ocean basin, and Gondwana and Baltica, the latter two continents having rejoined by then.

Finally, over a long period of time, and in several individual pulses of pressure from what is now Newfoundland to Texas, Gondwana fused with Laurentia as plate movement completed the formation of the supercontinent Pangaea. In the north, where plates had already joined, continued pressure caused some folding, faulting, and intrusion, but most of the major crumpling, including massive thrust faulting, occurred where the last ocean closed. (At various times,

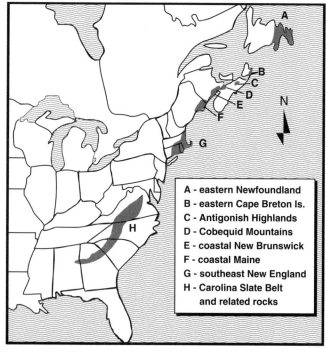

A - eastern Newfoundland
B - eastern Cape Breton Is.
C - Antigonish Highlands
D - Cobequid Mountains
E - coastal New Brunswick
F - coastal Maine
G - southeast New England
H - Carolina Slate Belt
    and related rocks

Figure 107. The areas indicated are remnants of old Avalonia (see Figure 106) left behind when Africa separated from North America.

other continents had moved around and joined to become other parts of Pangaea.) This episode began late in the Mississippian Epoch of the Carboniferous Period but climaxed during Pennsylvanian and Permian times, about 300 million to 250 million years ago. The ocean bottom crust moved under the eastern Appalachian crust, and Gondwana collided with the margins of Laurentia. This final collision is commonly called the Alleghenian Orogeny, from the mountain and plateau region of central and western Pennsylvania that was greatly affected by it. The Taconic, Acadian, and Alleghenian orogenies are considered to be three phases of the major Appalachian Orogeny.

The tectonic cycle continued into a rifting phase during the Triassic Period. Pangaea started to split, and the present Atlantic Ocean was born. Stresses in the crust promoted more volcanic activity, and large blocks of solid crust slid

Figure 108. Boulders, cobbles, and pebbles of a great variety of rock types form a beach in Scituate, Massachusetts. They were part of the till deposit from the last Ice Age glacier, but waves and currents have removed the sand and smaller grains.

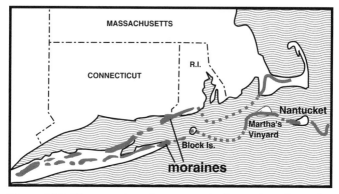

Figure 109. The remains of two terminal moraines from the last Ice Age glacier occur along the south coast and islands of New England.

down along normal faults. This block faulting formed basins that received lava and ash from volcanic eruptions, and thick deposits of silt, sand, and gravel from weathering and erosion of the neighboring highlands.

As rifting continued during the Mesozoic and Cenozoic eras, the continents did not separate at precisely the same sutures along which they had joined, so some parts of Avalonia traveled eastward while others were left behind in North America. Remnants of Avalonia can be recognized in Ireland, England, continental Europe, and North Africa. The North American remnants are shown and identified in Figure 107.

Aside from evidence of rifting, geologic remains of late Mesozoic and Cenozoic age in the Appalachian Province are few and hard to recognize, except for those from the Ice Age. The glaciers left abundant evidence from New York City northward, throughout New England and southeastern Canada. Till and outwash are obvious almost everywhere (Figure 108, Plate 37). Rocks from distant regions that melted out of the ice cap on mountaintops indicate how thick the ice was. The last receding glacier left behind numerous drumlins and eskers, mostly hidden in woods but recognizable when the leaves are off the hardwood trees (Plate 39). Some of the islands in Boston Harbor are remnants of drumlins, and large portions of two terminal moraines can be seen along the Atlantic coast (Figure 109). Cirques on some of the mountains show where valley glaciers originated during the Ice Age (Plate 40). Some hills exhibit cliffs that were formed when ice plucked and carried away weathered rock. Loose rock fragments on the hillsides facing in the direction of ice flow (the "downstream" facing sides) froze into the ice as the glacier moved over the hills. The glacier then carried the fragments as it flowed slowly away (Plate 40).

■ THE CENTRAL APPALACHIANS

The Central Appalachians consist of two mountain ranges and a flanking, relatively flat region, all with folded and faulted rock layers. The western range is in the Valley and Ridge Zone, the central range is the Blue Ridge, and the eastern portion is the Piedmont (Figure 110). The Valley and Ridge Zone has long parallel ridges and valleys carved in folded and faulted Paleozoic sedimentary rocks (Figure 111, Plate 17). The summits of the ridges are all approximately the same elevation, giving the ridges straight profiles. The Blue Ridge is a more "normal" range, with various folded and faulted Proterozoic and early Paleozoic rocks eroded into an irregular profile (Plate 18). The relatively flat Piedmont, east of the Blue Ridge, shows Proterozoic and Paleozoic rocks that have been greatly folded, faulted, and intruded, with few major hills and a deep soil. The Piedmont and the Coastal Plain between it and the ocean have similar subdued landscapes, but their bedrocks differ greatly. The Coastal Plain rocks have not been folded, faulted, intruded, or metamorphosed, and they formed much more recently, during the Mesozoic and Cenozoic Eras. They actually overlap the older rocks of the Piedmont, whose surface dips toward the ocean.

◆ The Valley and Ridge Zone

This zone runs a sinuous course from the lower Hudson Valley to northern Alabama, around Birmingham. It consists of broad valleys and long, simple parallel ridges. Its northern tip can be placed at the Hudson River, between Kingston and Newburgh, New York. The western margin of the Valley and Ridge Zone begins in the ridge formed by Kittatinny Moun-

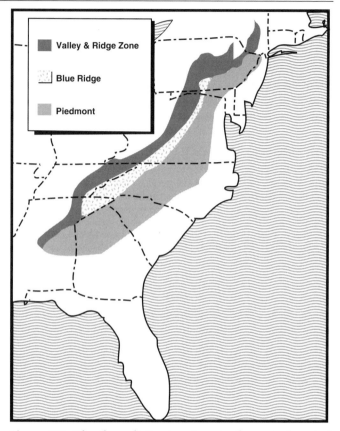

Figure 110. The three distinct regions in the central part of the Appalachian Province.

tain (New Jersey) and the Shawangunk Mountains (New York). It curves westward, then southward through central Pennsylvania, including the Allegheny Front at the edge of the Appalachian Plateau. It passes through the panhandle of Maryland just west of Cumberland, then across the "spout" of West Virginia, to closely follow the West Virginia–Virginia border to Bluefield. Then, on into Tennessee at about the point where that state, Kentucky, and Virginia come together, and past Harriman and Chattanooga to beyond Birmingham.

The eastern border of the Valley and Ridge Zone runs along

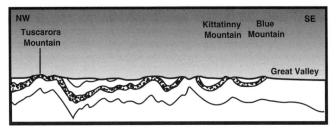

Figure 111. A cross section of the Appalachian Mountains in the Valley and Ridge Zone of Pennsylvania, showing folding. The folded layers were pushed along thrust faults deep below the surface. The harder rock layers, represented by the one emphasized in the diagram, make the mountains; valleys are in softer, more easily weathered rocks. This simplified sketch illustrates the scope of the folding. Most of the folds in central Pennsylvania are actually shaped like those in Figure 112.

the western edge of the irregular chain of mountains and hills from the Hudson and New Jersey Highlands through the Reading Hills in Pennsylvania and the Blue Ridge and Unaka Mountains (which include the Great Smokies).

The folded rocks in the Valley and Ridge Zone are mostly sandstones, shales, and limestones from the great Paleozoic inland sea that is the dominant feature of the Stable Interior, described in chapter 4. Early Mississippian terrestrial conglomerate, sandstone, and shale originated in the Valley and Ridge zone as sediments from the highlands created by the Acadian Orogeny. The Pennsylvanian terrestrial sedimentary rocks, including coal, stemmed from the upheaval of the Alleghenian Orogeny. The Valley and Ridge formations are continuous with those of the Stable Interior.

● The Great Valley

The Great Valley extends from the Hudson River southward through the Wallkill, Lehigh, Lebanon, Cumberland, Hagerstown, and Shenandoah valleys, and farther south in the valley of eastern Tennessee and the Coosa Valley of northeastern Alabama (Plate 17). The Great Valley forms the eastern portion of the Valley and Ridge Zone, bordered southeastward mostly by the Blue Ridge Mountains.

The bedrock formations in this valley region have been folded and faulted like their neighbors to the east and west, but erosion has rendered the landscape smoother than in the neighboring regions. The low ridges found there are not on the scale of those in the Valley and Ridge Zone's mountainous western portion (the true Appalachian Mountains).

The Great Valley's northern portion lies below the Shawangunk Mountains, Kittatinny Mountain, and the Blue Mountain Front (not to be confused with the Blue Ridge) on the northwest. To the southeast rise the Piedmont's Hudson Highlands, New Jersey Highlands, and Reading Hills. Between the Hudson and Susquehanna rivers, the bedrock is mostly Ordovician shale and slate, with a few sandstone and limestone beds. South of the Susquehanna, the formations consist of Cambrian and Ordovician quartzite, limestone, and dolomite. The formations all belong to a squashed and eroded synclinorium.

South of the Lehigh Valley, the Great Valley formations connect with the Appalachian Mountain rocks and structures to the west; they are just eroded more smoothly.

Karst topography has developed in the abundant limestone of the Great Valley—sink holes, disappearing creeks, springs, and caverns. Some well-known caverns in Virginia are Shenandoah Caverns (U.S. 11, near Mt. Jackson), Luray Caverns (U.S. 211, Luray), and Endless Caverns (U.S. 11, between New Market and Tenth Legion). Along I-81 and U.S. 11 down the Great Valley of Virginia, the town names of Lacey Spring and Mint Spring are indications of karst topography. A subterranean passage south of Lexington, Virginia, lost most of its roof to erosion and became the famous Natural Bridge. Some of the many Tennessee caverns are in the valley zone, such as Bristol Caverns, (U.S. 421, southeast of Bristol), and the towns of Sulphur Springs and Locust Springs show the presence of springs in that state.

Some Jurassic intrusive rocks occur in the Shenandoah Valley. Molten minerals intruded rocks in the Valley and Ridge Zone during the Tertiary Period of the Cenozoic Era, forming dikes and sill-like bodies. There are over 100 in the northern part of Virginia's Highland County. Such geologically recent intrusions lend credence to the theory that a live, unsolidified igneous body exists relatively close to the surface in the area of Warm Springs, Virginia, in the next county south of Highland.

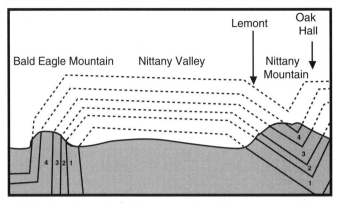

Figure 112. A simplified cross section of the overturned syncline and anticline in the vicinity of State College, Pennsylvania. The section is approximately 7.5 miles (12 km) long.

● The Valley and Ridge Mountains

The main western portion of the Valley and Ridge Zone is the Appalachian Mountain Range in the strict sense. It has the characteristic long parallel ridges with straight valleys in between, valleys in most cases narrower than the eastern Great Valley (Figure 111, Plate 17). The harder sandstones and limestones form the ridges and the softer shales erode into valleys.

The rock types of the mountain region are essentially the same as those of the neighboring Appalachian Plateau, and they tell the same story. The geology of the State College, Pennsylvania, area is a clear example of Appalachian geology in its simplest form (Figure 112).

The city of State College lies in the Nittany Valley, between two long parallel ridges, Nittany Mountain to the southeast and Bald Eagle Mountain to the northwest. The rocks of this area range from Cambrian to Devonian in age, and their compositions reflect a big part of Paleozoic Era history. The two oldest formations consist of Cambrian limestone and dolomite, from the sea that covered much of the present continent. Overlying the Cambrian rocks are three formations of Ordovician limestone and dolomite from the same sea. Next comes marine shale and sandstone from the later Ordovician, when silt and sand were being carried into the ocean, probably as a result of the uplifting and erosion of what is now New England, southeastern Canada, and the Piedmont during the Taconic Orogeny. Silurian marine sandstone and shale of the State College area indicate further erosion of the new land. The early Devonian rocks show a return to limestone along with sandstone, a sign that there must have been less erosion by slower rivers on a lowland, and silt did not get out into the ocean. The lessening erosion means the land mass had been worn down or lowered land was being covered by an encroaching sea. Later Devonian formations all consist of shale, which originated in the erosion of another new land mass, the land formed by the Acadian Orogeny.

Thus, the rock types of the State College area furnish evidence of what was happening to the north and east during the Taconic and Acadian orogenies, but the fact that the rocks are all marine shows that the local ocean was not disrupted by the orogenies.

Then came the Alleghenian Orogeny, and everything changed. Pressure from the southeast folded the Cambrian to Devonian ocean-bottom rocks into a syncline and an anticline and turned them into dry land. The docking of Gondwana squeezed the anticline against the continental core with such force that it overturned toward the northwest, and the rock in the center of the anticline broke, to form a reverse fault.

You can model the development of an anticline such as the one at State College with a sheet of paper. Put your left hand flat on a table to represent the unyielding continental core. Place the left edge of the paper against your left hand, but don't hold it down. Push the right edge toward the left edge with your right hand, keeping the right edge on the table (the horizontal pressure from the southeast). The paper forms an upward curve, which tends to slant toward the left as you push. That illustrates the typical Appalachian anticline structure, as in Figures 111 and 112. The left slant models an anticline overturned like that at State College.

At least some of the folds that were overturned toward the northwest were caused by underground thrust faults. Figures 113 and 114 illustrate thrust faults. Thrust faults start in layers of weak rock, usually of shale. If the layers above the shale do not crack, the fault remains below the surface, and pressure along the fault may push the layers above it into a fold that, as time goes by, may become overturned. South of Pennsylvania, thrust faults have reached the surface, but as

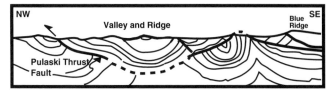

Figure 113. Thrust faulting of the Appalachians in the western point of Virginia. A large block of crust from the southeast moved along the Pulaski Thrust Fault; later, the fault itself was folded by continuing pressure. The southern extension of the Pulaski Fault is shown in Figure 116.

typical faults are traced northward, they change into anticlines. Apparently, the pressure in the south portion of the Appalachian Mountains was greater or lasted longer than that in Pennsylvania during the Alleghenian Orogeny.

As Figure 112 shows, the present Nittany Mountain is actually the remains of a syncline. That may seem illogical; you would think that ridges would come from anticlines. However, such a structure is typical of Valley and Ridge mountains. The summit of Nittany Mountain is resistant Ordovician sandstone. Bald Eagle Mountain eroded from the almost vertical series of Ordovician and Silurian formations on the northwest arm of the anticline. Its resistant core consists of a very dense and tough Silurian sandstone with quartz cement. In the Valley and Ridge Zone, sandstones typically form the ridges, and the particular ridge of Bald Eagle Mountain extends from near Altoona to Williamsport, a distance of about 140 miles (225 km).

A structure such as the one here described is so large that you can't see the whole fold in any one place, but if you follow a road that crosses the axis of a fold, you can extrapolate from outcrops and road cuts. For example, to reconstruct the structure of Nittany Mountain from scattered outcrops, you can leave U.S. 322 and drive to the small community of Oak Hall (see Figure 112). There you will find a large quarry south of the road, and a smaller long-abandoned one north of the road. In the latter, layers of limestone dip steeply downward toward the northwest; they are Ordovician limestones on the southeast limb of the syncline. Farther down the road toward Lemont, there are on the right a couple of exposures of Ordovician Reedsville Shale at the core of the syncline that outline a U-shape. Beyond Lemont, just before the Route 26

Figure 114. A road cut on I-81 north of Hazleton, Pennsylvania, with light Pennsylvanian sandstone overlying darker Mississippian sandstone and shale. The formations have been folded into a syncline and broken by two thrust faults (heavy lines). The fact that these rocks have been folded and faulted shows that the Alleghenian Orogeny began in this region after the beginning of the Pennsylvanian Epoch.

intersection, a small area of limestone outcrops, dipping down toward the southeast, just the opposite of the limestone at the old quarry. That is part of the northwest limb of the syncline. The series of exposures from Oak Hall to Route 26 illustrate the downfolded structure of the syncline.

There are few exposures in the Nittany Valley. Route 26 leads southwest to Business U.S. 322; on the latter road west of State College, you can see some Cambrian limestone and dolomite in the almost horizontal central portion of the anticline. The rocks dip only slightly because of the anticline's shape (Figure 112). Then, as the highway ascends Bald Eagle Mountain, you will pass various road cuts that expose the almost vertical Ordovician and Silurian rock layers of the ridge. Among the first is the Reedsville Shale, which you saw in the syncline between Oak Hall and Lemont. It originally curved upward from the syncline and is now sweeping downward along the overturned northwest limb of the anticline. As you drive up the hill, you see the younger formations that overlie the Reedsville, and at the top you will find the resistant quartz-cemented sandstone. Note in Figure 112 that the folds are not gradually curved in cross section; they are more abrupt V-shaped ones. These sharp folds are typical of structures in Pennsylvania.

By making field observations like these, you can discover

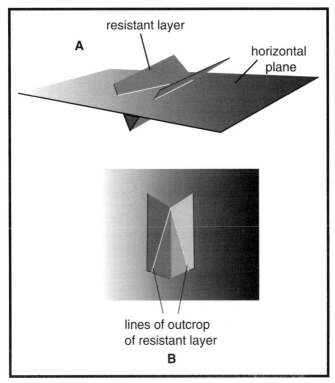

Figure 115. The origin of zigzag ridges in the Appalachian and Ouachita mountains, modeled by a folded sheet of paper. A, a syncline plunges to the left. B, the layers of a syncline will outcrop at the line where the syncline intersects the horizontal plane. The anticlines next to the syncline will make patterns that point toward the bottom of the page.

and reconstruct structures in most parts of the Appalachian Zone. In some places, highway construction has exposed structures (Plate 22). In the State College case, the portion of U.S. 322 that bypasses the city cuts right through the syncline of Nittany Mountain, and you can see the structure in an outstanding exposure.

The basic Appalachian structure of eroded parallel synclines and anticlines is further complicated in some places, where the folds themselves are tilted. When the dipping folds erode, the ridges join to form a zigzag pattern visible in aerial views (Plate 17). Again, you can model the structures with a piece of paper. The paper represents a resistant sandstone layer. Crease it down the middle and hold it in front of you sideways in a V-shape (Figure 115). First, hold it with the crease down, as in a syncline, then tilt it downward toward the right. Now, imagine a line formed by a horizontal plane intersecting the syncline. It would form a V-shape pointing toward the left. Hold the paper to model an anticline, and tilt it downward toward the right. The intersection line of a horizontal plane forms a V-shape pointing toward the right.

Going southward along the Appalachian Mountains, there is abundant evidence that the pressure from the east and southeast was more intense to the south than it was in Pennsylvania. Simple folds merge southward into synclinoria. Overturned anticlines grade southward into thrust faults, and thrust faults become more numerous from the Roanoke, Virginia, area southwestward. Under the continuous pressure, layers that were originally horizontal folded. Then, the folds overturned, with minor faulting. Thrust faults relieved some of the pressure, but the pressure continued, so that, in time, the thrust faults themselves were folded (Figure 113).

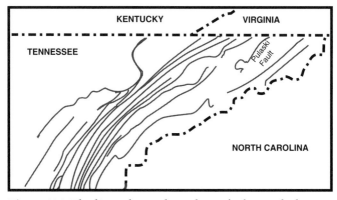

Figure 116. The lines show where thrust faults reach the surface in eastern Tennessee. Figure 113 includes a cross section of the Pulaski Thrust Fault in neighboring Virginia. Thrust faulting was less common farther north in the Valley and Ridge Zone.

The map in Figure 110 shows that the Valley and Ridge Zone trends toward the west in Pennsylvania, then it shifts toward the south. In the Roanoke area, the trend shifts again toward the southwest. The Valley and Ridge Zone was deformed by two different pulses of the Alleghenian Orogeny. In the northern part, folding and faulting occurred during the Pennsylvanian Epoch and the early part of the Permian Period. South of the Roanoke area, the folding and faulting started and ended earlier, possibly starting as early as Late Mississippian times. With the abundance of thrust faulting, the crust in the southern region was shortened more than that farther north. In one area of western Virginia, the Earth's crust was shortened about 15 miles (24 km). Figure 116 indicates how many major thrust faults shattered the formations in just one portion of the southern Valley and Ridge Zone.

In Knoxville, Tennessee, you can see a good example of thrust fault folding. Exit I-275 at Atlantic Avenue east and turn north on Central Avenue, and you will come to a road cut at Sharps Gap. The cut shows a sort of triangle of greatly broken and deformed Ordovician dolomite (Figure 117). It is bounded on the northwest by a thrust fault with Cambrian shale above it. The shale was thrust toward the northwest on top of the dolomite, probably when the dolomite was horizontal. But later, both formations were folded in such a way that the thrust fault now dips downward toward the northwest. The southeast boundary of the dolomite triangle is yet another thrust fault, along which Cambrian sandstone and

Figure 117. A simplified diagram of the road cut in Sharps Gap, Knoxville, Tennessee, that shows folded thrust faults, one cutting the other.

shale have been pushed approximately at right angles to the other two formations, breaking through them.

From Tennessee through Georgia, the Valley and Ridge geologic structure is characterized by broad folds in synclinoria, cut and distorted by the thrust faults. The Alabama portion is mostly an anticlinorium, with fewer thrust faults than in Georgia and Tennessee.

Throughout the Appalachian Valley and Ridge Zone, marine and terrestrial sedimentary rocks range in age from Cambrian to Pennsylvanian or Permian. Here and there, minor metamorphism has turned shale into slate. The terrestrial sediments reflect the Taconic, Acadian, and Alleghenian orogenies; the marine sediments originated in ocean basins east of the old Precambrian Laurentia.

As one might expect, caves can be found in the Valley and Ridge Zone's mountainous section. Two in Pennsylvania southwest of State College are Indian Caverns, on Route 45 near Spruce Creek, and Lincoln Caverns, on U.S. 22 near Huntingdon.

◆The Blue Ridge

The high hills and mountains of the Blue Ridge rise above the lowland from Pennsylvania to Georgia. The Blue Ridge starts out as a single ridge, South Mountain, that extends south from near Carlisle, Pennsylvania. South of the Pennsylvania-Maryland border, the Blue Ridge widens somewhat to continue as a long strip of hills sweeping down to Roanoke, Virginia (Plate 18). Beyond there, the mountainous zone becomes still wider, to include the Great Smoky Mountains to the west (Figure 110). From Roanoke to near Gainsville, Georgia, the edge of the Blue Ridge forms a prominent escarpment, the Blue Ridge Front, that overlooks the lower, mostly broadly rolling Piedmont country.

Generally, the Blue Ridge is a crushed and distorted anticlinorium of older rocks, thrust-faulted up and over younger rocks. The oldest rocks of the ridge are Proterozoic gneiss and igneous intrusives that are thought to be related to the Grenville Province of the Canadian Shield. They outcrop in South Mountain and down through most of the Virginia section, and then disappear, to reappear in westernmost North Carolina and along the North Carolina–Tennessee border halfway to Georgia.

Before the faulting and erosion, those ancient rocks were covered by Late Proterozoic and Early Cambrian sedimentary and volcanic rocks, some of which were metamorphosed more than others. They are now quartzite, metagraywacke, marble, slate, phyllite, schist, gneiss, amphibolite, and greenstone (metamorphosed basalt), along with rhyolite, basalt, limestone, dolomite, conglomerate, and sandstone. All in all, the rock series differs strikingly from that of the neighboring Valley and Ridge Zone. The older Proterozoic gneiss and intrusives generally occur along the eastern margin of the zone. The crest of South Mountain and its extension in the Blue Ridge of Virginia is mostly greenstone and quartzite.

In Virginia, near the corners of Tennessee and North Carolina, a Late Proterozoic formation contains what appears to be a tillite (solidified glacial till) formed during an ice age that would have been even earlier than the Pennsylvanian/Permian ice age mentioned in chapter 2. See Plate 43 for a New England rock of the same age with the characteristics of tillite. A rock type that accompanies the presumed tillite seems to have originated as ocean-bottom mud, with pebbles dropped into it from melting icebergs. The glacial interpretation may well be correct, because all the present continents show excellent evidence of at least two major ice ages in Late Proterozoic times.

The layered rocks of the Blue Ridge have been intensely folded and faulted. In some places, the rock type mylonite marks thrust faults. The mylonite developed from ground-up rock dust and fragments in the fault zones. The Great Smoky Mountains and others along the Tennessee–North Carolina border consist of great thrust sheets, pushed one on top of another.

Indeed, the southern portion of the Blue Ridge Zone was thrust westward over folded rocks of the Valley and Ridge Zone during the Alleghenian Orogeny. In the northwest part of the Great Smoky Mountains, smooth oval valleys called coves illustrate the thrusting. The coves are "windows" where Appalachian rocks show through the Blue Ridge rock series because of erosion of the Great Smoky Mountains thrust sheet that was pushed across the eastern border of the Valley and Ridge Zone. The bedrock on the cove floor consists of Ordovician shale and sandstone that belong to the Valley and Ridge Zone, with Blue Ridge Proterozoic metamorphosed sedimentary rocks all around. Cades Cove is in the northwestern corner of Great Smoky Mountains National Park, and two other coves, Wears Cove and Tucka-leeche Cove, are along U.S. 321/Tennessee 73 between Pine Grove and Maryville, just outside the park.

Caves in the Tennessee–North Carolina area include Tuck-aleeche Caverns, just south of Townsend, Tennessee, and Linville Caverns, along U.S. 221 between Ashford and Alta-mont, North Carolina.

This southern portion of the Blue Ridge Zone contains many mountain peaks over 6,000 feet (1,829 m) in elevation, including the highest mountain in the eastern United States, Mt. Mitchell, which rises to 6,684 feet (2,037 m), northeast of Asheville, North Carolina.

The metamorphism of the Blue Ridge Grenville gneiss occurred about 1 billion years ago. The other rocks were meta-morphosed during the major deformation of the Taconic Orogeny. Some faulting may date from the Acadian Orogeny, and some old faults may have been reactivated then. Apparently, the region was little affected by the Alleghenian Orogeny, except to be moved as a unit when the collision of Gondwana deformed the Valley and Ridge Zone.

◆The Piedmont

The landscape of the Piedmont (Figure 110) is significantly different from that of the Blue Ridge and Appalachian mountains; it is a gently rolling plain whose topography is here and there relieved by isolated hills or ridges of resistant rock (Plate 18). It slopes gradually from the Blue Ridge east to the equally flat Coastal Plain. The eastern border of the Piedmont is called the Fall Zone for the waterfalls and rapids that have formed where rivers and streams flow from the harder Piedmont rocks onto the softer ones of the Coastal Plain. Many cities and towns, including Baltimore, Richmond, and Columbia, grew along the Fall Zone, because that was the zone beyond which river boats could not go. Goods had to be transferred from boats to land transportation there.

In most of the Piedmont, the bedrock has weathered deeply into a very thick soil; in some places, the soil is as much as 100 feet (30.5 m) thick. Therefore you have to look carefully and creatively for outcrops and road cuts. Most of the hills occur in the southern portion and consist of erosion-resistant remnants of gneiss, granite, and other intrusive igneous rock. The best known is Stone Mountain, just east of Atlanta, Georgia, a mass of granite about 1.5 miles (2.5 km) long and 650 feet (198 m) high. The intrusions took place in three stages, getting younger from west to east. The first is dated at 595 million to 520 million years ago (a minor Late Proterozoic to Cambrian orogeny), the second at 440 million to 385 million years ago (possibly the Acadian Orogeny), and the last at 325 million to 265 million years ago (the Alleghenian Orogeny).

The Piedmont rocks resemble those of the Blue Ridge, being more or less metamorphosed sedimentary and volcanic rocks of Late Proterozoic, Cambrian, and Ordovician age. Piedmont rocks include, along with the intrusive igneous rocks and the gneiss, such types as quartzite, marble, slate, phyllite, amphibolite, and schist, as well as other less metamorphosed or unchanged sedimentary and volcanic rocks in some places.

The orogenies folded, faulted, and metamorphosed the Piedmont formations, although thrust faults are much less common there than they are in the mountain zones. It appears that, during the Alleghenian Orogeny, the Piedmont and Blue Ridge moved together northwestward along a "master sole thrust," or major crustal fault, in shales far below the surface. Most of the metamorphism occurred during the earlier Taconic Orogeny, although there may have been additional changes during the Acadian Orogeny.

The Piedmont's northeastern portion, from the Hudson River down to the Washington–Frederick, Maryland region, exhibits a confusing mixture of Late Proterozoic and early Paleozoic rocks. Some are metamorphosed, others are not, but all are greatly distorted. The far northeastern tip borders the Great Valley of the Valley and Ridge Zone, consisting of the Reading Hills of Pennsylvania, the New Jersey Highlands, and the Hudson Highlands of New York. Outcrops there show a complex assortment of Proterozoic, Cambrian, and Ordovician rocks, with a bit of Silurian and Devonian mixed in, faulted and tightly folded in the remnant of an anticlinorium. The Proterozoic rocks include granodiorite, gneiss, and marble/limestone, associated with Paleozoic limestone, shale, sandstone, and conglomerate.

The southeastern corner of Pennsylvania shows folded Cambrian and Ordovician limestone and shale replaced to the east by Proterozoic granodiorite and anorthosite with Cambrian sandstone and shale. Between Hanover and Lancaster, pressure warped the bedrock into a syncline of Cambrian and Ordovician limestone flanked by anticlines that contain Late Proterozoic to Early Cambrian quartzite and metamorphosed volcanic rocks.

Between Philadelphia and Westminster, Maryland, and down to Frederick, Maryland, folded and thrust-faulted metamorphic rocks that seem to underlie others in the Piedmont of Pennsylvania and Maryland are exposed, including schist, phyllite, quartzite, marble, and greenstone. There is some doubt about their age, but they probably date from the Late Proterozoic.

Approximately 20 miles (32 km) north and west of Baltimore and in the Doe Run–Avondale, Pennsylvania, area are eroded domes of gneiss, here and there overlain by schist, quartzite, and marble. The gneiss probably formed in the Late Proterozoic Era; the other rocks most likely belong to a Late Proterozoic to Early Cambrian series. The domelike structures started as overturned folded formations; later, the series of folds were broadly folded again into domes.

The same Late Proterozoic or Early Cambrian rocks extend down to the Washington-Frederick line. Directly west of Baltimore, and from there northeast into Pennsylvania, a group of granites and gabbros shows through the other rocks. Some are Cambrian or Ordovician in age; the ages of others have not been determined. Some outcrop between I-95 and U.S. 1.

South of the Washington-Frederick line, the geology becomes simpler. To the Roanoke River and beyond in Virginia, a large synclinorium borders the Blue Ridge, featuring phyllite, quartzite, schist, and greenstone; this zone may be a continuation of the rocks at Frederick.

To the southeast, an elongate zone of folds extends from between Fredericksburg and Washington to Cumberland–Buckingham, Virginia. In the center of a syncline there, a formation of Ordovician slate contains layers of such exceptional quality that it has been quarried continuously since 1726. The formation is named the Arvonia Formation, for the town of Arvonia, but the rock is widely known to architects and builders as Buckingham Slate, from the county where it is quarried. Cambrian and Ordovician quartzite, metagraywacke, and several varieties of schist accompany the slate in the zone of folds. The rock series also contains granite intrusions of Cambrian or Ordovician, Mississippian, and unknown age.

East and south of the Virginia structures, into North Carolina, stretches a mottled arrangement of metamorphic Late Proterozoic or Early Cambrian rocks of the kinds described above for the Piedmont in general, plus scattered bodies of

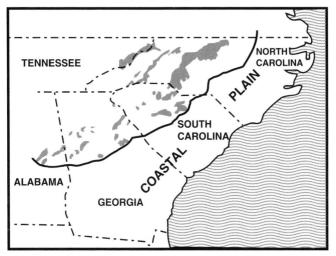

Figure 118. Igneous intrusions in the southern Piedmont and Blue Ridge regions.

granite and gabbro (Figure 118), results of the orogenies that affected the Piedmont. A large granite body of unknown age lies between Cumberland and Lunenburg, Virginia, and a Mississippian body lies from the vicinity of Wilsons on U.S. 460 to the state border south of Lawrenceville, with Virginia Route 46 running down the middle of it.

A belt of folded and faulted gneiss, amphibolite, and schist, with granite and gabbro-type intrusions borders the Blue Ridge in the Carolinas and through Georgia into Alabama. The rocks presumably range in age from Late Proterozoic to Ordovician. A folded zone of Cambrian to possibly Ordovician gneiss, schist, quartzite, and marble extends outward toward the Coastal Plain from that belt.

Bordering all those metamorphic rocks to the southeast is a strip called the Charlotte Belt, consisting mostly of granite and gneiss, roughly from Roxboro, North Carolina, through Greensboro and Charlotte, down to Union, South Carolina. I-85 goes along the Charlotte Belt from Graham to Gastonia, North Carolina. A narrow continuation of this zone crosses Georgia, centered approximately on a line from Washington to Thomaston. Rock ages range from Late Proterozoic and Early Cambrian to Pennsylvanian.

Southeast and south of the Charlotte Belt lies the Carolina Slate Belt, a long zone of latest Proterozoic to Cambrian rocks that extends approximately from southeast of Washington, Georgia, through the Carolinas to Lunenburg, Virginia. Between the Blue Ridge and this zone the rocks other than intrusive igneous ones are to a large extent highly metamorphosed schist and gneiss. The slate belt has a variety of less metamorphosed rocks; low-grade metamorphosed tuff, volcanic breccia, and lava flows overlain by metasediments such as argillite, metagraywacke, and quartzite. The formations were forced into a series of folds that dip down toward the southwest. The zone also contains numerous igneous intrusions like those of the Charlotte Belt. Volcanic rocks in this belt have been dated at 650 million to 550 million years old, which makes them very late Proterozoic. The suite of rocks in the Carolina Slate Belt and their age suggest that the region was originally part of the Avalonian subcontinent (Figures 106, 107).

◆Appalachian Zones and History

The Appalachian Province's three geographic zones generally correlate with phases of Paleozoic history.

The Valley and Ridge Cambrian to Pennsylvanian sedimentary rocks represent ocean sediments at the edge of the old Laurentia continent, and later terrestrial sediments from highlands to the east formed during the Paleozoic orogenies.

In the Blue Ridge and inner Piedmont, the rocks are metamorphic or intrusive igneous in nature. The metasediments and metavolcanics are almost entirely late Proterozoic to Ordovician in age. They represent volcanic material, marine sediments, and terrestrial sediments of the Late Proterozoic to early Paleozoic western island chain to the east of Laurentia (Figure 106). They also represent the ocean between the islands and Laurentia and presumably part of the ocean east of the islands. It is significant that the youngest rocks are Ordovician in age. Abundant evidence shows that they were folded, faulted, and metamorphosed during the Ordovician Taconic Orogeny, when the western island chain collided with Laurentia. They were intruded at various times.

Geologists believe that the eastern zone, the outer Piedmont including the Carolina Slate Belt, originated as part of Avalonia. Its rocks are presumed to have originally been volcanics and sediments from the land mass and the ocean basins on either side that were added to Laurentia during the Acadian Orogeny.

Finally, when Gondwana later collided with the expanded Laurentia, the Piedmont and Blue Ridge rocks were pushed up and onto some of the ones to the west of them, and the Valley and Ridge formations were folded and faulted.

◆The Triassic Basins

In the central and northern parts of the Appalachian Province, elongate bodies of younger rock have been downfaulted into the "normal" Proterozoic and Paleozoic rocks of the province (Figure 119). These are the Triassic Basins, remnants of the time when the supercontinent Pangaea began to

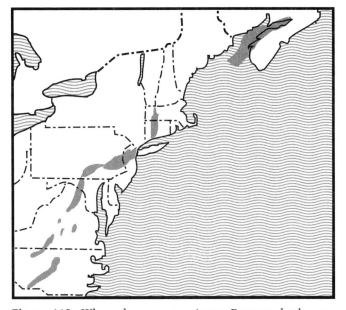

Figure 119. When the supercontinent Pangaea broke up, faulting due to stresses in Earth's crust formed the Triassic Basins in eastern North America. From Nova Scotia to North Carolina, fault blocks subsided.

break up, at the end of the Triassic Period and beginning of the Jurassic Period.

As Africa moved away from North America, the Piedmont Zone split; normal faults caused blocks of the crust to sink lower than their neighbors. Lava welled up from below, flowing out of volcanoes and fissures and forcing itself between layers of rock or sediment. Volcanic ash periodically clouded the sky. Rivers from the highlands carried sand and silt down into the basins formed by the down-dropped fault blocks. And dinosaurs walked across the sand and silt, leaving their footprints for us to see today (Plate 30). All this began near the end of the Triassic Period and continued into the Jurassic Period; the younger rocks in the basins are Jurassic in age (Plates 30, 42). Note also the Jurassic intrusions in the Valley and Ridge Zone mentioned above, which might have been caused by changes in the deeper crust as the continents separated.

The largest basin is the New York–Virginia Basin, which contains an estimated thickness of 3 miles (5 km) of sedimentary and volcanic rocks. The layers dip gradually toward the north and northwest from the southeast edge of the basin (Figure 120). Approximately the lower third of the sedimentary layers consists of reddish, purplish, and gray arkose, conglomerate, and shale in a redbed series. The rest is mostly

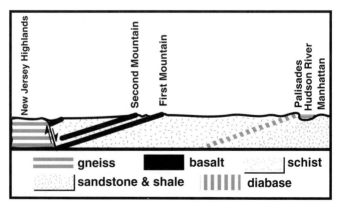

Figure 120. Cross section of the north end of the New York–Virginia Triassic Basin.

red shale and sandstone. The rocks are all river deposits of various kinds. The wide northeasternmost portion of the New York–Virginia Basin contains extensive lake deposits of dark gray and reddish shales.

A massive diabase sill occurs near the bottom of the series. Its outcrop forms the Hudson River palisades (Figure 120). Farther up in the series, basalt lava flows and tuff layers form

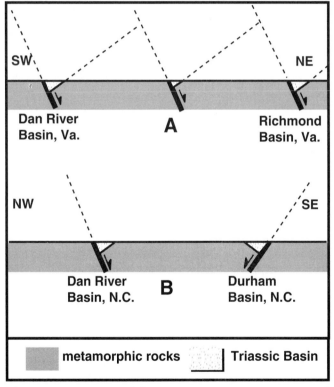

Figure 121. Sections of the Triassic Basins in Virginia and North Carolina. A shows a diagrammatic section of three basins downfaulted into the Piedmont rocks. The major faults are all parallel, indicating successive normal faulting. B shows that the major fault of the Durham Basin is opposite to those of the others.

ridges in the otherwise relatively flat countryside. First Mountain and Second Mountain in northern New Jersey are examples (Figure 120).

A series of normal faults forms the northwestern margin of the New York–Virginia Basin. It appears that the down-dropped crustal block tilted as its northwestern edge moved farther down than its southeastern edge, which makes the Triassic and Jurassic rock layers dip. Possibly, the crust on the northwest side of the faults moved upward. The fault zone on the other side of the basin has been eroded away, so no one knows the original width of the block or what happened at the southeastern edge.

The basins in Virginia south of the New York–Virginia one have similar rocks, and they all have faults on their northwest edges. It looks as if one large block had possibly broken into smaller ones (Figure 121A). The basin at the edge of the Coastal Plain in North Carolina has its faults along the southeast border, the opposite of the others (Figure 121B).

Two other basins have been exposed to the northeast: the Connecticut River valley of Connecticut and Massachusetts, and the Bay of Fundy region in the Maritime Provinces of Canada. There is also a series of basins out in the Atlantic continental shelf, east and south of the Piedmont, that are covered by Coastal Plain deposits. All of these basins testify to crustal stresses west of the primary rift as Pangaea broke up. It seems reasonable to expect similar evidence of stress on the eastern side of the rift, and, in fact, similar basins occur in Morocco, Portugal, England, Scotland, and east Greenland (Plate 30). All in all, the Triassic Basins offer abundant evidence of a rifting period in the cycle of crustal action and moving continents.

CENTRAL APPALACHIAN HIGHWAY SECTIONS

Newburgh to Port Jervis, New York: I-84 (Figure 122)
This stretch of I-84 traverses the Wallkill Valley, the northern end of the Valley and Ridge Zone's Great Valley. The highway crosses Ordovician rocks, mostly shale and sandstone from the inland sea.

Montgomery to I-84, New York: U.S. 6, NY 17
This route crosses the Hudson Highlands, at the northern end of the Piedmont. About half of it goes through the Prot-

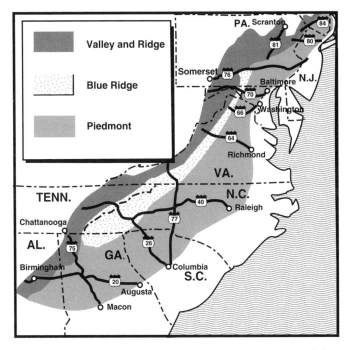

Figure 122. Geology and highways in the central part of the Appalachian Province.

erozoic metamorphic rocks of the Hudson Highlands. The gneiss, quartzite, and marble in this area are presumably related to the rocks of the Canadian Shield's Grenville Province and to the Proterozoic rocks in the Blue Ridge to the south. The route continues into the Ordovician rocks of the Wallkill Valley.

Pennsylvania, from I-78 to Scranton: I-81 (Figure 122) *This route traverses typical folds in the hard coal country at the northern end of the Appalachian Mountains.*

The route starts in the Lebanon Valley portion of the Great Valley, continues through Swatara Gap in Blue Mountain, and enters the mountainous part of the Valley and Ridge Zone. From there on, outcrops show rock layers tilted to various degrees on the limbs of eroded folds. Blue Mountain consists of Ordovician shale and Silurian quartzite and conglomerate, followed by Silurian shale and sandstone. The Silurian rocks accumulated as products of weathering and erosion in the highlands caused by the Taconic Orogeny.

The highway then follows the valley between Blue Mountain and Second Mountain, floored by Devonian shale. Second Mountain is held up by resistant sandstone and conglomerate that originated in the Pennsylvanian Epoch, from rivers and streams coming out of the highlands that resulted from the Alleghenian Orogeny. There are Taconic Orogeny sediments on the right in Blue Mountain and Alleghenian Orogeny sediments on the left in Second Mountain, evidence for two orogenies along one valley.

Between the Suedburg-Marstown exit and the Dorrance exit, a distance of about 56 miles (90 km), the highway crosses the Anthracite Synclinorium. Pennsylvanian coal layers lie in the synclines. Mississippian sandstone underlies the Pennsylvanian layers in Second Mountain. The Mississippian rocks show up again at the other end of the synclinorium. Figure 114 shows a road cut there. The coal is hard coal, or anthracite, which burns more efficiently than the soft coal of the Appalachian Plateaus. Evidence of former mining shows in the strip mines and piles of waste rock along the way. Pennsylvanian shale, sandstone, and conglomerate accompany the coal.

Mississippian shale and sandstone, flood plain and delta deposits from rivers of the Acadian Orogeny highlands, form the bedrock between the northern Hazleton exit and the Dorrance exit. From the Dorrance exit to the Pennsylvania Route 29 exit, the highway crosses an anticline. It first passes from the Mississippian rocks to Devonian terrestrial sandstone and shale, then to underlying Devonian marine sandstone and shale (younger to older). Then it goes back into the younger Devonian terrestrial beds and finally into the still younger Mississippian again. Younger to older to younger formations proves the anticline structure. (Figure 59 illustrates this relationship.) North of Route 29 the highway enters a syncline, with Pennsylvanian sandstone, shale, and coal flanked by Mississippian rocks. The highway follows the southeast flank of the syncline, mostly in Pennsylvanian rocks, toward Scranton. In the vicinity of the Pennsylvania Route 9 exit (northeast extension, Pennsylvania Turnpike), southwest of Scranton, I-81 leaves the syncline and enters the Allegheny Plateau.

Somerset to the Delaware River: Pennsylvania Turnpike (I-76) and I-276 (Figure 122) *This route affords a classic view of the relatively simple anticline/syncline structure of the northern Valley and Ridge Zone followed by the contrasting geologies of a Triassic Basin and the Piedmont. Through the Valley and Ridge Zone there is a variety of rocks from the Paleozoic inland seas, as well as terrestrial rocks from the orogenies.*

The Pennsylvania Turnpike leaves the Allegheny Plateau at the Allegheny Tunnel, east of Somerset. The tunnel penetrates a ridge of Mississippian and Pennsylvanian sandstone and shale that marks the northwest edge of a broad anticlinorium whose other edge is a few miles west of the Fort Littleton (U.S. 522) interchange. The valley beyond the first ridge has upper Devonian terrestrial sandstone and shale and underlying Devonian marine sandstone, shale, and limestone. Those beds continue to Wills Mountain, the next ridge, which forms the edge of a series of folds in Devonian, Silurian, Ordovician, and Cambrian sandstone, shale, and limestone within the anticlinorium. Wills Mountain shows narrow, steeply tilted layers of Devonian and Silurian shale, sandstone, and limestone. The Bedford interchange (U.S. 220) is a Silurian rock zone. Evitts Mountain and Tussey Mountain represent the flanks of an anticline with Cambrian marine limestone, sandstone, and shale in its center. Resistant Silurian and Ordovician sandstones support the mountains.

Between Everett and Rays Hill, the highway returns to the Devonian marine sandstone, shale, and limestone, followed by later Devonian terrestrial sandstone and shale at the southeast margin of the anticlinorium. The Breezewood interchange is in the latter rocks. Rays Hill is supported by Mississippian terrestrial sandstone and shale. These Devonian and Mississippian beds consist of river deposits from the Acadian Orogeny highlands.

The Mississippian rocks are downfolded here toward the northwest, so that just beyond the east end of the Sideling Hill tunnel the turnpike encounters the underlying Devonian sediments again. These formations introduce a whole new series of folds in mostly Ordovician and Silurian sandstone, shale, and limestone. Silurian sandstones, particularly the very resistant Tuscarora Sandstone, support the ridges of these eroded narrow folds. The Tuscarora Sandstone is also responsible for Tuscarora Mountain, Kittatinny Mountain, and Blue Mountain. The Willow Hill interchange occupies a zone of Ordovician limestone and dolomite, thrust-faulted up against the younger rocks to the northwest in an anticline between Tuscarora Mountain and Kittatinny Mountain. Kittatinny and Blue mountains form the limbs of a syncline in Silurian shales and sandstones (Figure 111).

The Great Valley begins at the east end of the Blue Mountain Tunnel. The remains of an overturned syncline in Ordovician shale with a little sandstone floor the valley. Since the shale weathers and erodes more easily than the limestones and resistant sandstones of the folded mountains, erosion has created a broad valley. From the Blue Mountain interchange to the Carlisle interchange the highway crosses the marine shale, which continues to about 10 miles (16 km) west of the Carlisle exit. At that point, limestone that underlies the shale appears.

About 10 miles (16 km) beyond Carlisle, a stretch of several shallow folds begins, with Ordovician limestone, dolomite, and shale broken by thrust faults. The deformation of the beds here may be a result of the Taconic Orogeny.

Approximately 5 miles (8 km) west of the Susquehanna River, the turnpike suddenly leaves the Appalachian Great Valley bedrock and enters the much younger rock in the Newark-Delaware Subbasin of the New York–Virginia Triassic Basin. Beyond the boundary fault there are no tight folds; the highway now traverses slightly tilted Triassic and Jurassic shale and sandstone, along with six diabase sills. The largest sill crosses the highway northeast of Middletown, and two more large ones occur in the Colebrook area. The Triassic basin ends about at the Berks-Lancaster county line just west of Morgantown, and the next zone is the Piedmont.

Piedmont country extends from the county line to the Delaware River. There, Proterozoic, Cambrian, and Ordovician formations have been folded and faulted, with some metamorphosed to a great extent and intruded. For about the first 5 miles (8 km), Cambrian limestone, dolomite, quartzite, and schist make up the bedrock. Then the rock changes to irregular bodies of gneiss, quartz monzonite, and granodiorite of Proterozoic age.

About 6 miles (9.5 km) beyond the Lionville exit (Pennsylvania 100), the bedrock is once again Cambrian rocks as described above, with some Ordovician limestone. These are in a narrow zone trending east northeast, which the turnpike follows as I-276. Extensions of the Triassic Basin occur east and west of Norristown. South of Southampton, the turnpike turns southeast and crosses a strip of gneisses to the

U.S. 1 exit. From there to the river, Ice Age sands and gravels mostly cover Proterozoic schist.

In the Philadelphia area, typical Piedmont rock can be studied along Wissahickon Creek in Fairmount Park. The park is just north of the U.S. 1 exit on I-76. A variable but mostly quartz and mica schist outcrops here that is famous among geologists; its age has been the subject of many studies and some controversy since the first serious study of it by Florence Bascomb in 1905, and it is still not known for sure. Experts now consider that the evidence suggests it was first deposited as marine sediment that may have become sandy shale during the Ordovician Period, and was metamorphosed during the Taconic Orogeny.

Hudson River to Delaware River, New Jersey: I-80 (Figure 122)

New Jersey contains part of the New York–Virginia Triassic Basin and a portion of the Piedmont. I-80 crosses them and enters the Great Valley.

I-80 starts on the Palisades of the Hudson. The Palisades are an eroded cross section of a diabase sill in the Triassic Basin that dips gradually northwestward (Figure 120). The sill is about 1,000 feet (305 m) thick, unusually thick for such an intrusion. It can be studied conveniently at Palisades State Park, Ross Dock, Fort Lee. The Palisades Diabase has been dated at 192 million to 186 million years ago, which puts it about on the border between Early and Middle Jurassic in age.

Almost immediately, the highway moves onto the outcrop area of the more or less red shales and sandstones that overlie the sill. These rocks continue westward to the hills of the New Jersey Highlands, in the vicinity of I-287 and Parsippany. Three major basalt lava flows that extend north and south in the I-80 area interrupt the succession of sediments. The first one occurs between the Paterson and Totowa exits. You can take a short side trip to see the basalt at the falls in the Great Falls District park in downtown Paterson. The outcrop area of the second flow crosses the highway between the Totowa and Mountain View (New Jersey 23) exits. The highway just touches the southern end of the third flow about a mile east of the I-280 intersection. The lava cooled to basalt at the beginning of the Jurassic Period.

The road enters the highlands just beyond the I-287 interchange, where there is a portion of the last ice cap's terminal moraine. The moraine extends to approximately the U.S. 46 exit. The rocks of the Highlands are typical Piedmont formations; Proterozoic to Devonian igneous, metamorphic, and sedimentary rocks, intensely folded and faulted. Along I-80, Proterozoic granite and gneiss form the bedrock. Strips of early Paleozoic sedimentary rocks and metasediments trend northeast-southwest through the Highlands; the highway crosses one of them between the Dover and Ledgewood exits. These Highlands rocks were presumably distorted during the Taconic Orogeny and again during the Alleghenian Orogeny.

About at the Allamuchy (Route 517) exit, the highway enters the Great Valley of the Appalachian Valley and Ridge Zone, with its Paleozoic unmetamorphosed to slightly metamorphosed sedimentary rocks. Even though the landscape is relatively flat, the rocks have been folded and faulted. The rocks are mostly Cambrian and Ordovician slate, shale, dolomite, and limestone, with some phyllite. Kittatinny Mountain and the Delaware River form the northwestern edge of the Great Valley.

Franklin, New Jersey, lies north of I-80, on New Jersey Route 23. There, and at Ogdensburg, there are old zinc mines in a strip of rocks that has been well known to mineral collectors for many years. The ores occur in the Proterozoic Franklin Limestone of the New Jersey Highlands, which is part of a down-dropped fault block that cuts into the Piedmont rocks and extends into the Great Valley, from Ogdensburg to Big Island, New York. Cambrian limestone and quartzite accompany the Franklin Limestone. The ore minerals are franklinite, willemite, and zincite, which occur in an attractive combination of green, reddish, and black crystals. The willemite is especially appealing because it is fluorescent. The minerals are on exhibit in the museum at the Franklin mines site.

Delaware: I-95

Coastal Plain sediments occupy almost all of Delaware, but the northern tip has Piedmont rocks above a line running just north of Newark and Wilmington. The rocks are masses of Proterozoic igneous and metamorphic types, such as anorthosite, gabbro, granite, amphibolite, and gneiss in a matrix of late Proterozoic to Cambrian quartzite, schist, and marble. I-95 crosses this zone.

In Bringhurst Woods Park, on the northeast outskirts of Wilmington, gneiss intruded by gabbro outcrops along Shellpot Creek. The gneiss's age is uncertain, but it may have been metamorphosed about 441 million years ago, which puts it in the vicinity of the Taconic Orogeny's end. The park can be reached from the U.S. 202 exit on I-95.

Baltimore to Cumberland, Maryland: I-70, I-68 (Figure 122) *This route traverses the whole Appalachian Province, through the Piedmont, the Blue Ridge, the Great Valley, and the folded rocks of the Appalachian Mountains.*

Baltimore lies at the edge of the Coastal Plain, where Cretaceous sandstone and shale lap up onto the contorted Piedmont rocks. From Baltimore to Frederick, I-70 passes over folded and faulted Late Proterozoic and Cambrian metasediments—quartzite, schist, and marble.

Between Frederick and Hagerstown the highway crosses a zone of Proterozoic metavolcanics, bordered on the southeast side by Cambrian marble and quartzite and on the northwest side by Cambrian and Ordovician dolomites. The metavolcanics make up part of the Blue Ridge Zone's northern end, which starts as South Mountain, a ridge that comes down into Maryland from Pennsylvania. About at the state line, South Mountain divides southward into two parallel ridges, the limbs of an anticlinorium. The southeast ridge forms Catoctin Mountain and the Bull Run Mountains, where it stops. The northwest ridge forms South Mountain, Short Hill, and the rest of the Blue Ridge. I-70 crosses the Monacacy River east of Frederick; beyond is Catoctin Mountain, then there is a valley, then South Mountain.

Beyond South Mountain the route enters the Hagerstown Valley, a part of the Great Valley. The valley contains eroded folds of Cambrian and Ordovician shale, limestone, and dolomite from the early Paleozoic sea. Between the Clear Spring and Big Pool exits, the highway arrives at the mountainous part of the Valley and Ridge Zone. From the Hagerstown Valley to Cumberland, the highway traverses a wide synclinorium, with a major syncline in the middle and smaller folds on the flanks. The synclinorium consists of Silurian sandstone and shale, Devonian limestone and shale, and some Mississippian terrestrial sandstones and conglomerates, the last being remains of the mountains formed during the Acadian Orogeny.

Three thrust faults complicate the folds between Clear Spring and Hancock. At Hancock, the road continues westward on I-68, as I-70 turns abruptly north. Six miles (9.5 km) west of Hancock is a stupendous road cut through Sideling Hill that clearly shows a syncline, with Mississippian conglomerate, sandstone, and shale (Plate 22). Most of the beds are river deposits associated with the highland created by the Acadian Orogeny, but the road cut shows some marine shale, indicating a brief incursion of the sea. The structure can be studied safely at the state visitor center there. The largest syncline in the series of folds occurs between Hancock and the vicinity of the Flintstone exit. The Allegheny Front, the edge of the Allegheny Plateau, crosses the highway just beyond Cumberland, and there the folds end.

Washington, D.C. to I-81, Virginia: I-66 (Figure 122) *This route crosses the Piedmont, with an included Triassic Basin, samples the Blue Ridge, and ends up in the Great Valley.*

Washington, like other cities south of it, lies at the edge of the Coastal Plain. Its bedrock is Cretaceous shale and sandstone, with a frosting of Ice Age sand. Neighboring Arlington, Virginia, is in the Piedmont. From Arlington to the Centreville exit (Virginia Route 28), the highway crosses folded and faulted Proterozoic and Cambrian phyllite, schist, quartzite, shale, and sandstone, with metavolcanics and some peridotite-family rocks from below the ancient sea floor. The formations occur in slices separated by three major thrust faults.

Just before the Centreville exit, I-66 passes into the Culpeper Subbasin of the New York–Virginia Triassic Basin, with Triassic sandstone and shale surmounted by Jurassic sandstone, shale, and three basalt lava flows. The lava flows occur in the area from directly north of Manassas to the U.S. 15 exit.

Between the U.S. 15 exit and the Virginia Route 55 exit, the highway reenters Piedmont rocks, with metavolcanics extending to the vicinity of Marshall.

From Marshall to the U.S. 522 (Riverton–Front Royal) exit, the Bull Run Mountains spur of the Blue Ridge shows mainly resistant Proterozoic metavolcanic greenstone, with older granite and gneiss. Two thrust faults occur along the northwest margin of the Blue Ridge, east of Front Royal.

Between the U.S. 522 exit and I-81, the highway traverses the Great Valley of the Valley and Ridge Zone; beyond are the long ridges of the Appalachian Mountains. In the Great Valley here, an eroded synclinorium contains Cambrian and Ordovician limestone, shale, and sandstone from the early Paleozoic seas.

Richmond, Virginia to the West Virginia border: I-64, U.S. 250 (Figure 122) *This route offers a complete section of the Appa-*

lachian Province: the Piedmont, the Blue Ridge, and the Valley and Ridge Zone. It also includes a bit of a Triassic Basin.

Richmond lies at the edge of the Coastal Plain, with Piedmont schist and gneiss intruded by Mississippian granite and here and there covered by fingers of Coastal Plain Tertiary shale and sandstone. About 10 miles (16 km) northwest of Richmond, I-64 crosses the narrow northern end of the Richmond Triassic Basin, with Triassic sandstone and shale.

From the Richmond Basin to the U.S. 522 (Gum Spring) exit, Piedmont Proterozoic gneisses of two different ages lead to a thrust fault. Beyond the fault, the bedrock consists of a broad series of folded Proterozoic, Cambrian, and Ordovician metasediments and metavolcanics, including various schists, phyllite, slate, and greenstone. This zone extends beyond Charlottesville. The Blue Ridge rises on the other side of Charlottesville, with Proterozoic greenstone, gneiss, and diorite. There are thrust faults on either side of the Blue Ridge and a third at Charlottesville.

At Waynesboro, the highway has descended into the Great Valley of the Valley and Ridge Zone, with eroded Cambrian sandstone, shale, limestone, and dolomite on either side of the Shenandoah River. The Great Valley extends from Waynesboro to about 2 miles (3 km) west of Churchville. Its formations of Cambrian and Ordovician sandstone, shale, limestone, and dolomite belong to a truncated synclinorium. There is a cross section of a thrust fault at Staunton in a road cut on U.S. 250 near the west end of the Gypsy Hill Park in that city. At the fault, Cambrian limestone has overridden Ordovician limestone. The fault zone is a breccia with broken limestone blocks of various sizes, some as big as automobiles. At the western edge of the Great Valley, beyond Churchville, yet another thrust fault crosses the highway.

Mountains of the Valley and Ridge Zone extend from Churchville to the West Virginia border. The mountains eroded from a synclinorium followed by an anticlinorium whose oldest rock outcrops around Hightown. The ridges and valleys have Ordovician limestone and shale, and Silurian and Devonian sandstone, shale, and limestone. The Allegheny Plateau begins a few miles into West Virginia.

Raleigh, North Carolina to Harriman, Tennessee: I-40 (Figure 122) *This highway section resembles others to the north in that it crosses the Piedmont with an included Triassic Basin, plus the Blue Ridge and the Valley and Ridge Zone.*

Other attractions are several masses of intrusive igneous rock and the Great Smoky Mountains, with their thrust-fault windows.

Typical Piedmont slates and metavolcanics form the bedrock for about 10 miles (16 km) west from the center of Raleigh. Then, the highway enters the Wadesboro–Deep River–Durham Triassic Basin, filled with Triassic sandstone and shale, which extends beyond Durham. Like other basins in the Piedmont, it has some Jurassic basalt lava flows, but their exposures do not reach I-40.

From just beyond Durham to Graham there is a strip of typical Piedmont slate, schist, and metavolcanics of late Proterozoic and Cambrian age. From Graham to Winston-Salem the highway crosses a zone of intrusive rocks that extends down into South Carolina. First, gabbro of unknown age accompanies Late Proterozoic or early Cambrian granite. Then, about halfway between Graham and Winston-Salem, the bedrock is Pennsylvanian granite. About 5 miles (8 km) east of Winston-Salem the highway returns to Late Proterozoic and Cambrian metasediments.

The country to Asheville is mostly underlain by Proterozoic gneiss and schist, but many other rock types occur. Gabbro and granite lie between Winston-Salem and the Mocksville (U.S. 601) exit. Beyond that for a couple of miles the highway crosses Cambrian granite, and farther west, Cambrian or Ordovician granite in the Morganton area. The North Carolina Route 18 exit is in the middle of the latter granite. About halfway between Morganton and Asheville, the rocks have been cut by a pair of thrust faults, the first major faults at the approach to the Blue Ridge Zone.

The metamorphosed Piedmont rocks continue to about 10 miles (16 km) beyond the I-26 Asheville exit. Then the highway curves directly across the Blue Ridge Mountains into Tennessee, through Proterozoic gneiss and schist into the resistant Cambrian sandstone-quartzite that forms much of the backbone of the whole Blue Ridge, and finally into some Cambrian and Ordovician limestone and dolomite in the region of Newport, Tennessee.

The southern end of the Blue Ridge Zone broadens significantly (Figure 110). The Great Smoky Mountains occupy the west part, with other mountains to the south and northeast. For a side trip to Great Smoky Mountains National Park, take U.S. 441, which traverses it from Cherokee, North Carolina to Gatlinburg, Tennessee. The main body of the mountain range consists of Proterozoic quartzite, conglomerate, and slate. Major thrust faults dislocate the formations in the north and east portions of the park and at the west end of it. At Cades Cove, erosion has opened a window through the Proterozoic rocks to show Ordovician limestone and dolomite that were overridden by the Proterozoic rocks in the thrust faulting.

Back on I-40, the Valley and Ridge Zone extends from Newport to Harriman. Between Newport and Knoxville the structure is basically that of two synclinoria separated by a thrust fault. Cambrian, Ordovician, and Silurian limestone, sandstone, and shale from the early Paleozoic sea outcrop in the area. Beyond Knoxville, the same set of rocks is repeated by one thrust fault after another. After Harriman, I-40 passes onto the nearly horizontal formations of the Cumberland Plateau.

Columbia, South Carolina to the West Virginia border: I-77 (Figure 122) *This southern Appalachian route passes through Piedmont metamorphic rocks and intrusives and a narrow strip of Blue Ridge sedimentary rocks into the Valley and Ridge Zone and some thrust-faulted sedimentary layers.*

Columbia is on the Fall Line between the Piedmont and the Coastal Plain. Coastal Plain Cretaceous sandstone and Tertiary shale overlap Proterozoic and Cambrian metasediments and metavolcanics. From Columbia to Charlotte, North Carolina, the usual complexly folded metasediments such as schist, phyllite, quartzite, and marble, along with metavolcanics and intrusive bodies, make up the bedrock. The intrusives belong to a northeast-trending belt of abundant intrusive rocks within the Piedmont matrix. Permian granite and then gabbro of uncertain age lie between the South Carolina Route 9 and U.S. 21 (Rock Hill) exits. From U.S. 21 to Charlotte you cross gabbro, then more Permian granite, then gabbro, then Ordovician granite.

The most productive gold mine in the eastern United States formerly operated about 35 miles (56 km) east of I-77, southeast of Lancaster, South Carolina, near the town of Kershaw. The Haile Mine, about 3 miles (5 km) northeast of Kershaw, was worked from about 1829 to 1942. The gold occurred in wide veins in metavolcanic rock, along with iron minerals and quartz; the site was probably an ancient hot spring, where hot solutions from below deposited minerals in cracks in the bedrock.

Gabbro occurs along I-77 north of Charlotte, followed by a long stretch of Proterozoic to Early Cambrian granite. In the

Mooresville area the bedrock switches to younger granite, Pennsylvanian or Permian in age, no doubt formed as part of the Alleghenian Orogeny. Then, between Mooresville and Statesville, the intrusive zone gives way to the Piedmont-type folded and metamorphosed sedimentary and volcanic rocks.

From about 10 miles (16 km) south of Statesville, North Carolina, to about the same distance north of the Hillsville, Virginia, exit (U.S. 58 and 221) the bedrock is a series of Proterozoic formations pushed horizontally from southeast to northwest in great overturned folds and metamorphosed in the process to gneiss, schist, and amphibolite. This is evidence of the Taconic Orogeny. Three thrust faults cut through the area, and the U.S. 21 exit that is 6 miles (9.5 km) north of the U.S. 421 exit lies between the first two of the three faults. All three cross the highway within a span of about 10 miles (16 km).

From Hillsville to the West Virginia border, I-77 goes through an unusually narrow strip of typical Blue Ridge rocks, so the Piedmont almost meets the Valley and Ridge Zone on this highway. The narrow strip begins about 7 miles (11 km) northwest of Hillsville and contains Blue Ridge sandstone, shale, and dolomite, that extends far northeastward along the inner margin of the Blue Ridge Zone.

These Blue Ridge rocks have been thrust over younger rocks. At Wytheville, you enter the Valley and Ridge Zone, with a series of partial synclines faulted into slices, forming narrow outcrop areas of tilted sandstone, shale, and limestone layers that trend northeastward. The first slice has Ordovician and Silurian formations, the second has Cambrian, Ordovician, Silurian, and Devonian thrust over Mississippian rocks, followed by Devonian, Silurian, and Ordovician rocks in a syncline. A last small fold with Ordovician, Silurian, and Devonian layers takes I-77 to the West Virginia border and beyond into the Allegheny Plateau.

Columbia, South Carolina to Asheville, North Carolina: I-26

(Figure 122) *This section of I-26 crosses a large area of the Piedmont and a variety of rock types.*

The trip starts where the Piedmont rocks disappear under Coastal Plain Cretaceous sandstone and Tertiary shale. The most common bedrock between Columbia and Asheville consists of metasediments such as gneiss, schist, slate, quartzite, and marble. Here and there, the highway crosses strips and blobs of metavolcanic rocks.

This mixture of Proterozoic and Cambrian metamorphic rocks contains a few intrusions. Silurian or Devonian granite occurs in the area of the Newberry exits, and just beyond the border in North Carolina the highway passes over a zone of Cambrian or Ordovician granite. Beyond that, Proterozoic to lower Cambrian gneiss and quartzite return. A series of thrust faults runs along the southeast border of the Blue Ridge and southeast of it. Two of the faults, close together, cross the countryside in the vicinity of the exit to Mountain Home, North Carolina. Between the faults the rock has been ground up by movement of the heavy masses and re-formed as a series of mylonitic rocks. Proterozoic gneiss and schist extend from the faults to Asheville.

Macon, Georgia to Chattanooga, Tennessee: I-75 (Figure 122)

The road from Macon to Atlanta and beyond traverses Piedmont rocks; those between Macon and Atlanta are mostly gneiss, schist, amphibolite, and quartzite of somewhat uncertain age, Proterozoic and/or Early Paleozoic. From Atlanta to Allatoona Lake the bedrock consists of gneiss, schist, metagraywacke, and quartzite of uncertain age.

At Allatoona Lake, the highway enters the southern equivalent of the Great Valley, beyond the end of the Blue Ridge. From Dalton to Chattanooga, a series of thrust faults has repeated the same formations of Cambrian and Ordovician limestone, dolomite, and some shale six times. Figure 116 shows the Tennessee thrust fault zone.

Augusta, Georgia to Birmingham, Alabama: I-20 (Figure 122)

I-20 traverses typical Piedmont rocks on its way to the folded formations in the Valley and Ridge Zone and into the coal and steel country around Birmingham.

Augusta lies at the edge of the Coastal Plain, where Cretaceous sandstone laps onto metavolcanics of the Piedmont. Between Augusta and Atlanta, zones of various gneisses, schist, quartzite, and amphibolite alternate with zones of granite. The highway crosses a mass of granite in the Siloam-Crawfordsville area. No one knows exactly how old these rocks are; the granite is Late Paleozoic and the others are probably Late Proterozoic or Early or Middle Paleozoic. A small body of Middle Paleozoic granite occurs in the western outskirts of Atlanta. From there almost to the Alabama line a tight mixture of gneiss and schist forms the bedrock, more formations of uncertain age.

Approximately 5 miles (8 km) east of the state line, Piedmont schist gives way to the Valley and Ridge Zone, with Pennsylvanian sandstone and shale and Mississippian limestone. A fault crosses just about at the state line and another at about the U.S. 431 exit in Alabama; the faults bracket a body of Devonian schist and other metasediments. West of the U.S. 431 exit, Cambrian and Ordovician limestone and dolomite form a broad anticline in the Coosa Valley, a southern cousin of the Great Valley. Devonian or Mississippian shale, then Pennsylvanian sandstone and shale follow in a syncline of the Coosa Coal Basin. The Coosa Valley rocks end at a thrust fault just beyond Logan Martin Lake.

Several thrust faults cut the region's structure; one separates the Coosa Coal Basin from the Cahoba Coal Basin's partial syncline to the west. The coal basins include formations of Carboniferous, Devonian, Ordovician, and Cambrian age, with a few small patches of Silurian. The rock suite consists of sandstones, shales, coal, and limestones, including rocks from the Paleozoic seas as well as terrestrial rocks from river systems of the lands formed by the orogenies. Birmingham grew in the western portion of the Cahoba Coal Basin, and beyond Birmingham, the last thrust fault forms the border of the Black Warrior Coal Basin, in the Cumberland Plateau.

The folded structure, in bringing a variety of old rock formations to the surface in eroded limbs of folds, made Birmingham's steel industry possible. The iron ore comes from a Silurian formation in which the iron mineral hematite replaced particles and fossil fragments and was deposited between grains of limestone. Other limestone formations furnish the lime used as a flux in the steel-making process. In the blast furnaces, a mixture called slag removes impurities from the liquid iron. The limestone flux makes the slag more fluid and lowers the melting temperature of the iron. And, of course, the coal mines in the area furnish fuel.

■ THE OUACHITA MOUNTAINS

The subcontinent Gondwana slowly but inexorably crunched into the eastern and southern parts of Laurentia as crustal plate motions completed the building of Pangaea (Figure 106). The portion of Gondwana that was to become Africa folded and faulted eastern formations to create the structures that eventually eroded into the Appalachian Mountains. The part destined to become South America similarly folded and faulted southern Laurentia; some of those distorted rocks there long afterward became the Ouachita Mountains of Arkansas and Oklahoma.

The Ouachita Mountains (see Figure 126 for location) appear to be a direct continuation of the central region of the Appalachian Province. The Ouachitas are separated from the Appalachians because the segment between Alabama and Arkansas sank and disappeared beneath Coastal Plain sediments of the Mississippi River region. The mountains themselves are ridges of resistant rock on the limbs of folds in a highly faulted anticlinorium (Figure 123, Plates 18, 22). The anticlinorium structure reacted to horizontal pressure unevenly, so that some folds dip, giving a zigzag pattern like those of Pennsylvania (Figure 124, Plate 17). The pressure overturned folds toward the continental center and created numerous thrust and other faults (Figures 123 and 125). The rocks range in age from Ordovician to Pennsylvanian. The most obvious difference between the Appalachians and the Ouachitas is the orientation—in the Ouachitas, the pressure was to a great extent from the south, so that most of the ridges trend essentially east and west, instead of northeast and southwest as in the Appalachian Mountains.

The primary ridge-maker is the Arkansas Novaculite, a Devonian and/or Mississippian formation that contains exceptionally hard and abrasive fine-grained layers formed almost entirely of silica (quartz). People have quarried those layers for many years to make grindstones and whetstones. Most of the novaculite consists of more normal shale and chert beds. Softer rocks have weathered and eroded away from the novaculite, leaving it as the backbone of major ridges.

Many of the ridges rise to over 2,500 feet (760 m) in elevation, and the maximum relief, from valley floor to moun-

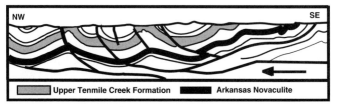

Figure 123. A simplified cross section of folds and faults in the Oklahoma Ouachita Mountains.

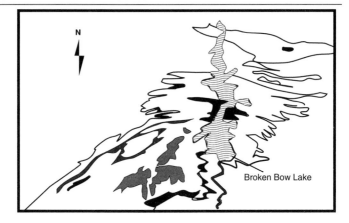

Figure 124. The Broken Bow Uplift in the Ouachita Mountains of Oklahoma. An anticlinorium has been bowed upward; its folds trending east–west make the zigzag pattern around the margin of the uplift. The oldest rocks in the anticlinorium are in the jagged "islands," lower center. These complexly folded Ordovician to Devonian formations are surrounded by relatively flat layers: Mississippian beds to the west, north, and east, Coastal Plain rocks to the south.

taintop, reaches about 1,100 feet (335 m) just west of the Oklahoma-Arkansas border. The relief decreases from there to the east and west.

◆Ouachita Rocks

The formations of the Ouachita Mountains consist mostly of shale and sandstone, along with a little Ordovician limestone and some chert. The Pennsylvanian Johns Valley Shale differs greatly from other formations in that it contains angular, unweathered blocks of sedimentary rock up to more than 300 feet (90 m) in diameter, surrounded by ordinary gray to tan shale. The big blocks occur in the lower part of the formation, but at all levels there are rock fragments of granule, pebble, and boulder size. The large blocks come from Upper Cambrian to Lower Pennsylvanian formations. Presumably, the Johns Valley Shale testifies to the beginning

Figure 125. A road cut on U.S. 259 in the Broken Bow Uplift. The tilting of the layers is evident. Note the normal fault in the right part of the illustration, indicated by the offset of the thinner beds.

of the Alleghenian Orogeny in the Ouachita region. According to the theory, a fault developed, and a mass of crust with many layers of sedimentary rock rose out of the ocean at the edge of Laurentia. Blocks and smaller fragments fell from the cliffs so formed and became embedded in nearshore ocean mud.

The Ouachita sedimentary rocks were formed from sediments deposited in an ocean along the south margin of Laurentia throughout most of the Paleozoic Era. Gondwana's collision happened at some time during the latter half of the Pennsylvanian Epoch and the beginning of the Permian Period, folding the formations and creating great thrust sheets that moved masses of rock many miles during the Alleghenian Orogeny. What direct evidence there may be for earlier orogenies is, unfortunately, buried under the Coastal Plain rocks south of the Ouachita Mountains.

OUACHITA MOUNTAIN HIGHWAY SECTIONS

In the Ouachita Mountain region, the interstate highways skirt the hills. The following routes are on smaller roads that traverse scenic, interesting, and geologically more diverse country.

Danville to Hot Springs, Arkansas: Route 27, U.S. 270 *This route crosses folds and faults in the main Ouachita structure, the Ouachita Anticlinorium.*

Between Danville and Mount Ida, Arkansas Route 27 crosses the northern half of the Ouachita Anticlinorium, which trends east northeast in the area. Mount Ida lies close to the structure's longitudinal axis. North of Danville, the Arkansas Valley parallels the mountains, with broad anticlines and synclines, not the severely contorted rocks of the mountains.

Pennsylvanian sandstone and shale extend from Danville to Aly. Markings on some slabs show that ocean currents flowed toward the west and southwest. Two or three miles (4 km) south of Aly, the road passes into Late Mississippian sandstone and shale, followed by a thrust fault in which earlier Mississippian shale, chert, and graywacke have been thrust northward over them.

The older rocks at the core of the structure begin to appear at and near the Arkansas Route 88 junction about 3 miles (5 km) south of Story. From north to south, the rocks are Mississippian shale; Devonian to Mississippian shale, chert, novaculite, and conglomerate; Silurian shale; Ordovician shale at the bridge over the arm of Lake Ouachita; and finally, Ordovician chert and shale.

From Mount Ida to Hot Springs, U.S. 270 enters the south flank of the anticlinorium. Mount Ida's bedrock is Ordovician chert and sandstone, and from there to Crystal Springs, Ordovician shale and sandstone outcrop. Quarries a quarter mile south of Crystal Springs show complexly folded chert and shale. Folded Ordovician sandstone, shale, limestone, and chert continue for about 5 miles (8 km) east of Crystal Springs. Then, the highway goes quickly back into the Devonian and/or Mississippian novaculite series of rocks and into Mississippian shale to Hot Springs.

Hot Springs to Hollis, Arkansas: Route 7 *This is a shorter trip through the Ouachita Anticlinorium than the one above.*

Hot Springs lies on Mississippian shale, next to outcrops of older rocks. Hot Springs Mountain, North Mountain, and Sugarloaf Mountain in Hot Springs National Park are all ridges supported by the Devonian and/or Mississippian novaculite formation. Action along a thrust fault between Hot

Springs Mountain and North Mountain moved novaculite northwestward over later Mississippian shale. North and Sugarloaf mountains emphasize the Devonian and Mississippian layers on the limbs of an anticline, with older Silurian shale/slate and Ordovician chert in between.

From Hot Springs to Hot Springs Village, the road traverses tight northeast-trending folds of the anticlinorium's core, from the novaculite through various older Ordovician shales and chert to the novaculite again in the vicinity of Hot Springs Village. Beyond the novaculite the road enters gently warped Mississippian shales and Pennsylvanian shales and sandstones in the Arkansas Valley.

Heavener to Broken Bow, Oklahoma: U.S. 59 and 259 *This route shows a good cross section of the Oklahoma Ouachita Mountains.*

From Heavener to Smithville the road crosses the western extension of the Arkansas Valley country, with broadly folded Carboniferous sandstones and shales. It consists of five thrust fault slices, with resistant sandstones forming ridges. The rocks originated as ocean-bottom sediments formed of material eroded from the Laurentia continent.

South of Smithville, U.S. 259 curves around the town of Sherwood. Mississippian shale in the area signals the edge of the much more intensely folded core of an anticlinorium. About where the highway goes south again, it enters a zone of greatly crumpled Devonian, Silurian, and Ordovician novaculite, shale, chert, and sandstone (Figures 124, 125, Plate 22). Nearby Broken Bow Lake extends through this zone. Between the road to the dam at Broken Bow Lake and the town of Broken Bow, the bedrock changes significantly, as almost horizontal Cretaceous sandstone, shale, and limestone of the Coastal Plain lap up on the folded and faulted Paleozoic rocks.

■ THE WESTERN WINDOWS

West and southwest of the Ouachita Mountains, remnants of mountain building during the Alleghenian Orogeny are almost entirely covered by the flat-lying sedimentary rocks of the Coastal Plain and the Central Lowland. However, four windows eroded through the covering formations reveal a curving zone of deformation that extends southwestward

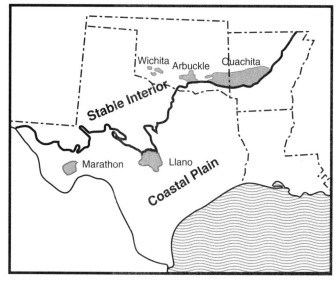

Figure 126. The Ouachita Mountains and the western windows (the Arbuckle Mountains, Wichita Mountains, Llano Uplift, and Marathon Region), are remnants of Appalachian-type structure west of the Mississippi River.

from the Ouachita Mountains to southwest Texas, along the margin of old Laurentia. The windows appear now as the Arbuckle and Wichita Mountains of Oklahoma and the Llano Uplift and Marathon Region of Texas (Figure 126). It seems certain that, along with the Ouachita Mountains, these windows belong to a continuation of the Valley and Ridge Zone of the central Appalachians. The rocks and structures of the Blue Ridge and Piedmont are thought to extend to the Gulf of Mexico, covered by Mesozoic and Cenozoic sedimentary rocks.

◆The Arbuckle Mountains

The Arbuckle Mountains are remnants of three anticlines trending northwest–southeast that have been modified by faulting (Figure 127). Resistant limestone causes ridges; ero-

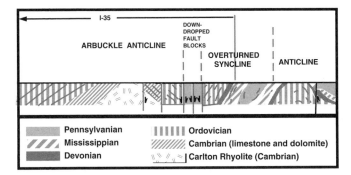

Figure 127. A north-south cross section of the western part of the Arbuckle Mountains, where I-35 crosses the Arbuckle Anticline. The overturned syncline indicates pressure from the left (south). Note that the anticline and the syncline have both been disrupted by vertical faults.

sion has carved the valleys in shale, with a maximum relief between valley floor and hilltop less than 500 feet (150 m) (Plate 19).

The rocks of the anticlines are Paleozoic, from Cambrian to Pennsylvanian in age. A large area in the eastern Arbuckles has granite bedrock, dated at 1.35 billion years old, and there are two small areas of granite in the western hills. The granites belong to a group of Proterozoic batholiths.

Almost all of the rock in these anticlines originated as sediments that accumulated in the ocean off the Laurentia shore. The series begins with Cambrian lava, a rhyolite porphyry, with sandstone on top of it. Upper Cambrian limestone overlies the sandstone, followed by younger Ordovician limestone, dolomite, sandstone, and shale (see Plate 23). Silurian, Devonian, and Mississippian deposition resulted in limestone and shale. Lower Pennsylvanian limestone, shale, sandstone, and conglomerate complete the series (Plate 32).

The Pennsylvanian sandstone and conglomerate suggest rising land and large rivers. As in the Ouachita Mountains, the formations of the Arbuckle Mountains were folded and faulted during the latter half of the Pennsylvanian Epoch as Gondwana collided with Laurentia. The Arbuckles show direct evidence of only the Alleghenian Orogeny.

◆The Wichita Mountains

This second window in southern Oklahoma shows several formations that also occur in the Arbuckle Mountains plus large areas of intrusive igneous rocks (Figure 126, Plate 19). The major hills are masses of Early Cambrian granite, diorite, and gabbro, pushed up through the Arbuckle-type sedimentary and volcanic rocks that were originally deposited on top of them. The upturned sedimentary and volcanic rocks outcrop in some places around the margins of the major hills and in smaller hills of the area; elsewhere, they have been eroded away. The granite (Plate 31) occurs as sills and stocks in all the hills to the west. The gabbro and diorite outcrop on Raggedy Mountain and on the slopes of Mt. Sheridan beneath a capping of granite (Figure 128).

During Early Cambrian times, gabbro, anorthosite, and diorite intruded a volcanic series of basalt, andesite, and tuff. Those rocks were in turn intruded by the granite. Then volcanoes erupted again, leaving another series of volcanic rocks consisting of rhyolite, tuff, and volcanic breccia, still Early Cambrian. All this volcanic and intrusive activity might have been caused by rifting as the supercontinent that gave birth to Laurentia broke up. An epoch of weathering and erosion followed. Later, the surface sank, and the shallow inland sea covered the area, so that Late Cambrian marine sediments settled on the earlier volcanics, first sand that became sandstone, then limey mud that became limestone. Later Paleozoic rocks, mostly below the surface, consist of Ordovician to Pennsylvanian sandstone, shale, and limestone from the inland sea, similar to those of the Arbuckle Mountains.

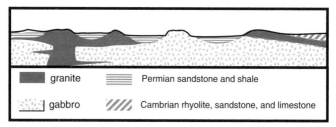

Figure 128. The Wichita Mountains in Oklahoma are hills of intrusive igneous rock surrounded mostly by flat-lying Permian sandstone and shale.

Horizontal Permian sandstone, shale, and conglomerate form the bedrock of the plains surrounding the Wichita Mountains. These typical Central Lowland rocks originated as sediments from highlands created by the Alleghenian Orogeny; evidence indicates that the Wichita uplift occurred during the Pennsylvanian Epoch and may have been related to the folding and faulting in the Arbuckle Mountains.

◆The Llano Uplift

Central Texas holds the third isolated fragment of the Appalachian Province (Figure 126). It is an upthrusted area on the margin between the Coastal Plain and the Central Lowland, a topographic basin of Proterozoic rocks surrounded by highlands that contain Paleozoic and Cretaceous formations.

The rocks in the basin represent the Proterozoic "basement" upon which Paleozoic marine sediments accumulated. Among the basement rocks, granites were intruded between 1.07 billion and 850 million years ago, and heat and pressure metamorphosed volcanic and marine sedimentary rocks into gneisses 1.12–1.05 billion and 970–910 million years ago (Plate 19). These ancient rocks, which seem to be an extension of the Canadian Shield's Grenville Province, were uplifted, eroded, and then lowered (or the sea level rose) during the Cambrian Period. They formed the base of the ocean floor for much of the Paleozoic Era.

Uplifting resumed in the latter part of the Paleozoic Era, from Late Mississippian into Middle Pennsylvanian times, then again at the end of the Pennsylvanian Epoch. The crustal dislocation was presumably part of the Alleghenian Orogeny. Now, remnants of the Paleozoic marine deposits outcrop around the margin of the uplift, including mostly Cambrian sandstone and limestone with smaller bits of Ordovician to Pennsylvanian rocks. Here and there along the east, south, and west margins, Cretaceous sandstone, shale, and limestone of the Coastal Plain are exposed.

◆The Marathon Region

In southwest Texas, another window, surrounded by Permian and Cretaceous rocks, exhibits results of the cataclys-

Figure 129. A simplified cross section of the Marathon Region of Texas, showing folding and faulting of Paleozoic formations in a structure similar to those of the Appalachian and Ouachita Mountains. To the northwest, beyond the Glass Mountains, the rocks were not affected. To the southeast, the contorted rocks extend beneath flat-lying formations of the Coastal Plain.

mic collision of Gondwana with Laurentia (Figure 126). The Marathon Region comprises, from southeast to northwest, a synclinorium broken by thrust faults into four slices and an anticlinorium that has been thrust northwestward over part of another anticlinorium (Figure 129). The complicated structure is similar to that of the Virginia-Tennessee region, indicating great horizontal pressure from the southeast.

The Marathon rock series contains Cambrian, Ordovician, possibly Devonian, and Pennsylvanian conglomerate, sandstone, arkose, shale, limestone, chert, and novaculite from the Paleozoic inland sea. Figure 130 shows a typical overturned fold. Gently dipping Early Permian formations cover folded Pennsylvanian layers with an unconformity in between, which indicates that the deformation of the Marathon rocks happened during the latter part of the Pennsylvanian Epoch as part of the Alleghenian Orogeny.

WESTERN WINDOWS HIGHWAY SECTIONS

Davis to Springer, Oklahoma: I-35 *This trip, between Texas Routes 7 and 53, follows a route in the western part of the Arbuckle Mountains through part of a syncline and then through an anticline disrupted by faults (Figure 127).*

Crustal forces during the Alleghenian Orogeny broke the

Arbuckle Anticline with two major down-dropped fault blocks and caused two other blocks to sink between the anticline and the syncline. Furthermore, the syncline dropped as a unit below the country on either side. I-35 crosses one limb of the syncline, the central lowered portion, and the anticline.

For about 2 miles (3 km) south of Route 7, the highway goes over only slightly distorted Pennsylvanian sandstone, shale, and limestone. For about the next 3 miles (5 km), I-35 skirts the edge of the syncline's southwest limb, then turns south to cross the anticline. The syncline's limb contains Devonian and Ordovician sandstone, shale, and limestone, with Pennsylvanian conglomerate overlying the older beds on an unconformity. Farther south is a down-faulted mass of Pennsylvanian conglomerate (Plate 32). At the scenic overlook, the highway has crossed over a border fault of the down-dropped block and reentered Ordovician limestone, in a formation that contains a rare limestone breccia. About a mile farther on, Cambrian limestone and dolomite outcrop in the central part of the anticline. In the middle of the sedimentary rock series, exposures show underlying sandstone and rhyolite. The rhyolite marks the core of the anticline in this area.

Cambrian to Ordovician Arbuckle Limestone outcrops at the southern scenic overlook, which is on the southern limb of the anticline (Plate 23). As the highway descends to the plain, it passes through Ordovician limestone, shale, and sandstone, then a thin strip of Devonian limestone. About 3 miles (5 km) north of Springer, the road leaves the hills and

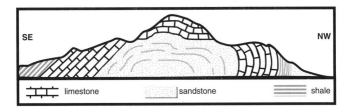

Figure 130. A diagram of the eroded end of an overturned anticline in the Marathon Region, showing the pressure was from the southeast. These exposed rocks are about 8 miles (13 km) south-southwest of Marathon, Texas.

enters Carboniferous rocks at the anticline's margin.

U.S. 77, which closely parallels I-35 through the hills, is marked with signs identifying all the individual formations. Also on U.S. 77 is Turner Falls Park, where Honey Creek falls 77 feet (232 m) over Cambrian limestone.

Lawton, Oklahoma through the Wichita Mountains: I-44, Route 49 *This route samples the extensive igneous rocks of the Wichita Mountains as they rise out of the surrounding Permian sedimentary rocks.*

From Lawton to the Oklahoma Route 49 exit, I-44 is on Permian conglomerate, a river deposit from the highlands of the Alleghenian Orogeny. Route 49 into the Wichita Mountains Wildlife Refuge encounters remnants of Cambrian granite, surrounded by the conglomerate. It is approximately 9 miles (14.5 km) to Mt. Scott Drive, which leads to the top of Mt. Scott, a hill carved by erosion from a sill of the Cambrian granite that overlies the gabbro in the area.

Route 49 continues westward along the scenic drive through the mountains to Route 54. The bedrock to Route 54 consists of granite and some gabbro in the hills, with Permian conglomerate in the flat areas around the intrusive rocks (Plate 19).

The gabbro and diorite that underlie the granite can be seen if you take a short side trip after you leave Mt. Scott. Turn north on Route 115, leave the wildlife refuge, and drive along the valley of Medicine Creek. The gabbro and diorite occur in the valley and form Little Mt. Sheridan on the left. There are many exposures of the rocks in the valley, including one at the bridge over Medicine Creek, which is about 5.5 miles (9 km) from Route 49. The dark rocks contain inclusions of quartzite, remnants of sandstone that the igneous material intruded.

West of Mason to Burnet, Texas: U.S. 377, Route 29 *This route traverses a cross section of the Proterozoic rocks in the Llano Uplift, from overlapping Cretaceous on one side to overlapping Cretaceous on the other side.*

The town of Mason is actually inside the Llano Uplift, built where masses of schist and gneiss join. The edge of the uplift crosses U.S. 377 about 7 miles (11 km) directly west of Mason, where Cretaceous shale and limestone overlie Proterozoic schist and granite. On U.S. 87, about 8 miles (13 km) north of Mason, steeply dipping layers of Pennsylvanian limestone outcrop.

From Mason for most of the way to Llano, the highway goes over gneiss, with an isolated body of granite about 10 to 12 miles (16 to 19 km) east of Mason. On the approach to Llano, the highway passes over patches of schist and granite. In the city of Llano, the Llano River has exposed gneiss and schist. Between Llano and the Buchanan Dam, the highway traverses gneiss, then granite. Beyond the dam, the bedrock becomes gneiss again, until about 3 miles (5 km) west of Burnet, when Cretaceous limestone suddenly appears, overlapping the gneiss.

U.S. 281 south from Burnet to Marble Falls skirts the eastern edge of the uplift mostly in Cretaceous rocks of the Coastal Plain. Just north of Marble Falls some Proterozoic granite shows up. Pennsylvanian limestone outcrops at Marble Falls, along the Llano River; it is the same formation that outcrops back north of Mason. A side trip off U.S. 281 to Inks Lake State Park will reveal a variety of granite, schist, and gneiss (Plate 19).

Sanderson to Alpine, Texas: U.S. 90 *This route crosses the folded and faulted Paleozoic rock layers of the Marathon Region, from east to west.*

West from Sanderson, the highway is on horizontal Cretaceous limestone of the Coastal Plain, where there are some road cuts in sand and gravel washed down from local mesas. About 35 miles (56 km) west of Sanderson, U.S. 90 enters the region of folded and faulted Paleozoic rocks, with steeply tilted Pennsylvanian limestone followed by almost vertical Pennsylvanian sandstone and shale. Along the way to Marathon, ridges formed by resistant layers of folded formations are visible, mostly Mississippian in age.

About 3 miles (5 km) east of Marathon the highway enters a Cambrian to Devonian zone. There, an east-west ridge capped by Devonian novaculite has been thrust over Pennsylvanian rocks. A major thrust fault crosses the highway about 3 miles (5 km) west of Marathon, where the older rocks have been thrust over the Pennsylvanian formations. The first hills west of Marathon have undistorted Permian limestone, gently dipping westward, that rests on top of the older folded rocks. The Permian rocks form the Glass Mountains to the north. At the U.S. 67 intersection, Permian limestone is visible in the cliff on the mountain south of the highway. About 2 miles (3 km) east of Alpine, overlapping Tertiary age rhyolite beds show the western edge of the Marathon Paleozoic region.

Marathon to Big Bend National Park, Texas: U.S. 385 *U.S. 385 crosses the two anticlinoria of the region.*

There is Ordovician limestone at Marathon. South of there are ridges of white novaculite, and about 3 miles (5 km) south of Marathon, the highway cuts through a ridge of Ordovician and Devonian chert and novaculite, then goes onto a strip of Mississippian shale and sandstone. At about the 12 mile (19 km) mark, novaculite appears again, nearly vertical. For the next 12 miles, the highway crosses folded and faulted Cambrian and Ordovician sandstone, shale, and limestone, with some Mississippian sandstone and shale at the end. After that, flat-lying Cretaceous limestone marks the edge of the Coastal Plain rocks, which, with some shale and sandstone, continue to the border of Big Bend National Park.

■ THE NORTHEASTERN APPALACHIANS: NEW ENGLAND

Most of New England is basically a relatively flat upland that rises gradually away from the ocean, with three mountain ranges and isolated peaks here and there. The ranges are the Green, Taconic, and White mountains; the best known and largest isolated peaks are Mt. Monadnock in New Hampshire and Mt. Katahdin in Maine (Plates 20, 31).

According to many experts, the southern edge of New England geology is in the Manhattan Prong, which ends at the tip of Manhattan Island, and the Reading Prong (the Hudson and New Jersey Highlands), which extends beyond the Hudson River to Reading, Pennsylvania. For the purposes of this book, is it convenient to cut off New England at the Hudson River in the New York City area. The New Jersey Highlands are described as part of the Piedmont, whose rocks resemble the New Jersey ones, and the eastern edge of New York State lies in the Appalachian Province (Figure 131).

The geologic history of New England essentially parallels that of the central and western portions of the Appalachian Province. Many New England rocks originated as volcanic material and as ocean bottom and land sediments from the western island chain, Avalonia, and the oceans east of the old continent of Laurentia (Figure 106). Those rocks were folded, faulted, metamorphosed to various degrees, and intruded during the same three orogenies that affected the central Appalachians, the Taconic, Acadian, and Alleghenian orogenies.

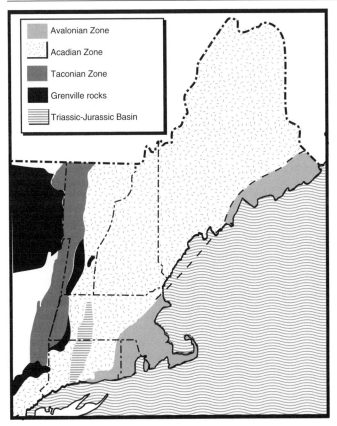

Figure 131. The geological zones of New England.

New England differs from the Appalachians to the south in that it has no Valley and Ridge Zone like that of the central Appalachians. Much of New England bedrock resembles that of the Piedmont, with Late Proterozoic and Early Paleozoic metamorphic and igneous rocks. The geology can be confusing, with miscellaneous rocks apparently tossed together without any order. The formations were originally organized, as they are in the Valley and Ridge Zone, but the orogenies did to New England rocks what they did to Piedmont rocks, as land masses crashed into Laurentia, deforming, intruding, and in some cases metamorphosing them, as well as breaking them up through faulting. Thrust faults, normal faults, reverse faults, vertical faults, and strike-slip faults abound (Plate 32). Massive sections of crust have been pushed upon one another, blocks have moved up and down, and blocks have slid horizontally. Faults have been folded and folds have been refolded, sometimes twice.

But one can discern order in all this chaos. In general, three Proterozoic and Paleozoic zones appear, which can be called the Taconian, Acadian, and Avalonian zones, along with a younger Triassic Basin cutting into the middle zone (Figure 131).

Since New England was covered by Ice Age glaciers, most of its bedrock is hidden by till and outwash. The ice also left a full complement of glacial phenomena—drumlins, eskers, moraines, kettle lakes, polished and striated pavements, and cirques in the mountains. The ice left its marks everywhere (Plates 37, 39, 40). Early in the last glacial stage, an individual ice cap formed in the mountains of Maine, New Hampshire, Vermont, and southeastern Québec, while the major continental glacier, the Laurentide Ice Cap (Figure 21), was growing in Canada. Eventually, the Laurentide sheet joined the local ice cap and continued flowing slowly to the Atlantic Ocean.

❖The Four New England Zones

The Taconian Zone, along the western border of New England, is a zone of Cambrian and Ordovician sedimentary rocks that have been folded and faulted in various ways. Some have been slightly to greatly metamorphosed. The degree of metamorphism increases from west to east, from shale to phyllite and slate, or from limestone to marble in Vermont; more intense metamorphism occurred in western Massachusetts and Connecticut. The deformation and metamorphism of these rocks resulted from the Taconic Orogeny, hence the zone's name. Figure 131 shows the areas of Grenville Proterozoic rocks at the eastern edge of the zone. They stand as upthrusted and exposed samples of the Proterozoic basement upon which the Taconian sediments accumulated.

The central, or Acadian, zone occupies most of New England. The Acadian Orogeny gave this zone most of its present features. It has moderately to highly metamorphosed rocks of Early Paleozoic age, including formations younger than those in the Taconian Zone. Here, there are also many intrusions of several ages, which the Taconian Zone does not have.

Embedded in the southern part of the Acadian Zone is the Triassic basin of the Connecticut River Valley (see Figures 119 and 131). It is a Triassic and Jurassic rift basin, with the major fault in the east and sandstones, shales, and volcanic rocks dipping toward it. The Connecticut Valley is obviously related to the Triassic basins to the south, formed in response to stretching of Earth's crust as parts of the supercontinent Pangaea separated.

The eastern zone, the Avalonian Zone, is the New England portion of the ancient subcontinent called Avalonia, which with associated ocean basins collided with the continent of Laurentia during the Acadian Orogeny (Figures 106 and 107). This eastern zone contains Cambrian to Carboniferous rocks, but its basic character differs greatly from those of the Taconian and Acadian zones. The characteristic rocks are Late Proterozoic and Cambrian sedimentary and volcanic rocks, with some metamorphics and some intrusives.

◆The Taconian Zone

The Taconian Zone in New England (Figure 131) extends from the Hudson River between Newburgh and Kingston, New York, up the Hudson River valley through the Champlain Valley, and includes western Massachusetts and Vermont. The Catskill Mountains, the Helderberg Escarpment, and the Adirondack Mountains border it on the west; the eastern edges of the Hudson Highlands, Taconic Mountains, and Green Mountains form the eastern border. The portion of the zone south of the Albany, New York, region has been described as part of the Stable Interior's Central Lowland, but it will also be discussed here because it is transitional between the Stable Interior and the Appalachian Province. The rocks on the west side of the valley are mostly Ordovician shale characteristic of the Stable Interior; most of those on the east side have become slate, characteristic of the Appalachian Province.

The Taconian Zone rocks formed from sediments and volcanic material in the Late Proterozoic, Cambrian, and Ordovician ocean basin that bordered the ancient continent of

Laurentia. The Taconic Orogeny's folding, faulting, and metamorphism happened when the western island chain collided with Laurentia. Thrust faulting was widespread. At one latitude in the Taconic Mountain area, 12 individual slices of Earth's crust separated by thrust faults have been identified. Some faults are truly immense. For instance, the Champlain Thrust Fault extends from southern Québec to the vicinity of Albany, New York. Along the fault, the upper block of Earth's crust moved an estimated 9 to 60 miles (15–100 km).

Along the eastern margin of the Taconian Zone, folding and faulting raised Proterozoic rocks that underlie the Cambrian ones enough so that erosion has exposed them. Proterozoic rocks form the major masses of the Hudson Highlands, the Berkshire Hills, and the southern Green Mountains (up to about halfway between the latitudes of Rutland and Middlebury). They consist of gneisses, schists, and marble from the Proterozoic Grenvillian Orogeny, metamorphosed about 1 billion years ago, and they come from part of the Laurentia continent's pre-Taconic Orogeny margin (Figure 131). The Grenville rocks have overridden the Paleozoic rocks to the west, and Paleozoics to the east have piled up against them. So the Proterozoic rocks are now surrounded by intensely folded, faulted, and metamorphosed early Paleozoic ones (Plate 24). The northern Green Mountains, along the border with the Acadian Zone, have Cambrian and Ordovician phyllite, quartzite, and an unusual form of metamorphosed limestone, plus greenstone.

In the Champlain lowlands, the Taconic Orogeny folded the formations of Cambrian and Ordovician limestone, dolomite, sandstone, and shale-argillite only slightly. However, the pressure broke the rocks by numerous normal faults, with the lowered sides on the east, and some reverse faults. The result is a complex patchwork with sections of formations instead of continuous bands. In western Vermont, erosion has exposed a "crumpled shale zone" associated with a major thrust fault along the eastern shore and islands of Lake Champlain.

In Vermont and Massachusetts west of the Green Mountains and the Berkshire Hills, thrust faults dominate the rock structure, forming many slices of tightly folded Cambrian and Ordovician metasedimentary rocks. The folds and faults trend more or less north and south, indicating a pressure from the east during the Taconic Orogeny. Across the Hudson River lowlands and into Vermont, Massachusetts, and eastern New York, the rocks grade from shale, sandstone, and limestone into phyllite, argillite, slate, schist, quartzite, and marble (Plates 26, 27). The rocks closer to the portion of the crust that carried the encroaching island chain were affected more than those farther away. In western Vermont

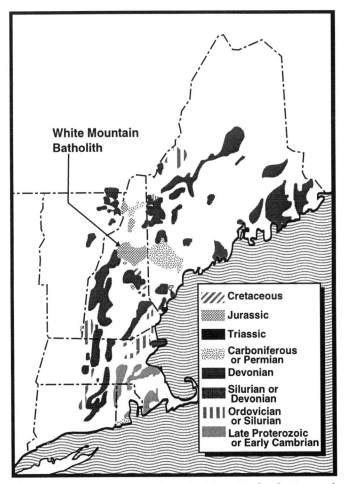

White Mountain Batholith

//////	Cretaceous
	Jurassic
■	Triassic
	Carboniferous or Permian
	Devonian
	Silurian or Devonian
|||||	Ordovician or Silurian
	Late Proterozoic or Early Cambrian

Figure 132. Igneous intrusions in New England. Most of them are Devonian, but intrusions of several other ages occur here and there.

and Massachusetts, metavolcanic rocks such as greenstone (Plate 44) also occur.

◆The Acadian Zone

Most of New England falls in the Acadian Zone (Figure 131), characterized by rocks that were formed from Late Proterozoic times through the Devonian Period and were folded, faulted, metamorphosed, and intruded during the Acadian Orogeny, when Avalonia collided with the growing continent of Laurentia. The rocks originated as volcanic lava and ash, with land and marine deposits from Laurentia, from Avalonia, and from the ocean basin between them.

The structure of the Acadian Zone, from west to east, is a complicated anticlinorium grading into an equally complicated synclinorium. Igneous intrusions of several ages punctured both of these major structures (Figure 132).

The earliest Acadian Zone rocks are the Late Proterozoic basement rocks, which consist mostly of gneiss, along with some metasediments such as quartzite (Plates 25, 26). They all originated as sedimentary and volcanic rocks. They are exposed in the core of the anticlinorium, in a series of eroded domes from Long Island Sound along the east edge of the Triassic Basin and extending north from just east of the Ver-

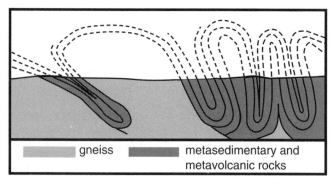

gneiss | metasedimentary and metavolcanic rocks

Figure 133. A simplified diagram of a typical Acadian Zone structure in west-central Massachusetts. Note how tight the vertical folds are.

mont–New Hampshire border to Berlin, New Hampshire.

Some of the domes in this chain of oldest Acadian rocks contain gneiss and amphibolite that started as Late Ordovician intrusive rocks during the Taconic Orogeny and were metamorphosed during the Acadian Orogeny. It is not unusual for intrusive rocks from one orogeny to be metamorphosed during a later orogeny.

The later Acadian Zone rocks are Early and Middle Paleozoic in age. They originated as sediments and volcanics during the Ordovician, Silurian, and Devonian Periods. Sedimentary rocks became phyllite, argillite, slate, schist, gneiss, marble, quartzite, and quartzite-conglomerate during the Acadian Orogeny. Volcanic rocks became gneiss, amphibolite, and greenstone. Here and there, some schists and gneisses were further modified by heat and pressure to become granulites. The Taconian and Acadian zones together contain a wealth of rock types.

The Acadian Zone shows much evidence of intense folding in the layered formations (Figures 133 and 134). Great thrust faults formed in the anticlinorium during the Acadian Orogeny, and blocks of Earth's crust were moved distances that must be measured in tens of miles. The pressure came from the east in Connecticut and Massachusetts and from the southeast in Vermont, New Hampshire, and Maine. Intrusions developed during the orogeny, with molten minerals cooling into igneous rocks from granite to gabbro.

● Southeastern New York

The Manhattan Prong, between the Hudson Highlands and the tip of Manhattan Island, has a series of rocks that were metamorphosed about 365 million years ago, which indicates they experienced the Acadian Orogeny and thus may be considered part of the Acadian Zone. The rock types include gneiss, quartzite, marble, and the Manhattan Schist that outcrops in Central Park. The quartzite, schist, and marble began their careers as Cambrian and Ordovician marine sediments. There is also the Peekskill Granite, which was intruded about the time of the Acadian Orogeny.

Some of the gneiss predates the Acadian Orogeny, and a body of gabbro and peridotite-type rocks in the Cortlandt–Croton Falls area was intruded during the Taconic Orogeny.

The New York City area occupies one of the most varied and therefore fascinating geological sites to be found among

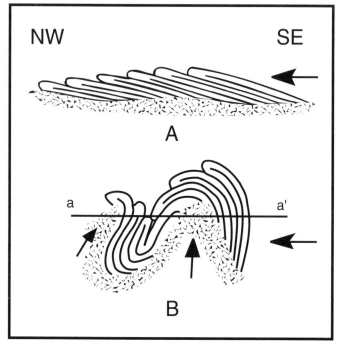

Figure 134. The development of a more complicated structure than that shown in Figure 133, with the anticlines uneroded. A shows tight overturned anticlines. B shows that with continuing pressure, part of the anticline series is forced downward into a syncline and deeper molten material is squeezed upward into neighboring anticlines. Imagine what a geologic map of the structure's surface would look like if the rocks were eroded to the plane of line a-a'.

North America's large cities. The Manhattan Prong belongs to the Acadian Zone. Across the river is the northern tip of the New York–Virginia Triassic Basin. And Brooklyn and Queens are underlain by Cretaceous clay, sand, and gravel of the Coastal Plain. The southeastern half (more or less) of Staten Island has the same Cretaceous material. The bedrock in the northwestern half of Staten Island (more or less), including the highest point, is a rare type called serpentinite that consists mostly of the light green, greasy-looking mineral serpentine. It is unusual in that it was part of the crust below the Ordovician ocean-bottom sediments that became limestone and shale. It was thrust upward in a fault block by the forces of the Taconic Orogeny.

Glacial till covers most of the bedrock in the area, for New York City and environs and all of Long Island are close to the southeastern extremity of the last Ice Age ice cap. Much of Long Island's topography is due to terminal moraines of the last great glacier.

● Vermont

Vermont bedrock east of the Green Mountains consists of metasediments and metavolcanics, mostly schist with some limestone-marble and some sandstone-quartzite, plus granite intrusions from the Acadian Orogeny. The schist in the eastern half of the state dates from the Silurian and Devonian periods; that to the west is older, having been formed during the Taconic Orogeny. In the northeastern part of the state, the rocks are less metamorphosed as you approach New Hampshire and Québec. The granite masses occur in the northeastern quarter of the state (Figure 132); the best known is the one that contains the Barre quarries.

● New Hampshire

The matrix of New Hampshire geology consists of Ordovician, Silurian, and Devonian metasediments and metavolcanics, which were thrown into overturned folds and faulted during the Acadian Orogeny. Phyllite, slate, schist, quartzite, metagraywacke, and marble make up the metasediments. The metavolcanics are amphibolite, greenstone, and a light-colored fine-grained gneiss. In the northern part of the state, the degree of metamorphism lessens, and shale, arkose, conglomerate, tuff, and rhyolite appear in the area bordering Maine, northeastern Vermont, and Québec. A line of gneiss domes that represents the core of the Acadian anticlinorium stretches inside the western border of the state.

Embedded in the New Hampshire geologic matrix are a large number of batholiths and stocks (Figure 132). They consist of granite, gabbro, and several varieties in between, intruded during the Ordovician, Devonian, Carboniferous, Jurassic, and Cretaceous periods.

The southern part of the White Mountain Range is a large batholithic complex. (Mt. Washington and the other peaks of the Presidential Range are north of the batholith, made up largely of Devonian schist.) The batholith consists primarily of varieties of granite (Plate 31), with some syenite; the rocks cooled in Middle to Late Jurassic times.

When the White Mountain Batholith's intrusive rock was molten, it seems to have engulfed volcanic rocks that are its extrusive equivalent in three places in the townships of Conway and Ossipee. The volcanic layers appear to have collapsed into emptying chambers as molten minerals moved to the surface. The volcanic eruptions lowered the liquid levels in the chambers and added to the weight of the rock above. Eventually, the roofs collapsed, and the volcanic rocks subsided into the emptying chambers. Trachyte (the lava equivalent of syenite), rhyolite, tuff, and breccia (Plate 43) form the volcanic rocks; they can be seen now on Moat Mountain and Mt. Kearsarge in the Conway area, and at West Ossipee. All this intrusive and volcanic activity presumably happened as part of the supercontinent Pangaea's breakup, which started in the Triassic and Jurassic periods and resulted also in the Triassic Basins.

● Maine

The geologic pattern in Maine resembles that in New Hampshire. A vast assemblage of Ordovician, Silurian, and Devonian sedimentary and volcanic rocks from Laurentia, Avalonia, and the ancient ocean was metamorphosed to varying degrees and thrown into overturned folds with northeast-southwest axes. Northeast-trending, high-angle faults broke the rock series during the Acadian Orogeny, and igneous intrusions formed before and during the orogeny (Plate 31). Later, in the Permian Period, strike-slip faults further disrupted the formations, presumably as part of the Alleghenian Orogeny. More intrusions occurred during and after the Alleghenian Orogeny. Some normal faulting occurred in the Triassic and Jurassic periods, as Europe and Africa split away from what was to become North America.

Some nonintrusive rocks in the western and central parts of the state are only slightly metamorphosed, including limestone, shale, sandstone, arkose, conglomerate, tuff, and rhyolite. Throughout the state, the metamorphics include phyllite, schist (Plate 24), quartzite, marble, amphibolite, greenstone, and a light colored, fine-grained gneiss that was originally volcanic.

The igneous intrusions (Figure 132) come in several varieties, such as granite, granodiorite, syenite, and gabbro. A few Ordovician and Early Silurian intrusions in the western part of the state presumably formed during the Taconic Orogeny, as indicated by their ages. After the orogeny, Late Silurian intrusions occurred. The majority of Maine's stocks, batholiths, and sills date from the Devonian Period, as molten minerals were forced or migrated upward while Avalonia was colliding with Laurentia during the Acadian Orogeny. Mississippian ones may also have been part of the Acadian Orogeny. Pennsylvanian and Permian intrusions in the southern tip of the state demonstrate stresses in Earth's crust when Gondwana docked with Laurentia in the Alleghenian Orogeny to form Pangaea. During the subsequent breakup of Pangaea, a small Triassic batholith at Ogunquit and west of Ogunquit, and even smaller bodies of Cretaceous age at Cape Neddick and near North Berwick cooled and solidified.

The list of rock types along the Maine coast outlines much of Paleozoic history. It suggests ancient oceans, islands, and subcontinents; it suggests volcanoes and fissure flows on land and in the sea; and it suggests crustal crumpling and mountain making during two orogenies. The list includes Proterozoic to Silurian sandstone and Carboniferous granite southwest of Casco Bay (Plate 26); phyllite, schist, marble, and gneiss (of uncertain age, from Proterozoic to Ordovician) in the Casco Bay area; schist, quartzite, marble, greenstone, and amphibolite (Cambrian and Ordovician) in the Penobscot Bay area; basalt, tuff, volcanic breccias, shale, argillite, and conglomerate (Silurian and Devonian); plus Devonian granite, diorite, and gabbro "down east" of there. (Some Avalonian Zone rocks are mixed with the Acadian coastal rocks; see the description of the Maine Avalonian Zone below.)

● Massachusetts and Connecticut

Except for the presence of the Connecticut Valley Triassic Basin, the portions of the Acadian Zone in Massachusetts and Connecticut generally contain the same folded, faulted, metamorphosed, and intruded Proterozoic and Paleozoic rocks that occur to the north in New Hampshire and Maine. East of the Triassic Basin lie the Acadian Zone anticlinorium and synclinorium that extend from the Connecticut coast up through New Hampshire and Maine. The intrusive rocks formed during the Acadian Orogeny vary from granite to gabbro. Most of the other rocks are metamorphic.

The oldest rock of the Acadian Zone occurs in the row of gneiss domes at the edge of the anticlinorium just east of the Triassic Basin. The domes are between the New Haven–Middletown line and the Connecticut River in Connecticut and roughly in the region bordered by Springfield, Northfield, Athol, and Monson in Massachusetts (Plate 25). There, Proterozoic gneiss metamorphosed from rhyolite, and associated quartzite formed from arkose. The gneiss and quartzite represent lava flows and sediments of the continents, islands, and ocean that existed when Laurentia, Gondwana, and Baltica separated before the Cambrian Period began.

The other nonintrusive rocks started as Ordovician, Silurian, and Devonian volcanic and marine sedimentary rocks, most of them later metamorphosed to varying degrees. Basalt became amphibolite, greenstone, and greenschist. Rhyolite became fine-grained, light-colored gneiss (Plate 44). Sedimentary rocks turned into phyllite, schist, gneiss, quartzite, marble, and granulite. As elsewhere in the Appalachian Province, the layered formations were severely deformed during the Acadian Orogeny. Those folded earlier in the Taconic Orogeny were refolded by pressure from the east; later ones were folded for the first time. Folds overturned, and some broke as thrust faults formed, moving immense blocks of rock many miles.

Intrusions that had occurred during the Taconic Orogeny were also metamorphosed. Taconian granite turned into gneiss and gabbro became amphibolite during the Acadian collision. The gneiss and amphibolite now outcrop in the zone of gneiss domes.

◆The Avalonian Zone

The Avalonian Zone of the Northeastern Appalachians, according to a recent theory, was originally part of the volcanic subcontinent called Avalonia, in the Late Proterozoic to Early Cambrian ocean (Figure 106, Plate 29). Its rocks originated as land and ocean sediments and as volcanic ash and lava. Avalonia rode along as the crustal plate it was on slowly collided with Laurentia in the Acadian Orogeny, part of the process that formed the supercontinent Pangaea.

Much later, when Pangaea broke up, part of what had been Avalonia was left behind. The New England remnants of the land and ocean now lie in southeastern Maine, eastern Massachusetts, Rhode Island, and contiguous Connecticut (Figure 131).

● Massachusetts and Rhode Island

The oldest Avalonian rocks in southern New England occur in northern and western suburbs of Boston. They belong to a formation that was originally sandstone and shale, believed to be 700–800 million years old, originally deposited toward the end of the Proterozoic Eon. Orogenic forces metamorphosed them to quartzite and greenschist. Some volcanic rocks that outcrop nearby may belong to the same rock series. These originally volcanic and sedimentary formations may be evidence for the rifting of the original Proterozoic supercontinent so long ago. The rifting promotes volcanic activity, and sand and silt accumulate in the basins formed by rifting.

Granite intrusions and volcanic rocks (rhyolite, tuff, volcanic breccia), dated in the vicinity of 600 million years ago, show a somewhat later period of crustal unrest. Following that igneous activity, sedimentary and volcanic rocks formed at the end of the Proterozoic Eon and the beginning of the Cambrian Period on and near the Avalonian subcontinent, which was at that time way out in the ocean, east of what is now New England. The sediments are now sandstone, argillite, conglomerate, limestone, and a controversial rock type that appears to be a tillite (Plate 43). Some geologists believe it is in fact a tillite, but others theorize that it had nothing to do with glaciers but originated when pebbles and cobbles somehow slid into ocean bottom mud. If it is tillite, it may be contemporaneous with the supposed Late Proterozoic tillite in Virginia's Blue Ridge. The Proterozoic and Early Cambrian members of this series occur in the Boston area (Plate 28); Early Cambrian ones also outcrop on islands in Narragansett Bay, Rhode Island (Figure 135).

This series of rocks contains a formation of Early Cambrian argillite that forms the bedrock in Braintree, Massachusetts, a southern suburb of Boston. Fossils found in the argillite are more closely related to European ones than they are to those of the same age in other parts of North America, which indicates that the animals did not live in the ocean

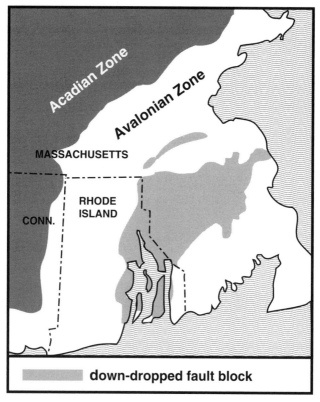

Figure 135. Down-dropped fault blocks of Pennsylvanian rocks in the Avalonian Zone of southeastern New England. The larger one is the Narragansett Basin; the smaller is the Norfolk Basin.

near Laurentia (since Laurentia became North America), but must have been nearer Avalonia. When Pangaea split in the Triassic and Jurassic Periods, some Avalonian fossils moved eastward to become part of Europe, while others remained in America. Plate 48 shows an example of such a fossil from Newfoundland.

An enigmatic sliver of volcanic rocks in the northeastern corner of Massachusetts may or may not be part of the Avalonian Zone. Shale interbedded with andesite and rhyolite lava flows and tuffs contains latest Silurian or early Devonian fossils. If these rocks do belong to Avalonia, they are atypical for Massachusetts Avalonian sequences in that they are younger than Cambrian.

The Narragansett and Norfolk basins (Figure 135) lie within the general Avalonian Zone boundary, but their rocks differ significantly from those of Avalonia. Just as the Connecticut Valley Triassic Basin resides as a foreign object in the Acadian Zone, so the Narragansett and Norfolk basins have rocks foreign to the Avalonian Zone. They contain Pennsylvanian-age sedimentary rocks, much younger than the typical Avalonian rocks. The sediments originated during the Alleghenian Orogeny, when Gondwana crushed the remnants of Avalonia. The folding and faulting of the crust during the collision caused highlands that weathered and eroded. The material that was deposited in the lowlands now forms Pennsylvanian conglomerate, sandstone, and shale, even some coal. After the Pennsylvanian Epoch ended, the basins were downfaulted into the older rocks that surround them today. Perhaps along with the Triassic Basins, they dropped down because of faults created as Pangaea split.

The dating of igneous intrusive rocks in Massachusetts and Rhode Island shows that there were, after the Cambrian Period, two main episodes of intrusion in the Avalonian Zone, which left mostly granite, with some diorite and gabbro. The earlier intrusions happened during the Ordovician Period, which at first blush seems to suggest the Taconic Orogeny. However, at that time, Avalonia was out in the ocean, and didn't collide with Laurentia. The second pulse of intrusions occurred in the Devonian Period, during the Acadian Orogeny, when Avalonia *did* collide with Laurentia.

A batholith of Permian granite along the southern coast of Rhode Island west of Narragansett Bay proves that there was igneous intrusion during the Alleghenian Orogeny. Although the most sensational and obvious effects of that orogeny are to be found in the Appalachian Province from Pennsylvania to Texas, it also had great effects in New England. Horizontal pressure threw the Narragansett Basin Pennsylvanian rocks into large-scale overturned folds, and gneisses formed in eastern Connecticut.

The Massachusetts Avalonian Zone has a famous remnant of the Ice Age in the form of Cape Cod. Erosion and ocean currents constructed the cape out of sand from the glacial till and outwash, furnishing many miles of beaches. Waves ate into the glacial material along the shore, and waves and currents moved sand away from the original shore to build the curved sandy peninsula of Cape Cod. Figure 165 illustrates the process. Some beaches to the north of Cape Cod show the glacial presence in a more subtle way. They are composed not of sand, but of boulders, cobbles, and pebbles deposited in a layer of till by the last glacier of the Ice Age as it melted. After the glacial ice disappeared, local conditions caused the Atlantic waves to wash away all the smaller grains along some stretches of coast to make sandy beaches next door to the rocky ones (Figure 108).

● Maine

There appear to be some bits of Avalonia in Maine. The islands in Penobscot Bay consist of two rock series. The older series contains Proterozoic schist, quartzite, marble, and amphibolite, whose age is about 600 million years or somewhat older. In age and character, the rocks seem to belong with the Late Proterozoic ones in the Boston area; they also closely resemble rocks in Newfoundland's Avalonian Zone. The younger series consists of slate, quartzite with conglomerate, and limestone and dolomite of early Paleozoic age. The exact age is uncertain, but they are definitely younger than Proterozoic and older than Devonian. These rocks may be extensions of the Cambrian ones in the Boston and Narragansett Bay areas.

The islands in Penobscot Bay are parts of a fault block of Avalonian rocks surrounded by different formations ranging in age from Cambrian to Devonian. The fault on the west is a major strike-slip one that can be traced across the Gulf of Maine to the vicinity of Newburyport, Massachusetts, where it separates related early Paleozoic rocks from those of the Acadian Zone. Thus, the Penobscot Bay block seems to be an extension of the Rhode Island–eastern Massachusetts block.

The rest of the Avalonian boundary in Maine can be extrapolated from Penobscot Bay to New Brunswick, but orogenic forces and erosion have obscured the details; the region shown in Figure 131 has for bedrock a complicated assortment of Avalonian and Acadian rocks. Those that may be as old as Proterozoic, plus Cambrian and Ordovician ones, belong to the Avalonian series, while the Silurian and Devonian rocks belong to the Acadian series.

The coast of Maine also has an interesting set of large and

small terminal moraines from the last Ice Age ice cap, formed during temporary stable periods as the glacier retreated. The moraines run along the coast more or less parallel to it, from Eastport southwest for over 60 miles (100 km), in a zone about 25 miles (40 km) wide. The large ones consist of sand, gravel, till, and silt in ridges as much as 60 feet (18 m) high, 300 feet (90 m) wide, and 10 miles (16 km) long. The largest small ones measure about 10 feet (3 m) high, 30 feet (9 m) wide, and half a mile (1 km) long and are made up of till. The various moraines occur as clusters of many curved ridges, evenly spaced and parallel to each other. The glacier built most of these moraine segments below the present sea level, for the weight of the ice had bowed the crust downward slightly. Since the end of the Ice Age, the crust has rebounded and lifted the moraines upward.

◆The Connecticut Valley Triassic Basin

The Connecticut River has carved its valley into relatively soft rocks of New England's Triassic Basin, which extends from just south of the Massachusetts-Vermont border southward to Long Island Sound (Figure 131). The basin's large areas of flat topography make it look strikingly different from the hills to the east and west. The basin's softer sedimentary rocks have eroded flat, leaving only a few long ridges of resistant volcanic rocks (Plate 20). The tougher metamorphics stand above the basin floor in hills on either side.

The basin contains Late Triassic and Early Jurassic sandstone, arkose, and shale interbedded with basalt lava flows and tuff. Normal faults occurred on the basin's east side as it subsided, and rivers carried sediments into the lowland that was being created. The sediments were left where the rivers flowed out of hills at the edges of the valley, as well as on flood plains and lake bottoms in the valley's interior. Since the east side of the basin was continually being lowered, the layers of sediment tilted down toward the east (Plate 30). Meanwhile, there were three pulses of volcanism at the end of the Triassic Period, resulting in three sets of surface flows. One of the pulses also intruded sills, and another intruded dikes. Dikes were also formed during a fourth smaller incident of igneous activity in Early Jurassic times.

As in the New York–Virginia Basin, the sedimentary rocks vary in color from black through several shades of gray, pur-

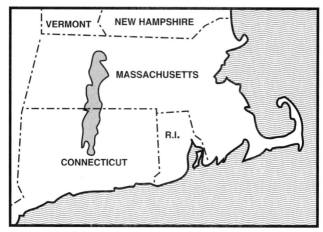

Figure 136. Glacial Lake Hitchcock, formed in the Connecticut River valley by a dam of outwash material south of Hartford as the last Ice Age glacier retreated.

ple, and red (Plate 42). Dinosaur footprints can be seen on some sandstone bedding planes. There are many footprint trails under cover at the Dinosaur State Park near Rocky Hill, Connecticut (Plate 30).

When the last Ice Age ice cap retreated northward, a dam of glacial till and outwash accumulated across the valley. Behind the dam, a meltwater lake was born, called Glacial Lake Hitchcock, in the portion of the valley that is now central Connecticut. As the ice melted back, the lake spread up the valley, from Rocky Hill, Connecticut, to Turners Falls, Massachusetts (Figure 136). The lake-bottom silt is visible here and there today, along with coarser deposits of deltas formed by rivers that flowed into the lake.

NEW ENGLAND HIGHWAY SECTIONS

New York City to Boston, Massachusetts: I-95 (Figure 137) *This section of I-95 passes through Acadian Zone metamorphic rocks, crosses the Triassic Basin, and ends in a variety of*

Avalonian Zone rocks. In the Avalonian Zone, it crosses the down-dropped Pennsylvanian fault blocks.

Early Paleozoic schist and gneiss form the bedrock from the Bronx into Connecticut. Another body of gneiss extends from about the state border eastward almost as far as the Norwalk exit. Then schist appears again. From there to the western outskirts of New Haven, the highway continues through schist with some gneiss and quartzite. All are typical rocks of the Acadian Zone, products of the Acadian Orogeny, when Avalonia collided with Laurentia.

New Haven is at the edge of the Connecticut Valley Triassic Basin's southern tip, and for a few miles, the bedrock is sandstone and shale. Then, beyond the greater New Haven area, the rock changes back to Acadian Zone gneiss, with some schist. The highway enters the Avalonian Zone as it crosses the Connecticut River and continues in that zone all the way to Boston. From the Connecticut River to Rhode Island, most of the rock is Late Proterozoic or Early Paleozoic gneiss, with smaller amounts of schist and quartzite. Ten miles (16 km) or so into Rhode Island, the bedrock changes to an area of granite and metavolcanic rocks, still Late Proterozoic or Early Paleozoic.

An even bigger change occurs in the vicinity of the I-295 exit as I-95 enters the Narragansett Basin, the large down-dropped fault block that is filled with folded Pennsylvanian sedimentary rocks. There, the sandstone, shale, and conglomerate, from sediments eroded off highlands of the Appalachian Orogeny, are a far cry from Proterozoic granite and gneiss in rock types and age. The Pennsylvanian rocks continue through Providence and into Massachusetts. Just beyond the Foxboro-Mansfield exit, the road crosses the northern boundary fault of the Pennsylvanian block and returns to "normal" Avalonian rock—late Proterozoic granite. The granite continues until I-95 joins Massachusetts Route 128, except for a brief encounter with an outlier of the Narragansett Basin just before Route 128—the narrow, northeast-trending, down-dropped Norfolk Basin, with Pennsylvanian conglomerate and sandstone (Figure 135). Then more granite in another fault block (the Boston Basin) along I-95/Route 128 northward, followed by Late Proterozoic rhyolite and conglomerate in the towns of Needham and Wellesley. The northern fault of that block crosses the highway close to the intersection with I-90, the Massachusetts Turnpike, which leads to Boston and its immediate suburbs. There, the bedrock is Late Proterozoic conglomerate, basalt, and argillite.

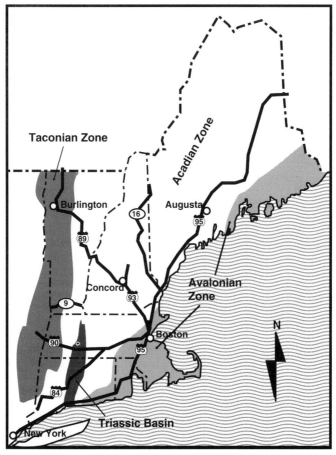

Figure 137. Geology and highways in the New England part of the Northern Appalachians.

Newburgh, New York to I-90 in Massachusetts: I-84 (Figure 137)

This route goes from the Taconian Zone into the Acadian Zone and traverses the local Triassic and Jurassic basin for many miles. The contrasts are evident as the highway passes from sedimentary rocks to metamorphics to sedimentary and volcanic rocks to metamorphics again.

From the Hudson River to beyond the Connecticut border, the bedrock along I-84 is Taconian Zone Cambrian and Ordovician shale, sandstone, and limestone or dolomite that become more highly metamorphosed the farther east you go, as a result of the Taconic Orogeny, turning into slate, schist, quartzite, and marble. The indistinct border between the Taconian and Acadian zones runs east of the state line and parallel to it. From the border to the Connecticut River Triassic Basin, the bedrock consists of greatly metamorphosed Cambrian and Ordovician sediments and volcanics, which have turned into schist and gneiss because of high heat and pressure during the Acadian Orogeny.

The highway descends into the Triassic Basin at the I-691 exit. From there through Hartford to Vernon, it crosses the flat valley of the basin, ornamented by ridges of basalt. Most of the Triassic and Jurassic bedrock is gray to red sandstone, arkose, and shale. Dinosaur State Park near Rocky Hill, off I-91 south of Hartford, is well worth a side trip (Plate 30).

About at the Route 30 Vernon exit, the route reenters the hills of the Acadian Zone, with bedrock mostly of Paleozoic gneiss and then schist. Gneisses flank a narrow strip of Silurian to Devonian schist-quartzite-marble formations in the Rockville-Tolland area. Beyond that, an extensive body of Ordovician schist continues into Massachusetts and all the way to I-90.

Hudson River, New York to Boston, Massachusetts: I-90 (Figure 137)

This part of I-90 passes through all four zones of New England geology, from the Taconian Zone in New York and western Massachusetts through the Acadian Zone and Triassic Basin to the Avalonian Zone in the Boston Basin. From west to east, the highway encounters sedimentary, then metamorphic, then sedimentary and volcanic, then mostly metamorphic and igneous rocks.

Near the Hudson River, the rocks are flat-lying Ordovician shales and sandstones of the Central Lowland. As you proceed east into the Taconian Zone, the layers become tilted and the rocks become metamorphosed. The shale turns into slate before the Massachusetts line. In Massachusetts, the westernmost rocks along I-90 consist of Cambrian and Ordovician phyllite, marble (Plate 26), metagraywacke, and schist. They were greatly folded and thrust-faulted into slices during the Taconic Orogeny. Approximately 2 miles (3 km) beyond the Stockbridge exit, the highway passes into one of the zones of Grenville rocks (Figure 131). The Proterozoic Grenville rocks underlie the younger Cambrian and Ordovician ones from here to the Adirondacks. In this zone, Grenville schist and gneiss have been thrust up and over the younger rocks to the west, and the younger rocks above them have been eroded away to expose them. After about 13 miles (21 km), the highway returns to the Paleozoic Era rocks at the western edge of the Acadian Zone. The rocks from there to the Triassic Basin are mostly Ordovician and Devonian gneiss, schist, greenstone, and amphibolite.

The western edge of the Connecticut Valley Triassic Basin shows as the abrupt end of a series of road cuts through gnarled and tortured rocks, with a long view over the flat lowland ahead. The schist ends at the boundary and is replaced with smoothly eroded Triassic and Jurassic formations of the basin. In the middle of the basin, resistant basalt forms a prominent ridge trending north and south (Plate 20). The basin contains basalt, tuff, arkose, shale, and sandstone. Rocks of the ridge and to the east are Jurassic; those to the west of the ridge are Triassic. The layers in the basin are tilted down to the east, so the older ones outcrop to the west. Between the I-91 interchange at the Connecticut River and the Chicopee (Route 33) exit, the highway goes across the flood plain of the Connecticut River and up a slope onto the delta of a river that flowed at the end of the Ice Age into the Glacial Lake Hitchcock (Figure 136). The advent of hilly country marks the basin's eastern side fault, about 9 miles (14.5 km) east of the Connecticut River.

The hills contain typical Acadian Zone rocks, and for miles the bedrocks consist of folded and faulted Ordovician, Silurian, and Devonian metasediments and metavolcanics such as gneiss, schist, phyllite, slate, marble, and amphibolite. There is also some Silurian granite west of Auburn; Massachusetts Route 56 (no exit) is near the west end of the outcrop area. The metamorphics were formed, folded, and faulted during the Acadian Orogeny.

The boundary fault between the Acadian Zone and the Avalonian Zone crosses I-90 on a line going north-northeast from the town of Grafton. The Avalonian Zone's Proterozoic bedrock on the way to the I-95 interchange includes a quartzite-schist-amphibolite formation, some gneiss, greenschist of volcanic origin, and a body of granite. Beyond the I-95 interchange, the turnpike enters the down-dropped fault block of the Boston Basin, with its Late Proterozoic argillite, conglomerate, and basalt.

Boston, Massachusetts to Burlington, Vermont: I-93, I-89 (Figure 137) *This route gives a more northern view of New England geology's three basic zones, starting in the Avalonian Zone and ending in the Taconian Zone. It covers a variety of igneous and metamorphic rocks.*

I-93 leaves downtown Boston on Late Proterozoic argillite from the ocean near ancient Avalonia. In the near suburbs the rock switches to Avalonian granite, then a formation of diorite and gabbro, also Late Proterozoic. The I-95 interchange lies about in the middle of the diorite-gabbro outcrop area.

I-93 crosses the Avalonian Zone boundary fault approximately 4 miles (6.5 km) north of the I-95 interchange and enters Silurian granite. Granites continue far into the Acadian Zone; after the Silurian variety comes a large body aged either Ordovician or Silurian. Just before the Merrimack River, another major fault passes under the highway; the granite ends and is replaced by a Silurian metasedimentary formation consisting of argillite, quartzite, and schist, which continues into New Hampshire.

The rock along I-93 between Salem and Derry, New Hampshire, consists mostly of Ordovician or Silurian phyllite, schist, and gneiss in the Acadian Zone's major synclinorium, which contains metamorphosed sediments from the ocean basin that separated Laurentia and Avalonia. Between the New Hampshire Route 28 exit and Manchester, the road crosses into an elongate northeast-trending Devonian granite batholith, whose margin is near the Hooksett exit. From Hooksett to the I-89 exit south of Concord, the bedrock is mostly Silurian or Devonian gneiss and a mixture of igneous and metamorphic textures called migmatite.

Concord lies in the middle of a small body of Devonian granite, and as the route turns northwestward on I-89, it traverses the southern end of the granite mass. Soon, metamorphics reappear—schist and quartzite of Devonian age, more remnants of the Paleozoic ocean. Just before the New Hampshire Route 109 exit, the highway goes onto a large mass of monzonite, which is an intrusive igneous rock related to granite. The monzonite is unusual in that in many places it is a porphyry containing extra-large grains (phenocrysts) of feldspar in a groundmass of smaller grains (Figure 6). It also contains inclusions of the schist into which it intruded during the Acadian Orogeny. The monzonite, which is Devonian, extends as far as the Grantham–New Hampshire Route 10 exit. It belongs to the largest single mass of igneous rock (some of it gneissic) in New Hampshire, which extends from the latitude of Keene all the way up to the White Mountains.

The region from Grantham to White River Junction is part of the strip of gneiss domes that extends from the Connecticut shore into Maine. The domes are along the core of the major anticlinorium in the Acadian Zone. The highway passes through a small body of Devonian gneiss, and then Ordovician and Silurian schist, marble, quartzite, and greenschist, on the way to White River Junction. Lebanon, on the way to White River Junction, marks the southern tip of a small Ordovician gneiss dome. The rock at the I-91 interchange is greenschist, originally Ordovician basaltic volcanic rocks.

The Connecticut River flows through a synclinorium; just beyond it, the bedrock switches to a new set of Silurian and Early Devonian schists and related rocks in a north–south-trending anticlinorium that continues to Montpelier. The schists are in some places gneissic, and in other places they grade into phyllites, according to the degree of metamorphism to which they were subjected.

West of Montpelier, in the vicinity of Bolton, I-89 enters the Taconian Zone, with no rocks younger than Ordovician. The metamorphism to the east is Acadian and that to the west is Taconian. From Montpelier to Burlington is a series of six major thrust-fault slices similar to those in western Massachusetts and eastern Tennessee. Presumably, most of them originated in the Taconic Orogeny, but some may have been reactivated during the Acadian Orogeny. The rocks in the slices were folded by the same horizontal pressure that faulted them. Schists and phyllites west of Montpelier are Proterozoic to Ordovician in age. Cambrian schist extends from Waterbury to the Vermont Route 2A exit.

Burlington is situated on a synclinal structure in Cambrian and Ordovician quartzite and marble that lies to the west of Route 2A. The metamorphism decreases westward, and the bedrock along the Lake Champlain shore and over to New York grades to Ordovician limestone, shale, and argillite, with Cambrian sandstone.

Brattleboro to Bennington, Vermont: Route 9 (Figure 137) *This route traverses rocks and structures similar to those of the Vermont section of I-89, but one part of it cuts through the southern Green Mountains, where the rocks differ from those of the range's northern portion. Here, the core of the* *mountains consists of Proterozoic rocks metamorphosed during the Grenvillian Orogeny approximately a billion years ago. These rocks are related to those of the Canadian Shield's Grenville Province, which includes the Adirondacks.*

From Brattleboro to Wilmington, the highway traverses tight folds and many faults in the western anticlinorium of the Acadian Zone. The rocks are schist, phyllite, greenschist, and amphibolite, Devonian in age from Brattleboro to Marlboro and Cambrian and Ordovician from Marlboro to Wilmington.

The highway passes through Wilmington on the east flank of the Green Mountains and Bennington on the west flank. In between, the mountains consist of a large body of Late Proterozoic gneiss, with a band of Lower Cambrian schist and quartzite on either side. Bennington is at the west edge of the western Lower Cambrian band. To the west, as at Burlington, Taconian Zone Cambrian and Ordovician rocks grade from metamorphic in Vermont to sedimentary in New York.

Portsmouth to Berlin, New Hampshire: Spaulding Turnpike/New Hampshire 16 (Figure 137) *Along New Hampshire Route 16 are typical New England Acadian Zone rocks, but the route is different from other routes in that it crosses the White Mountain Batholith, which was formed as Pangaea broke up, and thus contains some of the youngest bedrock in New England.*

In the Portsmouth area, bedrock consists of Paleozoic slate, phyllite, quartzite, and schist, with some volcanic material (I-95 south of Portsmouth cuts through quartzite, and on Route 16, basalt occurs near the first Dover exit). The rocks are possibly Ordovician or Silurian in age, derived from the early Paleozoic ocean basin sediments that accumulated between Laurentia and Avalonia. About halfway from Dover to Rochester the highway enters the outcrop area of an elongate gneiss-granite body; Rochester is just beyond the other edge, back in metamorphosed sedimentary rocks.

From the Rochester area to the Union–Milton Mills area, the bedrock is Silurian or Devonian schist, which in some places is modified into a migmatite. About 2 miles (3 km) north of the Union–Milton Mills exit (the end of the Spaulding Turnpike) there is an outcrop that is just the tip of a Cretaceous granite intrusion, one of a scattered row of Cretaceous intrusions from the breakup of Pangaea that extends up to Montréal. Acadian Zone granite and schist continue beyond the New Hampshire Route 109 intersection. At West Ossipee and the junction with New Hampshire Route 25 west, the hills to the west of the highway are part of a complicated mixture of rock that is round in outline and about 9 miles (14.5 km) in diameter. It contains granite, with volcanic rocks that apparently collapsed into an emptied molten mineral chamber in the Jurassic Period. It is one of the cases described above in the section on the Acadian Zone in New Hampshire.

From West Ossipee to about halfway between Chocorua and Conway the bedrock belongs to the Silurian or Devonian schist complex. Then the highway enters the White Mountain Batholith, with its suite of various Jurassic granites. Mt. Chocorua, rising to the west of the highway, is part of the batholith. From before Conway to beyond Glen, the mountains on either side of Route 16 eroded from the batholith and its associated volcanic rocks. The rocks in the White Mountain Batholith and the structure at West Ossipee resulted from the breaking of the crust as Pangaea divided into smaller continents.

Most of the way from Glen to Gorham is over Silurian and Devonian schist and quartzite, in a zone that includes Mt. Washington. The view of that mountain from Pinkham Notch reveals three large concavities on its eastern face; they are glacial cirques, carved by valley glaciers that developed before the last ice cap of the Ice Age covered the area.

About 3 miles (5 km) north of the Mt. Washington Auto Road, the highway passes the edge of an Ordovician granite intrusion. Gorham lies beyond its other edge. The highway heads west out of Gorham; just after it turns north again it crosses another batholithic complex of Ordovician granite. Berlin is at the northeast end of the granite mass.

I-90, Weston, Massachusetts to Houlton, Maine: I-95 (Figure 137)

This section of I-95 starts with a small sample of Avalonian Zone igneous and metamorphic rocks and then follows the trend of folded, faulted, and intruded rocks in the Acadian Zone. In Maine, rocks of the Acadian Zone's major synclinorium vary from schist to much less metamorphosed sedimentary rocks in great folds.

The bedrock at the I-90/I-95 interchange in Weston is Avalonian Zone Proterozoic granite. From there along I-95 into southern New Hampshire is an almost bewildering patchwork of rocks, due to the folding, faulting, metamorphism, and intrusion of the Avalonian formations, first during the Ordovician Period out in the ocean, later during the Acadian Orogeny, and finally during the Alleghenian Orogeny.

Just beyond the end of the granite outcrop area, a major thrust fault crosses I-95. The fault is marked by a tall outcrop of mylonite on the southbound side of the highway, a bit less than a mile north of the U.S. 20 overpass. In between the granite and the mylonite, part of a small zone of gabbro/diorite outcrops. Beyond the fault stretches Proterozoic diorite and gabbro, along with metamorphosed volcanic rocks, to just beyond the U.S. 1 exit in Danvers. Within those Proterozoic rocks, however, are two areas of Devonian granite, one in Reading and the other centered where Massachusetts Route 128 separates from I-95.

After the U.S. 1 exit in Danvers, the next major outcrop is Silurian diorite. It starts about 2 miles (3 km) north of the exit, where you cross the boundry fault between the Avalonian Zone and the Acadian Zone. The Acadian Zone Silurian diorite body continues to about 2 miles south of the Massachusetts Route 113 exit, but in the middle of this intrusive mass is an area of Proterozoic or Ordovician gneiss and amphibolite. Beyond the Silurian diorite, a separate body of Ordovician or Silurian diorite extends to a point between where I-495 joins I-95 and the New Hampshire border, about 1 mile south of the border. From there, granite bedrock continues for about a mile and a half (2.4 km) into New Hampshire.

At this point the geology gets simpler. The highway enters the outcrop area of a possibly Ordovician formation of metasediments—here, quartzite and phyllite, that originated as ocean bottom sand and silt between Laurentia and Avalonia. Those rocks continue into Maine (Plate 26). About 2 miles (3 km) north of the York exit, the highway goes onto Triassic granite, one of the several intrusions that occurred in New England during the breakup of Pangaea. After about 5 miles (8 km) of granite, the road is back on argillite and phyllite of the formation that outcrops before the Triassic granite. Outcrops along the seashore near here show that the rock layers were thrown into tight, overturned folds. Beyond the Triassic granite, the highway goes right along the edge of a large Devonian or Mississippian granite intrusion, so granite occurs to the west and argillite and phyllite to the east up to about 5 miles (8 km) north of the Wells exit, where the granite disappears. Carboniferous granite intrudes between the Kennebunk and Saco rivers.

In the Portland area, the phyllite grades into schist, and

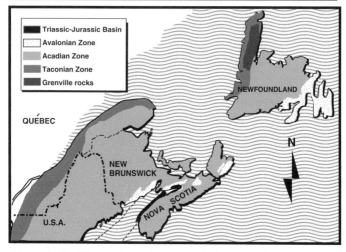

Figure 138. The geological zones of the Canadian Northern Appalachians.

just north of Portland, a formation possibly of Silurian or Devonian age contains more schist. Schist of an age between Proterozoic and Ordovician continues into the Gardiner area. From Gardiner through Bangor and Orono, the rocks are a scrambled mixture of Silurian with some possibly Ordovician phyllite, slate, shale, sandstone, conglomerate, and limestone. Rocks of this nature and age continue on to the Canadian border at Houlton. These formations originated in the ocean between Laurentia and Avalonia, after the Taconic Orogeny. They occur in broad, northeast-trending anticlines and synclines within the overall synclinorium of the Acadian Zone, which the forces of the Acadian Orogeny created.

■ NORTHEAST APPALACHIANS: ATLANTIC PROVINCES, SOUTHERN QUÉBEC

The major geologic zones of New England continue up into Canada, showing similar rocks and structures in their northern extensions (Figure 138, Plates 20,21). The northwestern Taconian Zone, a central Acadian Zone, and a more or less eastern Avalonian Zone are readily distinguishable.

The oldest rocks in the Taconian Zone are the Grenville-age Proterozoic ones on Newfoundland's Great Northern Peninsula, which resemble in age and composition the Grenville rocks of Vermont's Green Mountains, Massachusetts's Berkshires, and New York's Hudson Highlands and Adirondack Mountains (Figure 131). All of these Grenville rocks occur at the eastern edge of the zone and were metamorphosed approximately 1 billion years ago. The Northern Peninsula rocks, like those of the Adirondacks and other areas, are extensions of the Canadian Shield.

The other Taconian rocks are Proterozoic, Cambrian, and Ordovician sedimentaries and volcanics that were deformed and in some cases metamorphosed during the Taconic Orogeny, when the western island chain docked against the old continent of Laurentia (Figure 106). To a great extent, the rocks formed originally as sediments, lava flows, and volcanic ash of Laurentia and the adjacent ocean.

The Acadian Zone contains the region for which the orogeny was named. Acadia comprises former French colonies of present-day Nova Scotia, Cape Breton Island, Prince Edward Island, and coastal areas of New Brunswick. This zone has Cambrian and Ordovician rocks as in the Taconian Zone plus Silurian through Pennsylvanian ones. The Cambrian through Devonian rocks were deformed and slightly to strongly metamorphosed during the Acadian Orogeny, when the Avalonia subcontinent collided with the growing Laurentia. The Silurian and Devonian rocks were originally formed from sediments of the ocean between Laurentia and Avalonia. Large areas of these Acadian rocks are now covered by Carboniferous rocks deposited after the Acadian Orogeny. Further modifications occurred during the Alleghenian Orogeny.

The Avalonian Zone is represented by remnants of Avalonian rocks among the later ones that were affected by the Acadian Orogeny. Faulting and erosion have left rocks that resemble the Avalonian rocks of New England: Late Proterozoic originally marine and terrestrial sediments, lava flows, and ash that were covered by Cambrian and Ordovician sediments. The resulting rocks were folded, faulted, and metamorphosed before the Cambrian Period, during a minor orogeny out in the ocean, far from Laurentia. The Avalon Peninsula of Newfoundland, for which the zone is named, along with the neighboring part of the island, contains the largest area of these rocks in Canada.

Mostly under water in the Bay of Fundy are the remains of a large Triassic Basin that extends along the ocean floor beyond the mouth of the bay. Like the other Triassic Basins mentioned in this book, it formed when the crust stretched and broke as Pangaea split up, starting near the end of the Triassic Period.

The Appalachian Province in Canada was greatly affected by the glaciers of the Ice Age. The last glacial stage began with the development of three major local ice caps in the northern Appalachian region. One covered the highlands of Maine, New Hampshire, Vermont, and adjoining Québec. A second ice cap formed in the highlands of New Brunswick and spread out to cover Nova Scotia, and a third covered almost all of Newfoundland. There was a lesser ice cap on the Gaspé Peninsula and possibly a small one on Cape Breton Island. As the Laurentide ice sheet advanced from its Labrador center (Figure 21), it seems to have flowed over the mountains of Québec and New England and fused with the local ice in the Maritime Provinces and Newfoundland to completely cover the area. The glaciers scoured valleys, carved cirques in the mountains, and polished and striated bedrock surfaces. When they finally retreated, they left drumlins in Newfoundland and Nova Scotia, eskers on Prince Edward Island, boulders and pebbles from the Canadian Shield in New Brunswick and Prince Edward Island, and moraines in many places, including offshore Nova Scotia (when sea level was lower). And, of course, the glaciers left a layer of till to be found now under the topsoil.

◆The Taconian Zone

The Taconian Zone (Figure 138) covers southern Québec and western Newfoundland. In Québec, Taconian formations occupy the central portion of the Eastern Townships and the Gaspé Peninsula's northern third. Cambrian and Ordovician rocks grade from northwest to southeast from unmetamorphosed varieties connected with those in the Central Lowland's Northeast Corridor to highly metamorphosed types closer to the Acadian Zone. This series of sedimentary and volcanic rocks was folded and thrust over the Ordovician rocks of the Central Lowland during the Taconic Orogeny. The Taconian marine rocks consist of shale, sandstone, graywacke, conglomerate, limestone, slate, phyllite, argillite, schist, greenstone, and amphibolite.

The great thrust fault at the northwestern edge of the Taconian Zone extends from Lake Champlain to Québec City and on down the river to the Gulf of St. Lawrence. As in Vermont and Massachusetts, there are other thrust faults in the zone, and massive blocks of Earth's crust moved significant distances during the Taconic Orogeny. In the process, the rock layers were strongly folded, overturned toward the northwest, and, in many areas, metamorphosed. The general structure southeast of the boundary fault is that of three anticlinoria in line. They have weathered and eroded into the Sutton, Notre Dame, and Schickschock Mountains, and they form a continuation of the Green Mountain Anticlinorium in Vermont.

In western Newfoundland, Taconian rocks consist of Cambrian and Ordovician shale, sandstone, limestone, limestone conglomerate, graywacke, slate, argillite, quartzite, schist, marble, andesite, basalt, and greenschist. They were folded and thrust faulted, with large chunks moved westward as much as 25 to 60 miles (40 to 97 km) during the Taconic Orogeny. Some of the coastal rocks grade into the undistorted and unmetamorphosed ones of the Central Lowland in the Anticosti Basin. Other Cambrian and Ordovician rocks were thrust on top of them. Schist, gneiss, quartzite, and anorthosite form the Grenvillian rocks of the Great Northern Peninsula, which were intruded by granites between 960 million and 830 million years ago; they are all much older than the Proterozoic rocks in the Avalonian Zone.

Faulting during the Taconic upheaval resulted in a series of bodies containing basalt, gabbro, and peridotite being thrown up from below sedimentary rocks of the Ordovician ocean bottom. They are rare samples of what Earth's crust was like below the sediments in the latest Cambrian and/or early Ordovician open ocean. There is a string of them in Québec's Eastern Townships from Mansonville to between Thetford-Mines and Ste-Marie. The asbestos mines at Asbestos and Thetford-Mines are in metamorphosed portions of the deep crustal rocks, and there are associated copper and iron deposits. A single mass of these rocks occurs on the Gaspé Peninsula in the Shickshock Mountains east-southeast of Ste.-Anne-des-Monts. Four more occur in western Newfoundland, on either side of the Bay of Islands, where the Table Mountain, North Arm Mountain, Blow-Me-Down Mountain, and Lewis Hill vicinities consist largely of gabbro and peridotite.

The Taconic Orogeny included intrusions. In northern Newfoundland, on the peninsula between White Bay and Notre Dame Bay, a presumed Ordovician granodiorite batholithic complex mixes in with other rocks. In the Madeleine River area of the Gaspé Peninsula, a Devonian intrusion of granite and diorite into Ordovician rocks promoted the formation of copper ores.

◆The Acadian Zone

As in New England, the rocks that define the Acadian Zone (Figure 138) are those of Silurian and Devonian age that were folded, faulted, metamorphosed, and intruded (along with older ones) during the Acadian Orogeny. Ordovician and possibly Cambrian rocks occur in western New Brunswick and in Nova Scotia southeast of a line roughly from Windsor to Chedabucto Bay. In central Nova Scotia, the beds mix with Devonian intrusives. Rocks include argillite, quartzite, schist, gneiss, and volcanics in New Brunswick, sandstone, quartzite, shale, and argillite in Nova Scotia. Two northeast-southwest strips in central Newfoundland have Ordovician shale, slate, graywacke, limestone, and conglomerate.

● Silurian and Devonian Rocks

The Acadian Zone's Silurian to Middle Devonian sediments and metasediments are almost all marine (Plates 46, 48). They make up a series with shale, sandstone, conglomerate, graywacke, and limestone, as well as schist, gneiss, phyllite, slate, quartzite, and marble. Rhyolite, andesite, and basalt were metamorphosed to gneiss and amphibolite. All the Acadian rocks so far mentioned were deformed during the Acadian Orogeny, with fold and fault axes trending more or less northeast; the faults include thrust, normal, and strike-slip varieties. Folded Ordovician rocks from Bathurst, New Brunswick, southwestward contain well-known mining areas, with ores of iron, zinc, lead, and copper.

Acadian Orogeny intrusions resulted in large batholiths of granite, gabbro, and other rock varieties in southwestern Nova Scotia, western and southwestern New Brunswick (Figure 139), and throughout the Acadian Zone of Newfoundland (Figure 140). Small batholiths and stocks formed

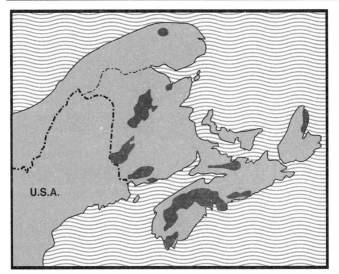

Figure 139. Devonian igneous intrusions in New Brunswick, Nova Scotia, and the Gaspé Peninsula.

in southern Québec and in southeastern Newfoundland.

In the later Devonian, during and after the Acadian Orogeny, terrestrial sediments accumulated on the newly uplifted highlands. The most famous are those now found along the north shore of Chaleur Bay on the Gaspé Peninsula (Plate 21). Among the shale, sandstone, and conglomerate layers, some beds contain fossils of freshwater fish and impressions of leaves from fernlike plants. In southern New Brunswick and near Moncton there are terrestrial redbeds of sandstone, shale, and conglomerate. A couple of small areas of Devonian redbed sandstone, shale, and conglomerate occur on the northwest coast of Fortune Bay, in southern Newfoundland.

● Carboniferous and Permian Rocks

The typical Acadian rocks of New Brunswick and Nova Scotia disappear under remnants of a Carboniferous and Permian cover. Mississippian rocks show up around the large Pennsylvanian area in New Brunswick (Figure 141) and in the northern half of Nova Scotia. Western Newfoundland

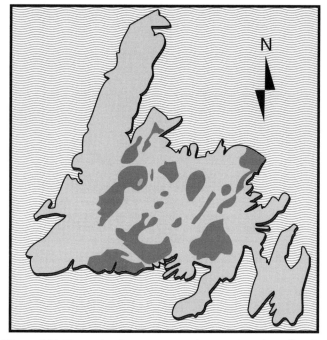

Figure 140. Devonian igneous intrusions in Newfoundland.

also has two Mississippian regions, one south of White Bay and the other along the southeast coast of St. George's Bay and inland. All the Mississippian rocks indicate a return to shallow seas after the Acadian Orogeny, with sandstone, limestone, and red shale, plus gypsum and anhydrite beds.

The Alleghenian Orogeny did not strongly deform the Atlantic Provinces as it did the regions to the south, but it left broad open folds with reverse and strike-slip faults and uplifted areas (Plate 21). The result was a combination of uplands and lowlands. The lowlands received terrestrial sediments like those of the Pennsylvanian Epoch in the United States—conglomerate, sandstone, shale, and coal (Plates 21, 48). The rocks show evidence of river channels, flood plains, and lowland coal swamps. A geologic map of the Maritime Provinces (Figure 141), shows a vast shallow syncline in New Brunswick, the adjoining portion of Nova Scotia, and

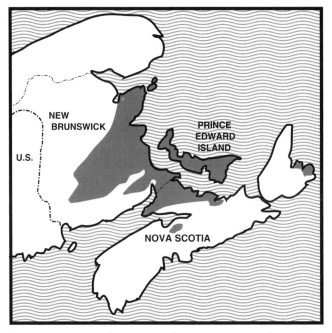

Figure 141. Pennsylvanian- and Permian-age rocks in New Brunswick, Nova Scotia, and Prince Edward Island.

Prince Edward Island. The syncline plunges toward the northeast, and its Pennsylvanian rocks are flanked as on two sides of a triangle by older rocks. Since the syncline plunges northeastward, it is logical to assume that the youngest rocks in it would be on Prince Edward Island, and there the youngest rocks are, in fact, Permian in age. The synclinal structure does not extend to Newfoundland.

The Pennsylvanian deposits have supported major mining operations at coal fields near Sydney, Inverness, Pictou, and Springhill in Nova Scotia, and Minto in New Brunswick.

◆The Avalonian Zone

The vicissitudes of geologic history have left the Canadian Avalonian Zone distributed differently from the continuous fringe it forms in New England and the central Appalachians. It is scattered from coastal New Brunswick through north-central Nova Scotia to eastern Newfoundland (Figure 138). Its rocks, like the Avalonian ones of New England, are basically sedimentary and volcanic types that were originally land and ocean sediments, lava flows, and volcanic ash falls. They indicate land masses with volcanoes (Avalonia) situated out in the ocean (Figure 106). During the Acadian Orogeny, the Avalonia region collided with Laurentia.

● Newfoundland

Eastern Newfoundland contains the largest piece of Avalonia in Canada, on the Avalon Peninsula and the contiguous "mainland" up to a line running approximately from Wellington on Hare Bay to Terrenceville at the base of the Burin Peninsula, then along the coast westward to Hermitage Bay. The rocks range in age from Late Proterozoic to Ordovician, though most of them are Proterozoic or Cambrian. As is typical for the Avalonian Zone, in spite of the great age of the rocks, the degree of metamorphism is light. According to one theory, the present Avalonian surface rocks rode along on top of those that actually collided with Laurentia and were not seriously changed in composition by the collision.

On the eastern half of the Avalon Peninsula, the bedrock consists of tightly folded and thrust-faulted Late Proterozoic shale, slate, arkose, and graywacke, with rhyolite and basalt-type volcanic rocks surrounding a central up-faulted area. The central area consists of rhyolite to basalt volcanics, with a batholith of granite, monzonite (a relative of granite), and diorite. The batholith is in the vicinity of 574 million years old, making it Late Proterozoic. Geologists believe that the batholith is related to the volcanic rocks; that they are intrusive and extrusive offspring of the same mixture of molten minerals. A strip of Cambrian marine sedimentary rocks, mostly shale, overlies the batholith on the shore of Conception Bay (Plate 28), and Bell Island in the bay has Ordovician shale and sandstone, including beds of the mineral hematite that have been mined as iron ore.

The bedrock on both sides of Trinity Bay and southwestward consists mostly of the same Proterozoic sedimentary and volcanic rock series, overlain by Proterozoic quartzite and erosion remnants of Cambrian shale, sandstone, conglomerate, and limestone. This seems to be a continuous series, with no unconformity between the Proterozoic and Cambrian rocks. The whole series was folded into a northeast-trending synclinorium, with some strike-slip faults, probably during the Acadian Orogeny.

West of a line roughly from Sunnyside to Eastport, a region of fault blocks carries the same folded Proterozoic, Cambrian, and Ordovician sedimentary and volcanics already described. Most of the rocks are volcanics. This region differs from the others in that it contains several Devonian batholiths, granite to diorite in composition. The batholiths' age suggests that the Acadian Orogeny deformed the region.

● New Brunswick and Nova Scotia

In the Maritime Provinces, the westernmost block of Avalonian rocks lies north and west of Saint John, New Brunswick. It extends roughly from between Kennebecasis Bay and Bellisle Bay southwestward to East Head. An extension of that block goes eastward from Saint John almost to Shepody Bay. The two central blocks are in northern Nova Scotia. One lies north of the Minas Basin and Cobequid Bay, from Cape Chignecto to about 4 miles (6–7 km) into Pictou County (the Cobequid Mountains). The other forms the Antigonish Highlands in central and eastern Pictou County and western Antigonish County. The eastern block is in the eastern portion of Cape Breton Island.

Proterozoic gneiss approximately 1 billion years old forms the oldest Avalonian rocks in New Brunswick. The gneiss may come from the time of the Grenvillian Orogeny that affected eastern Québec, the Adirondack area, and points south. Two different series of later Proterozoic sedimentary and metasedimentary rocks overlie the gneiss. They include sandstone, graywacke, conglomerate, dolomite, shale, argillite, and schist, with some basalt. A volcanic epoch, with basalt flows, tuffs, and volcanic breccias followed the deposition of the sediments. The last Proterozoic activity was intrusive, with granite and diorite intruding the older rocks.

All the above rocks are overlain by Early Cambrian basalt, rhyolite, and tuff, with associated breccias and shales, plus diorite and gabbro intrusives. The youngest rocks in the region consist of Late Cambrian and Early Ordovician conglomerate, sandstone, quartzite, and shale, along with some basalt.

In Nova Scotia, Late Proterozoic quartzite, graywacke,

slate, and marble overlie a Grenvillian-age gneiss in the Cobequid Mountains and Cape Breton. The next younger series, still Proterozoic, occurs in the three Nova Scotian blocks as marine shale, with basalt, andesite, and rhyolite flows plus tuff and volcanic breccia. Intrusions of granite and diorite came from the molten mineral reservoir that furnished material for the volcanoes.

Nova Scotia also has Early Paleozoic rocks. Cambrian and Ordovician conglomerate, sandstone, graywacke, and shale, along with some volcanics, form the bedrock in southern Cape Breton Island and in the Antigonish Highlands. Silurian and Devonian conglomerate, sandstone, shale, argillite, and limestone, with volcanics, also occur. These Silurian and Devonian formations also overlie the Proterozoic rocks of the Cobequid Mountains. Silurian and Devonian rocks are associated with the Acadian Zone, and one might not think these belong to the Avalonian Zone; however, the fossils in them are more closely related to European ones than to those of the same age now found elsewhere in North America. That suggests that the Nova Scotian and European animals lived in the same ancestral ocean, in the vicinity of Avalonia.

◆The Triassic Basin

The Triassic Basin in the Maritime Provinces lies mostly under water in the Bay of Fundy, but parts of its edges are visible in New Brunswick and Nova Scotia. Like the Triassic Basins to the south (Figure 119), its rocks consist of terrestrial sandstone and shale plus volcanics. The basin resulted from downfaulting as the supercontinent Pangaea started to break apart. Rivers from bordering uplands carried sediments into the valley created by the faulting, and volcanic eruptions associated with the faulting left lava and ash. The major fault was along the New Brunswick side, and rock layers tilted downward to the northwest, like those in the New York–Virginia Basin but opposite to those in the Connecticut River Basin, and folded into a syncline. The long ridge along the northwestern shore of Nova Scotia from Capes Split and Blomidon to Whipple Point on Briar Island is the outcrop of a series of tilted layers of Jurassic basalt. The basalt outcrops again to the east, north of Cobequid Bay around the town of Five Islands and north of Bass River. Layers of Triassic conglomerate, sandstone, and shale underlie the basalt and out-crop to the southeast of it. They have eroded into the Annap-olis River valley and St. Mary Bay. These sedimentary beds continue across Cobequid Bay and form the bedrock from the vicinity of Five Islands to the Salmon River, and along the south shore almost to Walton.

The Jurassic layers above the basalt lie on the bottom of the Bay of Fundy, but there is a fragment of a second basalt flow on Grand Manan Island. Of the sedimentary layers above this second basalt, only a few tiny bits remain on the New Brunswick shore, at Point Lepreau, St. Martins and Quaco, and Salisbury Bay.

ATLANTIC PROVINCES AND SOUTHERN QUEBEC HIGHWAY SECTIONS

Québec City to the Maine border: Routes 73, 173 (Figure 142)

This route crosses the Taconian Zone and penetrates the Ac-adian Zone for several miles. One of the rock masses from below the Ordovician ocean bottom can be seen from the road.

The major thrust fault on which the Taconian Zone rocks overrode the Central Lowland formations goes right through Québec City, so the route starts at the northern edge of the zone. The bedrock from Québec City to Vallée Jonction con-sists of folded Cambrian and Ordovician marine sedimentary rocks, that become more or less metamorphosed toward the east because of the Taconic Orogeny. There is limestone, sandstone, and graywacke, with shale, slate, argillite, phyl-lite, and schist. The highway crosses a northeast-trending anticlinorium into the Notre Dame Mountains. Most of the Cambrian rocks outcrop in the center of the anticlinorium, in the region of Scott-Jonction, Ste.-Marie, and Vallée Jonc-tion; the ones on the flanks date mostly from the Ordovician Period.

The deep crustal rocks of the asbestos-mining country, the rocks that were part of the lower oceanic crust in the Ordov-ician Period, outcrop in the vicinity of Thetford-Mines, which is about 28 miles (44 km) southwest of Vallée Jonc-tion. The rock is exposed on Route 267 southeast of Thet-ford-Mines, southeast of the road to the Mt. Adstock ski area. Black peridotite and the relatively soft dark greenish

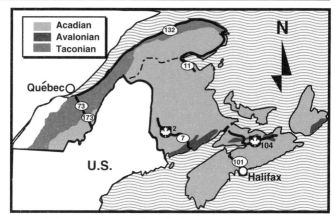

Figure 142. Highways and geologic zones in New Brunswick and Nova Scotia.

rock called serpentinite, which contains the asbestos, are visible here. There is also some volcanic material and fractured sandstone. The Black Lake Lookout, south of Thetford-Mines, overlooks an asbestos mine that occupies the site of Black Lake, which was drained for mining.

Continuing southeastward, Route 173 enters the Acadian Zone and traverses more Ordovician rocks, to the vicinity of St.-Georges-de-Beauce. There, the highway enters a major synclinorium, with Lower Devonian marine sedimentary and metasedimentary rocks that extend to the Maine border, folded and metamorphosed by the Acadian Orogeny.

Around the Gaspé Peninsula, Québec: Route 132 (Figure 142)

This scenic route follows the shore eastward from the base of the peninsula through Acadian Zone rocks and back westward through those of the Taconian Zone. There are many exposures of marine sedimentary and metasedimentary rocks, with some volcanics, from the ancient sea east of Laurentia, as well as sediments from the land created by the Acadian Orogeny.

The route starts in the Late Devonian beds of the Acadian Zone around Miguasha, which are terrestrial sandstone and shale, with evidence of currents from the east, and fossils of land plants and freshwater fish. The sediments came from the eroded highlands uplifted during the Acadian Orogeny.

To the east, in the Black Cape area, there is along the shore of Chaleur Bay a long series of folded Silurian marine sedimentary and volcanic rocks: conglomerate, sandstone, shale, limestone, andesite, basalt, and volcanic breccia. About 2 miles (3 km) northeast of L'Anse-aux-Gascons, Silurian limestone, sandstone, and shale lie with an unconformity on Ordovician schist. Along the shore about a mile north of Newport, Ordovician quartzite and slate-phyllite outcrop, with a variety of greenschist that is probably metamorphosed andesite. Along most of the coastal area to approximately this point, the unconformity atop the Silurian and Ordovician rocks is overlain by Mississippian sandstone and conglomerate, originally sediments washed from the Acadian Orogeny highlands. The older rocks show up mostly along the shoreline. Then there is a gap in the Mississippian rocks until about halfway to Cap-d'Espoir, where they again lie atop the older formations.

The Percé-area bedrock consists of folded and faulted Ordovician and Devonian limestone, sandstone, and shale, with a frosting of the Mississippian conglomerate and sandstone. Percé rock exhibits almost vertical layers of Lower Devonian limestone. Mont-Sainte-Anne and Bonaventure Island are formed by the Mississippian rocks.

The Ordovician and Devonian rocks extend northward along the coast. The shore cliffs at the northeast outskirts of the town of Gaspé consist of Devonian sandstone, shale, and conglomerate that appear to be river deposits on land, as opposed to the usual marine sediments of that age elsewhere along the coast. Marine Devonian limestone and shale extend to Cap-des-Rosiers.

At Cap-des-Rosiers, the highway enters the Taconian Zone, where it stays for the rest of the trip. The area shows greatly distorted beds of Ordovician shale, argillite, graywacke, and limestone, with a conglomerate of limestone pieces. The Taconian Zone is famous for its thrust faults, and there is one about 2.5 miles (4 km) northwest of Cap-des-Rosiers and another at Pointe-Jaune. Ordovician limestone, sandstone, graywacke, shale, argillite, slate, breccia, and limestone conglomerate continue along the St. Lawrence shore almost all the way to Québec City. At Tourelle, vertical layers of sandstone, shale, and argillite outcrop. The sandstone has weathered and eroded into pillars, giving the town its name: tourelle is French for "turret." Cambrian limestone, limestone conglomerate, sandstone, and shale are exposed in the La Pocatière area and at L'Isletsur-Mer. To the west, to Québec City, the bedrock consists of the Ordovician series described above.

Saint John to Woodstock, New Brunswick: Route 7, Trans-Canada Highway 2 (Figure 142) *This route starts in Avalonia and continues deep into the Acadian Zone. It crosses the point of the dipping syncline in the Acadian Zone that is filled with Pennsylvanian rocks, with the older ones on either side (Figure 141).*

The geology of Saint John exhibits in one city some of the oldest and some of the youngest rocks in the province. Among the oldest, at the north end of the toll bridge at the head of the harbor, an assortment of Avalonian Proterozoic and Lower Cambrian layers lies in the limb of a fold along Main Street. The rocks include limestone, conglomerate, sandstone, and shale, along with some volcanics. These Avalonian rocks were moved east to New Brunswick during the Acadian Orogeny. Among the youngest rocks, in the seaside park at Duck Cove, tilted Pennsylvanian sandstone and shale outcrop; the rocks originated as terrestrial deposits, as indicated by stream-type cross bedding, current ripple marks, and land plant fossils. These rocks are evidence of the Alleghenian Orogeny.

From just southwest of Saint John to Nerepis, the highway crosses an Avalonian synclinorium in Proterozoic and Lower Cambrian rocks, containing quartzite, argillite, limestone, and granite. The road from Nerepis to Welsford is in the Acadian Zone, with Silurian marine volcanic rocks and then Devonian granite. Silurian volcanics and Silurian and Devonian limestone, shale, argillite, sandstone, and conglomerate continue northward beyond Welsford.

Approximately 30 miles (50 km) south of Fredericton a formation of Mississippian lava occurs that appears to be rhyolite but is of a slightly different chemical composition. It forms the earliest of the Carboniferous rocks that overlie the older ones that were deformed in the Acadian Orogeny. Approximately the next 4 miles (6.5 km) are on Mississippian rocks in a series that contains sandstone, shale, limestone, gypsum, and anhydrite. These rocks indicate a marine near-shore environment, and the gypsum and anhydrite tell of evaporating seawater in shallow embayments. Next come Pennsylvanian conglomerate, sandstone, and shale (without significant coal here) to Fredericton. At the Fredericton airport are sand dunes as much as 65 feet (20 m) high, created by wind action from Ice Age sand deposits.

On Trans-Canada Highway 2 west of Fredericton, at the exit to the Mactaquac Dam, Silurian graywacke and slate underlie the Pennsylvanian rocks. Near the powerhouse, the truncated ends of steeply dipping to vertical Silurian layers are exposed, overlain by almost horizontal Pennsylvanian conglomerate and arkosic sandstone.

Silverwood approximately marks the border between the Pennsylvanian bedrock and the Silurian and Devonian rocks on the northwestern side of the great synclinal structure, the same series that occurs north of Welsford. In the past, Silurian and Ordovician rocks extended almost all the way to what is now Woodstock. However, erosion has uncovered a large batholith of Devonian granite in the middle of the outcrop area between Prince William and a couple of miles east of Temple. Also, around Prince William, a small outlier of the Pennsylvanian rocks covers the older ones. Between Temple and Woodstock the bedrock changes to a Silurian then Ordovician series of argillite, quartzite, schist, and gneiss, with volcanics between Temple and Meductic.

Dalhousie to Bathurst, New Brunswick: Route 11, or Routes 11 and 134 along the coast (Figure 142) *This trip introduces a variety of rocks on the northwest flank of New Brunswick's central synclinal structure. The older formations in the region were distorted by the Acadian Orogeny; the Ordovician ones were also affected earlier by Silurian deformation.*

The bedrock of the Dalhousie-Charlo region is folded and faulted Devonian limestone, shale, basalt, and tuff. From Charlo to Pointe Verte is a mixture of the above with Silurian limestone, shale, and sandstone; the Silurian rocks occur in the vicinities of Dickie and Belledune. There is also a bit of Pennsylvanian sandstone and conglomerate in the Nash Creek–Jacquet River area. At Quinn Point, near Nash Creek, folded and faulted Silurian rocks are overlain by flat-lying Pennsylvanian conglomerate. Belledune Point has tilted layers of Silurian sandstone, shale, and conglomerate, with basalt prophyry lava flows.

Silurian rocks in the Pointe Verte area become mixed with Ordovician and Silurian graywacke, slate, basalt, gabbro, and peridotite between there and Bathurst. Bathurst lies at the north end of a Devonian granite batholith, on which Pennsylvanian shale and sandstone of central New Brunswick overlap. Southwest of Bathurst, the bedrock contains large deposits of Ordovician iron ore, in iron formation interbedded with layers of schist and phyllite-argillite. The iron formation contains a combination of pyrite, chalcopyrite, and pyrrhotite, along with the zinc mineral sphalerite and the lead mineral galena. Large-scale mining has gone on for years at Bathurst Mines and Heath Steele. The closest mine to Bathurst is about 12 miles (19 km) to the southwest.

Halifax to Wolfville, Nova Scotia: Routes 102, 101 (Figure 142)
This route travels through some typical Ordovician, Silurian, and Mississippian rocks of the Acadian Zone into the Triassic Basin.

Halifax is situated on tightly folded Ordovician rocks at the edge of the great Devonian batholithic complex that forms the backbone of the southwestern half of Nova Scotia (Figure 139). At Point Pleasant Park in Halifax, Ordovician slate and shale are exposed. These slaty rocks, with quartzite and other metasediments, form the bedrock most of the way from Halifax to St. Croix. A lobe of the great batholith crosses the highway between Pockwock Lake and Five Mile Lake.

At St. Croix are Mississippian shale and limestone layers of a series of folds that have been overturned to the north. The shale originated in silt and clay from the weathering and erosion of highlands formed in the Acadian Orogeny; both kinds of rock were deposited in the sea that invaded the edge of Laurentia between the Acadian and Alleghenian orogenies. There is a sample of these rocks at Windsor, on the east side of the Avon River.

Highway 101 traverses Mississippian sandstone and shale up to the Horton Landing area, where it skirts the edge of the Triassic Basin. If you take a side trip to the northeast shore of Oak Island, 3 miles (5 km) east-northeast of Grand Pre, you can see the horizontal conglomeratic base of the Triassic rock series on top of folded Mississippian shale and sandstone. The rocks outcrop at the tidal flat on the west shore of the Avon Estuary. The maximum difference between high and low tide here is 43.6 feet (13.3 m); be aware that the tide comes in very quickly. The highway continues southwestward from Wolfville down the Annapolis River valley in sandstone and shale of the Triassic Basin.

Wolfville is situated at the junction of Triassic, Mississippian, and Ordovician and Silurian rocks. Southwest of Wolfville is a formation containing quartzite that some geologists have theorized is Upper Ordovician glacial outwash, modified by seashore waves. The rock no doubt gave its name to the town of White Rock, near which it is quarried.

Antigonish to Amherst, Nova Scotia: Trans-Canada Highway 104

(Figure 142) *Highway 104 from Antigonish to Amherst displays the greatest geological variety in Nova Scotia, through two parts of the Avalonian Zone as well as a large section of the Acadian Zone and a bit of the Triassic Basin.*

The bedrock at Antigonish consists of Mississippian shale, sandstone, limestone, and gypsum of the Acadian Zone, marine rocks that lie on top of the earlier Paleozoic ones modified by the Acadian Orogeny. Just beyond James River, the highway enters one of the central segments in the Canadian Avalonian Zone (Figure 138), the Antigonish Highlands. The highlands evolved from a complexly faulted island of Proterozoic to Lower Devonian sedimentary and volcanic rocks surrounded by Carboniferous formations. Between James River and Marshy Hope, the highway traverses a rock series that includes Late Proterozoic, Silurian, and Devonian basalt, andesite, and rhyolite, with tuff and breccia, as well as shale, graywacke, and sandstone-quartzite. From beyond Marshy Hope to just beyond French River, Ordovician and Silurian rocks occur, including shale and sandstone.

After French River, the highway enters an area of Pennsylvanian sandstone and shale, here with coal, that covers the older rocks. The Pennsylvanian rocks continue along the highway most of the way to Truro, in some places dipping steeply to the northeast. The edge of the Triassic Basin is near Valley, with gently tilted conglomerate, sandstone, and shale. East of Glenholme, basalt overlies the basin's sedimentary rocks.

At Glenholme, the highway turns abruptly north, and in approximately 3 miles (5 km) it leaves the Triassic bedrock and passes briefly through about 3 miles of Pennsylvanian rocks back into the Avalonian Zone, in the Cobequid Mountains. First is Proterozoic quartzite, slate, and marble, followed by Mississippian granite and diorite, and then Devonian lava, sandstone, and shale. The true Avalonian rocks are along the southern portion of the mountains. Beyond them are rocks of the Acadian Zone. In the mountains, quarries near Folly Lake have exposed what were originally Proterozoic sandstone and shale, folded and metamorphosed into argillite, schist, and quartzite at the edge of the granite intrusion. Near the northern edge of the highland region is Wentworth Station, and about at the Route 246 intersection, the highway reenters the Pennsylvanian sandstone, conglomerate, and shale. Those rocks continue the rest of the way to Amherst, covering the distorted Acadian rocks below.

Pennsylvanian coal deposits occur around Springhill and in a strip eastward from Joggins through Maccan almost to Highway 104. Mines in the region have operated at Springhill and River Hebert. Along the coast at Joggins, river channel and flood plain deposits of sandstone and shale outcrop, gently dipping toward the southeast on the limb of a syncline (Plate 21). In these beds, paleontologists have found the remains of small amphibians and reptiles in fossil tree stumps.

St. John's to Corner Brook, Newfoundland: Trans-Canada Highway 1, Route 60 (Figure 143) *This route crosses all three zones of the Northeastern Appalachians in their northern extremities, including a good sampling of the area for which the Avalonian Zone was named.*

This crossing of the island begins on the Avalonian Zone rocks of the Avalon Peninsula. The route shows rocks that originated as sediments and volcanic material in the ancient Avalonian region and in the ocean between Avalonia and Laurentia in Late Proterozoic and early and middle Paleozoic times, punctured by Devonian intrusives and modified by the Acadian Orogeny.

St. John's lies on the western limb of a syncline trending north to south. The range of hills that includes Signal Hill consists of sandstone and conglomerate, which dip under the ocean to reappear at Cape Spear. West of the hills at St. John's are thick layers of shale, underlain by volcanic rocks in the western limb. All of these rocks are Proterozoic. Most of the city rests on the shale.

To see the Cambrian rocks that were formed after the Proterozoic ones of St. John's, follow Highway 1 to Donovan's and pick up Route 60, which follows the coast. The Cambrian rocks occur from Topsail to the Seal Cove area, where they overlie, with an unconformity, the Proterozoic rocks. The Cambrian ones are mostly marine shale, with some sandstone, conglomerate, and limestone. At the Manuels River, the contact between Proterozoic granite and Cambrian shale shows just downstream from the bridge, with a 20-foot (6 m) layer of conglomerate in between. The shale contains European fossils, one of the primary characteristics of the Avalonian Zone (Plate 48). They show that parts of Avalonia have become parts of Europe during the breakup of Pangaea. From Seal Cove to Highway 1 via Holyrood and Route 90, the bedrock consists of Proterozoic granite and diorite, then volcanic rocks.

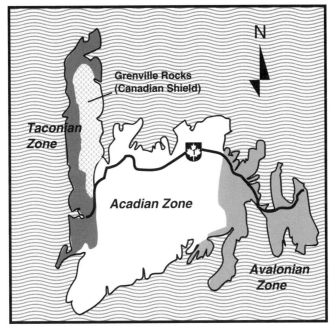

Figure 143. The Trans-Canada Highway and geologic zones in Newfoundland.

From the Route 90 intersection to halfway between the Routes 70 and 71 exits on Highway 1, the rocks are mostly Proterozoic shale, slate, graywacke, and conglomerate on the eastern side of the northeast-trending Trinity Bay Synclinorium. From there, the road passes into younger Proterozoic shale, slate, arkose, and conglomerate, with some volcanic rocks. Those formations extend all the way to about where the highway crosses the latitude of Chance Cove, on the other side of the synclinorium. The volcanic rocks are in the western part of the outcrop area. They show up in the vicinity of the intersection with Routes 203 and 201, where rhyolite is exposed. Some Cambrian rocks—mostly shale, with some sandstone and limestone—occur in the center of the synclinorium, between Routes 80 and 202. Between the Chance Cove latitude and near Clarenville, the highway is back in the older Proterozoic rocks it crossed between Routes 70 and 71, on the western limb of the synclinorium.

From Clarenville to the vicinity of Glovertown, the highway follows and crosses a northeast-trending synclinal structure of Proterozoic rocks; the older on the flanks, and the younger in the center. The rocks include shale, slate, phyllite, graywacke, arkose, conglomerate, and volcanics. Just beyond the junction of the road to Traytown, gneiss, amphibolite, and granite outcrop that geologists consider to correlate with the Proterozoic Grenvillian Orogeny, much older than the Avalonian rocks.

Between Glovertown and Gambo, the road crosses a mass of Devonian Granite that signals the eastern margin of the Acadian Zone. From Gambo to Glenwood, Middle Ordovician or older schist, slate, graywacke, and quartzite form the bedrock. Rocks from Glenwood to Windsor consist of Silurian shale, sandstone, graywacke, conglomerate, limestone, and volcanics. Of particular interest is red and green, mica-rich sandstone exposed about half a mile northwest of the eastern interchange to Bishops Falls. It shows cross bedding, ripple marks, mud cracks, and raindrop imprints, all of which indicate a terrestrial environment, a rarity in Newfoundland's world of marine rocks. A large batholithic complex, mostly Devonian in age, breaks into the middle of this zone of Silurian rocks. The highway runs along the edge of this complex of intrusive igneous rocks, which vary from gabbro to syenite.

Between Windsor and the Catamaran Brook Provincial Park, the bedrocks are Ordovician argillite, chert, graywacke, slate, some limestone, and some volcanics, with evidence of tight upright folds. North of the turnoff to Catamaran Provincial Park, the highway enters a Devonian batholith of granite and diorite and follows those rocks up to South Brook. There, it enters another Silurian zone that extends along the highway to Sandy Lake. The rocks include shale, sandstone, graywacke, conglomerate, limestone, and volcanics. In this zone, another set of terrestrial rocks outcrops about 3.5 miles (5.6 km) southeast of the Route 390–Springdale exit. They are redbeds, including shale, arkosic sandstone, and conglomerate with pebbles of volcanic rocks.

The highway parallels the shore of Sandy Lake through the edge of a Devonian granite batholith, and where it curves around the northern tip of the lake it enters the northern Carboniferous basin. At the margin, where the basin meets the batholith, less than a mile east of the Route 420–Hampden exit, is a road cut in latest Proterozoic to earliest Cambrian schist. The schist appears to have started as a graywacke and was metamorphosed possibly before the Ordovician Period and certainly during the later Acadian Orogeny.

The Carboniferous basin contains Mississippian shale, sandstone, conglomerate, limestone, gypsum, and anhydrite. The sandstone and conglomerate are sediments that rivers carried to the sea from the highlands formed in the Acadian Orogeny. The highway follows the Mississippian rocks to South Brook. There, it enters the Taconian Zone. To Steady Brook, the bedrock is schist, gneiss, quartzite, and marble of latest Proterozoic to earliest Cambrian age; this is the series that underlies the Cambrian and Ordovician rocks to the west. From Steady Brook to Corner Brook, the highway crosses a narrow fault block composed of folded Ordovician shale, sandstone, quartzite, limestone, and dolomite.

Near the edge of Corner Brook the western thrust fault of the fault block crosses the highway. The city rests on a larger block of Cambrian and Ordovician sedimentary, volcanic, and intrusive rocks that was moved from the east during the Taconic Orogeny. Typical rocks include shale, slate, phyllite, sandstone, graywacke, conglomerate, limestone, and basaltic types. This large fault block also contains the deep intrusive rocks from below the Ordovician sea floor that are described in the introduction to the Taconian Zone.

6

THE COASTAL PLAIN

The Appalachian Province stands like a rough, curved backbone in the eastern United States, breaking the generally flat surface of the regions to the east, west, and south. Toward the interior of the continent, the land surface descends gradually across nearly horizontal formations to the Mississippi Valley. The same is true in the other direction, to the ocean, where the land dips into the water and becomes the continental shelf. The landward portion of the latter surface is the Coastal Plain.

During the Paleozoic Era, the orogenies described in chapter 5 helped to create the supercontinent Pangaea. The last of them (Figure 106) was so powerful and extensive that it crumpled layers of rock thousands of feet thick into folds and fault blocks that formed a mountain range presumably rivaling today's Rocky Mountains, extending from what is now southeast Canada to Mexico. By the beginning of the Mesozoic Era, highlands in the present Rocky Mountain region furnished more rugged topography to Pangaea. As Pangaea split during the Mesozoic Era, Africa and South America moved away from Europe, North America, and Greenland, opening the Atlantic Ocean (Figure 27). Since the Triassic Period, river systems have been carrying sediments from the highlands into the slowly widening ocean. The result is thousands of feet of sedimentary rock and unconsolidated (not solidly cemented) sediments in deepening troughs and basins off the eastern and Gulf coasts of North America and extending from the edge of the continental shelf through the Coastal Plain to the margin of the Appalachian Province.

The Coastal Plain is a region of unfolded sedimentary rock layers and unconsolidated sediments that appear to be horizontal but actually slope gently toward the ocean at an angle of about 1°. (Note: the cross sections in the figures for this chapter all have large vertical exaggerations.) The Coastal Plain sediments accumulated during the Mesozoic and Cenozoic eras. They overlie the folded, faulted, and eroded Paleozoic rocks of the Appalachian Province (Figure 144). In contrast to the neighboring Appalachian Province rocks, the formations are younger and are not folded or significantly intruded, and some consist of or contain unconsolidated gravel, sand, and silt. There are fewer faults than in the Appalachian Province, and no thrust faults.

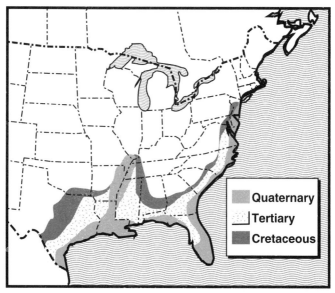

Figure 144. A simplified geologic map of the Coastal Plain.

■ THE LAY OF THE LAND

The Coastal Plain offers generally flat or rolling country, rising here and there in escarpments, or long ridges or rows of hills that represent outcroppings of the more resistant rock layers (Plates 33, 35). The drowned river valleys along the coast of Maine and Canada, carved at the end of the Ice Age when sea level was lower, make the shoreline very irregular in places. Yet much of the shoreline south of New England is smoother and is bordered by long, narrow, sandy islands. Extensive beaches along the coastline make parts of the coast a far cry from the rugged, rocky shores of New England and the Maritime Provinces, with their small beaches in coves and river mouths.

■THE REGIONS OF THE COASTAL PLAIN

The Coastal Plain formations make up the bedrock of eastern North America from Mexico to New England (Fig. 144). They pinch out to the north in southeastern Massachusetts; they extend farthest into the continent in the Mississippi Valley; and south of Texas, they border the Gulf of Mexico from the Rio Grande through the Yucatán Peninsula. From Mexico to the Cape Cod area in Massachusetts, the Coastal Plain formations continue from the land into the ocean to form the bedrock of the continental shelf; beyond Cape Cod, Mesozoic and Tertiary sediments lie underwater to as far north as the continental shelf off Labrador.

Because there is little variation in its topography, the Coastal Plain may seem to be all the same from Texas to Massachusetts. However, there are eight distinct regions of this interesting province, each with its individual features.

◆The Northern End

The northernmost outcrops of Coastal Plain sediments occur in southeastern New England. There are a few tiny spots in the Marshfield-Duxbury area of Massachusetts, just north of Cape Cod, where they are mixed with glacial material from the Ice Age. Tertiary sediments are found below the glacial material on Cape Cod, and there are well-known and spectacular cliffs of Cretaceous and Tertiary sand, silt, and clay on the island of Martha's Vineyard, off Cape Cod (Plate 36).

The Ice Age glacial deposits of Long Island rest on an escarpment of Cretaceous and Tertiary sediments that is part of the Coastal Plain's western edge; the scarp continues northeastward and forms the nuclei of Block Island, Martha's Vineyard, and Nantucket. The Cretaceous material extends southward through Staten Island, where the bedrock is partly Cretaceous and partly Paleozoic. From Staten Island, the edge of the Coastal Plain continues to Trenton, New Jersey; between the edge and the coast, Tertiary sediments overlie the Cretaceous formations.

◆The Region of Shoreline and Inland Bays

The Chesapeake Bay area occupies a broad downwarping of the rocks in which the lower portions of the Delaware and Susquehanna river valleys have been drowned. The formations of the peninsula between Delaware Bay and Chesapeake Bay are almost entirely Quaternary in age, flanked by older deposits as would be expected in the middle of a structural basin. Chesapeake Bay is essentially the drowned lower portion of the Susquehanna River and its tributaries, the James, Mattaponi, Rappahannock, Potomac, and Patuxent rivers. This region of drowned river mouths continues along the North Carolina coast, where the Roanoke, Pamlico, and Neuse River mouths form large bays, and other rivers form smaller bays. Just offshore, from Sandy Hook in New Jersey to Cape Fear in North Carolina, a series of elongate barrier islands parallels the coast.

Flat areas inland are mostly modern river flood plains and alluvial deposits of earlier Quaternary age. Older formations west of the Quaternary deposits are described below. The countryside south of Chesapeake Bay in this region contains numerous large oval, elliptical, or otherwise irregular depressions called bays that look more or less like meteoroid craters. The name comes from the fact that a typical bay is filled with bay trees. There is no evidence that the bays are in fact meteoroid craters, and no one knows how they originated. The inland bays continue down through the next region to the south.

◆The Sea Island Region

The Sea Island Region extends down the coast roughly from Cape Fear, North Carolina, to the base of the Florida peninsula. Here, there are only a couple of barrier islands, and the river mouths have been drowned much less than those to the north. The northern part of the South Carolina coast is smooth, but from Georgetown, South Carolina, to Jacksonville, Florida, irregularly shaped islands called sea islands follow the coast. Some of the sea islands are flat and covered with sand or with plant-filled marshes. Others have ridges of sand.

Inland in this region, the variously called Sand Hills or Red Hills occur in a belt about 20 to 40 miles (32 to 64 km) wide. They consist of resistant Cretaceous and early Tertiary rocks that rise as high as 300 feet (91 m) above the lowland, with the relief decreasing toward the coast. The belt of hills starts at about Columbia, South Carolina, and extends past Augusta, Georgia, to south of Macon. It then goes beyond the Sea Island Region and curves southwest, west, and northwest through Alabama, roughly past Eufaula, Troy, and Greenville, and south of Demopolis, and continuing northward into central Mississippi (Figure 145).

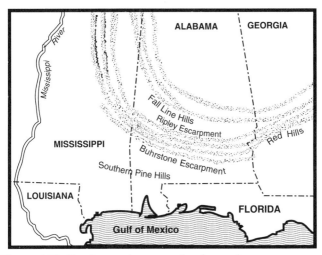

Figure 145. Erosion-resistant rocks form the escarpments and hills of the eastern Gulf Coastal Plain.

◆The Fall Zone

At the inland edge of the Coastal Plain from Washington, D.C., to Alabama is the Fall Zone: rapids and waterfalls mark the points where harder Piedmont rocks meet the softer Coastal Plain rocks. During the 18th and early 19th centuries, goods were shipped up the rivers to the Fall Zone, where they had to be unloaded from the boats and sent on by land. Products from the hinterlands were carried from the fall line back down the rivers to the more populated areas. Because of this activity, and because the waterfalls were good sources of power for mills, many towns sprang up along the Fall Zone. Some developed into the cities that mark the border between the Coastal Plain and the Piedmont: Richmond, Raleigh, Columbia, Augusta, Macon, and Columbus.

◆The Florida Section

The Florida section of the Coastal Plain is dominated by the Ocala Arch, which forms the core of the state. The highest point of the arch lies in the northwestern part of the peninsula, where Early Tertiary marine Ocala Limestone outcrops at about 150 feet (45 m) above sea level. From there, the crest tilts gradually downward, so that the Early Tertiary limestone disappears under younger rocks south of Bushnell. The younger formations also dip away from the Ocala Limestone on the flanks of the arch in northern Florida.

Since the Tertiary limestone occupies much of the northern half of Florida, the countryside exhibits many karst features—caves, underground passages, sinkholes, and springs. Florida karst country differs from that of other states in that the caverns and passages developed mostly below the groundwater level, so they are filled with water. Thus sinkholes show up as ponds and lakes, and there are vast underground systems, some discharging water into the ocean.

Some caves in Florida are at least partly above water level, such as Ocala Caverns, south of Ocala (not open to the public), and the cave at Florida Caverns State Park, in the panhandle, just north of Marianna. Falling Waters State Recreation Area is also in the panhandle, south of Chipley. There, waterfalls tumble into sinkholes, unless the season is dry. Ichetucknee Springs, at the state park north of Fort White, pour out an estimated 233 million gallons of water per day. At Wakulla Springs, south of Tallahassee, bones of elephant-like mastodons and other animals that fell into the springs during the Ice Age have been found. Silver Springs, east of Ocala off Florida Route 40, is reputed to be the largest limestone artesian spring in the world, producing an average of about 800 million gallons a day.

South-central Florida, north of the Everglades and Big Cypress Swamp and extending northward east of the Ocala Arch, is characterized by flatlands of Ice Age sediments, mostly sand. The sandy formation continues southward under the Everglades, Big Cypress Swamp, and Lake Okeechobee, where freshwater limestone and limey clay accumulated during the Ice Age.

Long barrier islands, like those north of Cape Fear, North Carolina, protect the coast south of Jacksonville. The islands along Florida's west coast more closely resemble the shorter, less regular sea islands of Georgia and South Carolina.

The Florida Keys are a chain made up of two different kinds of islands. North and east of Big Pine Key is a strip of long, narrow islands that look like a broken ridge poking up out of the ocean. Those keys are the remains of a coral reef that formed during the Ice Age, when the sea level was 25 feet (8 m) higher than it is today. The reef consists of massive coral heads surrounded by smaller coral colonies and shell fragments in a limey matrix. The southern end of the reef is the southern point of Big Pine Key; northward, it stretches past Sands Key to Miami, mostly just below the ocean surface. The southern keys, which are irregularly shaped islands (including most of Big Pine Key), consist of a rock made up of tiny (1 mm or less) spherical particles of the mineral calcite in a calcareous groundmass. The particles grow in layers of calcite from the seawater, accreting around tiny shell fragments or quartz grains. Waves and currents keep the particles intermittently moving, so that the calcite can accrete all around them, giving them their spherical, or nearly spherical, shape. This deposit formed in shoal waters during the Ice Age, at about the same time the reef was growing. The formation continues alongside the reef to become the bedrock in Miami.

◆The Eastern Gulf Plain

The eastern Gulf plain continues westward from the Florida peninsula, with more and thicker Cretaceous and Early Tertiary formations. The rocks become thicker toward the west into the downwarped structure of the Mississippi Basin. Past central Alabama, where the Coastal Plain formations lap up onto those of the Central Lowland, the Fall Zone is less noticeable.

Much of the region consists of flat lowland with concentric escarpments and rows of hills where resistant formations break the surface (Figure 145, Plate 33). The Fall Line Hills, formed of Cretaceous sandstone, run just seaward of the Fall Zone. Next is the Ripley Escarpment, formed of Cretaceous sandy shale, followed by the Red Hills, of Early Tertiary sandstone. Then comes the Buhrstone Escarpment, held up by Early Tertiary shale and sandstone. The Southern Pine Hills, in southern Mississippi, southwestern Alabama, and extreme northwestern Florida, have for bedrock later Tertiary sandstone and conglomerate. Along the coast is a narrow strip of Ice Age river deposits, mostly flood plain sand and silt.

◆The Mississippi Embayment

The Mississippi Valley occupies a structural basin, or trough, caused by faulting as Pangaea broke, beginning in the Mesozoic Era. Downwarped Paleozoic formations in the fault block underlie Cretaceous and Tertiary rocks that filled the subsiding trough, up as far as the southern tip of Illinois. The margins of the trough include portions of Alabama, Tennessee, Kentucky, Missouri, and Arkansas (Figure 144). From the Mississippi River delta to Illinois, the trough is the Mississippi Embayment. In the northern prong of the embayment, Tertiary sediments lie under a mostly complete covering of Ice Age deposits of sand. Gravel, silt, and wind-blown dust (loess) from the meltwater-swollen Ice Age Mississippi River and its tributaries are also visible.

Officially, the Mississippi alluvial plain extends from Cape Girardeau, Missouri, to the head of the Atchafalaya River in Louisiana. South and east of there is the delta country. The alluvial plain varies from about 125 miles (200 km) wide at the latitude of Little Rock, Arkansas, to only about 25 miles (40 km) wide just north of Natchez, Mississippi. The plain forms the floor of a valley whose walls rise sharply as much as 200 feet (60 m).

Cyclic deposits of silt and clay alternating with sand and gravel accumulated in the lower part of the valley's alluvial plain during alternating glacial and interglacial stages of the Ice Age. When an ice cap held sway, the rivers eroded and deposited gravel and sand along the valley. Silt and clay were carried farther, to the delta and the ocean. Then, when the ice disappeared during an interglacial stage, the sea level rose and drowned the lower valley. While interglacial sand and gravel were deposited along the shoreline and inland, contemporary silt and clay settled down on top of the older sand and gravel in the temporary marine bay that had been the valley. As the next glacial stage evolved, the ocean retreated, and the rivers once again deposited sand and gravel in the

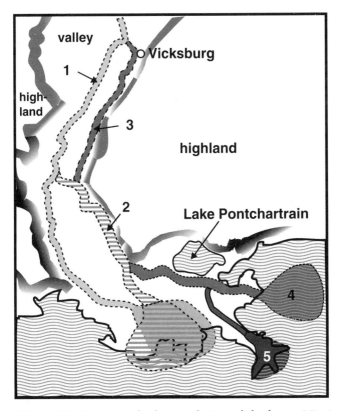

Figure 146. Stages in the late evolution of the lower Mississippi River. 1, the river's path before A.D. 300; 2, a diversion eastward in the middle of the old path, A.D. 300–400; 3, a diversion upstream near the position of the present Vicksburg, A.D. 1000–1100; 4, a downstream diversion, A.D.1100–1200; 5, the final diversion, A.D. 1500–1600, forming the modern river and delta.

lower valley, forming new layers on top of the silt and clay. Since the end of the Ice Age, the Mississippi River has been cutting down through the old deposits in much of the valley. Figure 146 shows how the Mississippi River and Delta have changed in modern times. The lower Mississippi Valley is almost as wide as the distance between Lafayette and Baton Rouge, about 45 miles (72 km). The delta begins approximately at I-10, its greatest width is roughly from Lake Pontchartrain to Vermillion Bay, and the distance from I-10 to the

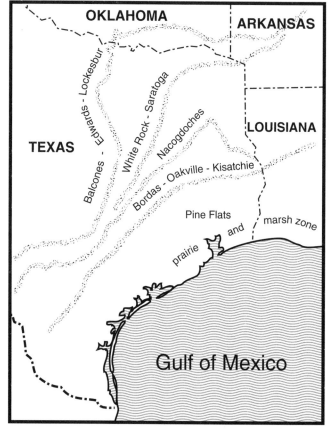

Figure 147. The major escarpments of the western Gulf Coastal Plain.

farthest point on the delta is about 175 miles (280 km); it contains a truly amazing amount of sediment from one river system.

◆The Western Gulf Plain

The Western Gulf Coastal Plain is generally similar to the eastern Gulf Plain, with flat lowlands and occasional long escarpments or rows of hills. The hills and ridges are similar in origin and structure to those east of the Mississippi River valley, and they have similar resistant rocks. From northwest to southeast, the four major scarps are the Balcones-Edwards-Lockesburg (Cretaceous), the White Rock–Saratoga (Cretaceous), the Nacogdoches (Tertiary), and the Bordas-Oakville-Kisatchie (Tertiary) (Figure 147).

As in the eastern Gulf region, a strip of Quaternary sediments along the coast overlaps the Tertiary formations that dip underneath the Gulf of Mexico. The nine rivers that flow over the Texas Coastal Plain have all contributed gravel, sand, silt, and clay. Their deltas and flood plains have joined since the end of the Ice Age to form the soil and underlying material of the pine flats and the prairie and marsh zone.

An "outer coastline" of almost continuous sand bars and dune-covered islands runs from Galveston to the Rio Grande, including the famous Padre Island. East of Galveston, in Texas and Louisiana, old ridges of sand extend along the shore, with tidal marshes, mud flats, and freshwater and brackish lakes in between.

■THE ROCKS OF THE COASTAL PLAIN

The Coastal Plain differs greatly from the rest of eastern North America in that many of its formations consist of or contain poorly consolidated rocks or completely unconsolidated sediments (sand, gravel, silt, and clay). With only a few tiny exceptions in Texas, all the formations are sedimentary; some sediments were deposited by rivers on land, the rest by the ocean that periodically invaded the Coastal Plain region (Plates 34–36). The exceptions are several small Cretaceous intrusive igneous rock masses, several in the vicinity of Uvalde, Texas, and one south of Austin.

A special characteristic of Coastal Plain sediments is the presence of large quantities of the mineral glauconite, especially in the sands. Glauconite is a greenish mineral related to muscovite mica, and, like the mica, it cleaves in one direction. It gives the color to those sands called greensands in the Coastal Plain. It occurs in sands, silts, and clays in the form of tiny pellets that, according to a modern theory, originated as fecal pellets from oceanic filter-feeding organisms. After the pellets sank to the ocean bottom, the matter inside them turned to glauconite. Glauconite pellets occur here and there in rocks from Cambrian to Quaternary in age, but the pellets are especially common throughout both Cretaceous and Tertiary sediments on the Coastal Plain.

◆Quaternary Sediments

The Quaternary sediments are almost all unconsolidated. From Staten Island, New York, and a small adjacent area of New Jersey north to the Cape Cod area, glacial till and outwash cover most of the bedrock in and near the terminal moraines of the last Ice Age glacier (Figure 109). In the rest of the Coastal Plain, the youngest material consists of Ice Age (Pleistocene) and recent (Holocene) river, marsh, and beach deposits of gravel, sand, silt, and clay, along with the windblown dust called loess.

◆Tertiary Sediments

Some Tertiary sediments are unconsolidated, and others are now solid rocks (Plate 36). There are channel, flood plain, and delta deposits from rivers, as well as nearshore to deeper ocean sediments and rocks. The coarser-grained layers consist of conglomerate, sandstone, gravel, and sand. Fine-grained sediments and rocks include shale, silt, clay, and marl. Marl is a clay that contains a lot of calcium carbonate; thus it is a limey, or calcareous, clay. The Tertiary solid rocks include limestone and the unusual rock type chalk. The chalk occurs in an early Tertiary formation in Mississippi.

Various sediments on the Coastal Plain grade into one another. Some sands are silty, clayey, or calcareous, and some clays and silts are sandy. Many layers of Coastal Plain sediments contain fossil shells of clams, oysters, and snails, and possibly some other marine fossils, such as shark teeth. A few of the layers contain fossil leaf impressions. There are also deposits of lignite, which is a mass of more or less degraded plant material that is in an early stage of the process that produces coal.

◆Cretaceous Sediments

Cretaceous formations of the Coastal Plain resemble the Tertiary ones in composition and include conglomerate, sandstone, shale, limestone, sand, silt, clay, and marl. Chalk occurs in two Cretaceous formations of the Coastal Plain. The Austin Chalk formation of Texas (Plate 35) and Mississippi contains chalk as well as other types of rock, and there are other chalks in Alabama and Mississippi (Plate 35). Some Cretaceous layers contain abundant marine fossils (Plate 35), and a few have fossils of land plants.

◆Jurassic and Triassic Sediments

The oldest rocks that outcrop in the Coastal Plain are Cretaceous in age, but Jurassic limestone, dolomite, and rock salt are hidden below them. Under the Jurassic rocks are Triassic terrestrial sandstones and shales, including redbeds, that rest on the Piedmont-type basement of the Coastal Plain and continental shelf. These older, deeper rocks were discovered during the drilling of oil wells.

■THE HISTORY OF THE COASTAL PLAIN

The Coastal Plain reflects the movement of crustal plates, which has caused supercontinents to form, break apart, and reform again over millions of years. The last supercontinent was Pangaea, formed during the Paleozoic Era. Core samples from the Coastal Plain have shown that as Pangaea eroded during the Permian and Triassic Periods, the Piedmont surface developed a hilly topography with a relief of about 200 to 500 feet (60 to 150 m). (What is now the Piedmont was much broader then than it is today, and was in the middle of Pangaea.) By the end of the Triassic Period, Coastal Plain sediments were being deposited on the Piedmont surface.

◆Block Faulting: Arches and Basins

Pangaea started to slowly break into smaller units during the Triassic Period. As Earth's crust stretched where Africa and South America were splitting away from North America, Greenland, and Europe, it cracked to form major faults, and some fault blocks of the crust sank, forming the Triassic and

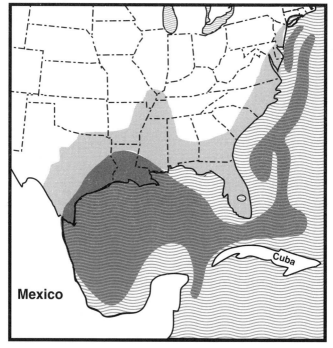

Figure 148. Regions of subsiding basins along the Atlantic continental shelf and in the Gulf of Mexico.

Jurassic basins described in chapter 5, on the Appalachian Province (Figure 119).

While the ancestors of today's continents moved apart from each other to give birth to the Atlantic Ocean, the zone of fault-block troughs and basins extended eastward and southward to what is now the continental shelf (Figures 148, 149, and 150). The Gulf of Mexico Basin does not appear to be faulted as massively as the troughs and basins off the Atlantic coast (Figure 151). As the crustal blocks have slowly dropped during the Mesozoic and Cenozoic eras, the old eroded Piedmont surface has tilted oceanward, and rivers from the North American continent have delivered massive amounts of sediments onto the Piedmont surface and into the troughs and basins. Deposition kept pace with sinking;

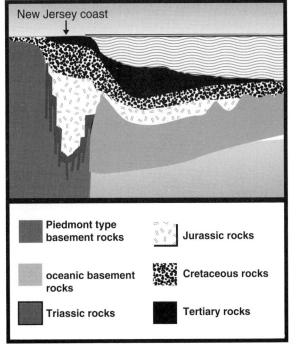

Figure 149. A cross section of the New Jersey continental margin, showing the fault-block basin.

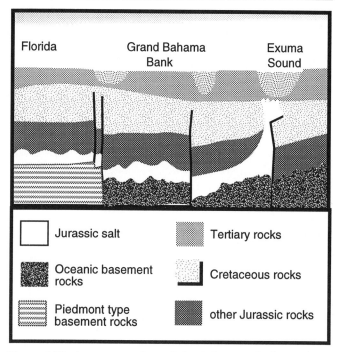

Figure 150. A cross section of the Bahama Platform, from Florida to the Bahama Islands.

as the troughs and basins have sunk, so much sediment has settled into them that the ocean in those areas has never been very deep. The sediments now underlie the shallow continental shelf. The amount of material deposited since the Triassic Period is staggering; under the continental shelf in the Gulf of Mexico Basin, the sediments are more than 6 miles (more than 10 km) thick.

The amount of sinking varied from area to area because of a complex faulting pattern. As the developing North America moved away from the Mid-Atlantic Ridge, it also rotated clockwise, and the stresses in the crust caused four sets of faults, running north-south, northeast-southwest, east-west, and northwest-southeast. Some of the faults are normal up-and-down faults, which cause down-dropped blocks, and

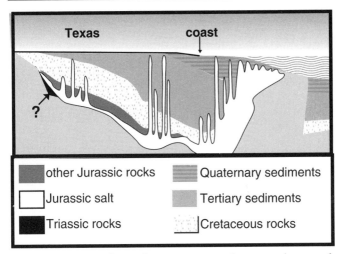

Figure 151. A north-south cross section from south-central Texas into the Gulf of Mexico Basin. The deepest part of the subsiding basin is near the present coastline. In some places, salt has been forced up toward the surface.

others are strike-slip ones that distort the major blocks by breaking them into smaller, offset chunks.

As all this faulting has happened, some areas have sunk more than others. Figure 152 shows the up-and-down, arch and embayment structure. In embayments, the basement rocks have sunk lowest and the Coastal Plain sediments have become thickest; in arches, or higher ridges of the basement, the sediments are arched upward and are thinner (Figure 153). So, while the Coastal Plain sediments generally dip toward the ocean from the plain, they also have large-scale ripples on axes at right angles to the coast.

◆The Moving Shoreline

The layers of rocks and unconsolidated sediments that underlie the Coastal Plain were not simply deposited one on top of another. The sea level rose and fell several times, so that at some times the shore was farther inland then it is today and at other times farther out. The record of these transgressions and regressions can be read in the rock layers, in the details of how the formations were deposited on one another. Figure 154 depicts the process.

When sea level rises, the shore advances inland (transgression), and the ocean bottom deposits extend farther landward than the deposits below them. When the relative sea level falls (regression), the newer land deposits will extend farther seaward than those below them. In Figure 154, the shoreline of the oldest deposit is in the middle. During a period of regression, the shoreline moved seaward, and later, during a transgression, the sea level rose, so the shore moved landward. Triassic and Jurassic rocks do not outcrop on the Coastal Plain because their landward portions were covered by a transgressive, landward-moving ocean and have not yet been uncovered by erosion. Figure 154 illustrates this process: assume that formations 1 and 2 are Jurassic and have

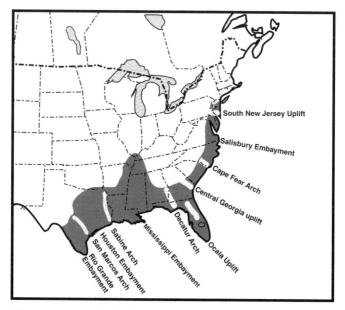

Figure 152. Coastal Plain rocks not only tilt downward toward the ocean but are also gently warped into arches and basins or troughs. The latter form embayments.

been hidden by transgressive formations 3 and 4 during the Cretaceous Period.

In the Coastal Plain's history, there have been several transgressions and regressions. Early Cretaceous deposits accumulated on land, which means the shore was seaward of where they now outcrop. But the greatest transgression of all Mesozoic and Cenozoic history occurred during Middle Cretaceous times. The sea advanced over the Great Plains from

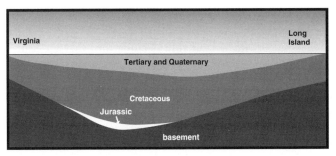

Figure 153. A northeast-southwest cross section of the Salisbury Embayment parallel to the coastline from Norfolk, Virginia, to Long Island, New York. The section indicates the arches on either side of the embayment.

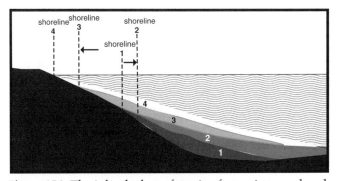

Figure 154. The inland edges of marine formations are beach deposits, and the positions of ancient beaches can indicate changes in sea level. In the figure, 1 is earliest and 4 is latest. 1 and 2 show a seaward movement of the coastline as the sea level lowered. 3 and 4 show an advance of the sea over the land as the sea level rose.

the Gulf of Mexico to the Arctic Ocean, into the present Rocky Mountain region, across the Mississippi Embayment, and across all of Florida and southern Georgia and Alabama. It also inundated much of northern South America and Europe. Then, during the last third of the Cretaceous Period, there was a regression and a second transgression.

As the Cretaceous Period gave way to the Tertiary Period, there was a time of minor back-and-forth movement, followed in the Early Tertiary by another major transgression. The sea retreated during the Middle Tertiary, and another transgression followed in the Late Tertiary.

Today, the Coastal Plain seems to be in transgression mode. The drowned river valleys along the coast are prime evidence. It has been estimated that along the Delaware coast between 7,000 and 3,000 years ago, sea level rose about 1 foot (.3 m) per century, and for the last 3,000 years it has risen about half a foot per century. Studies of Colonial historical sites give some evidence of the rising sea level. For instance, at the site of the Colonial Saugus Iron Works in Saugus, Massachusetts, the sea level is estimated to have risen about 3 feet (1 m) between the late 1600s and the 1950s.

What causes sea level to fluctuate? Geologists have theorized that as the continents around the Atlantic Ocean and elsewhere have separated, there have been variations in the plate tectonic process, in the amount of faulting, and in the amount of activity at midocean ridges (Figure 24). When the ocean basins subside more and the midocean ridges grow less, the sea level goes down as the ocean bottom goes down. When the lava flows more abundantly, to add more rock to the midocean ridges, and at the same time the basins subside less, the extra rock on the ocean bottom raises the sea floor in some areas and makes the basins smaller, and the sea level rises.

■ FEATURES OF THE COASTAL PLAIN

The Coastal Plain has significant geologic features apart from its gently tilted sedimentary rocks and sediments, features associated with both ancient and modern geologic history. Faults and dikes have resulted from movements in Earth's crust. This is the only place in America where salt domes are found, and the shoreline reflects the modern interaction of land and sea.

◆Faults

As the fault blocks under the continental shelf have subsided, the inner portions of the Coastal Plain formations
have been put under stress, and many faults have formed
more or less parallel to the shoreline. Faults cut the formations from the Cape Fear Arch (Figure 152) through Texas
(Plate 35). There are high-angle reverse faults in Virginia that
parallel the inner edge of the Coastal Plain. The Balcones and
Mexia fault zones in Texas (Figure 155) form the two most
extensive zones of faults that reach the surface. In the Balcones Zone, the blocks have moved down like stair steps
(Figure 156), forming a series of terraces named by early
Spaniards with their word for balconies. In each case, the
down-thrown side is toward the Gulf of Mexico. In the
Mexia Zone, the faults have produced down-dropped blocks
(Figure 156).

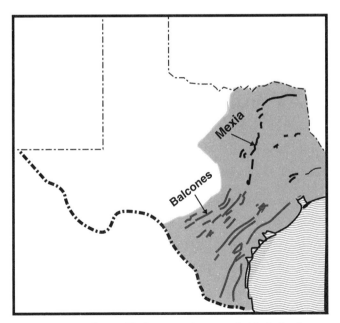

Figure 155. As the Gulf of Mexico Basin subsided, the formations cracked to form major fault zones in the Texas Coastal
Plain, including the Balcones and Mexia Zones.

Balcones

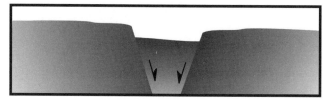

Mexia

Figure 156. Diagrammatic cross sections illustrate faulting in the Balcones and Mexia fault zones of Texas.

The faulting has gone on for a long time (Figure 157). Many old faults were covered by later sediments and don't show on the surface. There were at least nine discrete episodes of faulting in the Gulf coastal area from the Late Cretaceous to the Ice Age, and the faulting has continued up to modern times. In the Houston area, faults have occasionally revealed themselves in the dislocation of road pavement as much as 2 feet or more (.6 m) within a horizontal distance of about 200 feet (60 m).

◆Dikes

In your travels on the Coastal Plain, you may come upon dikes of basalt and other igneous rocks cutting through the masses of sedimentary material. They most commonly stretch north-south and northwest-southeast (Figure 158). The dikes show where the crust cracked as the continents separated and the troughs and basins developed, mostly during the Jurassic Period. As the cracks formed, they were filled with molten minerals from below to create the dikes.

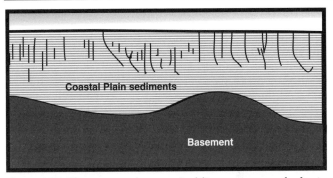

Figure 157. The more or less vertical lines represent faults in Coastal Plain sediments of Texas. Most do not reach the surface, indicating that they were covered by later sediments.

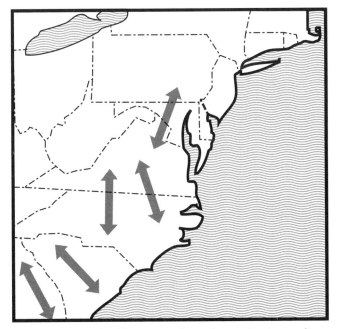

Figure 158. The double arrows show the predominant directions of dikes in the southeastern United States. The dikes represent cracks in the crust that resulted from the stress of continental separation during the Mesozoic Era.

◆Salt Domes

Salt domes are produced by underground columns of rock salt that occur on land in North America only in the Gulf coastal region, but salt columns, vertical walls, and ridges occur throughout the continental shelf all the way from off Texas to off Newfoundland (Figures 159 and 160; also see Figures 150 and 151). The columns, which are circular or elliptical in section, walls, and ridges are masses of salt that have been forced up through thousands of feet of sediments from a Jurassic deposit at the bottom of the Coastal Plain rock series.

The tips of some columns have risen close enough to the surface to produce visible (though eroded) domes in the surface rocks. A typical salt body rises 10,000 feet (3,050 m)

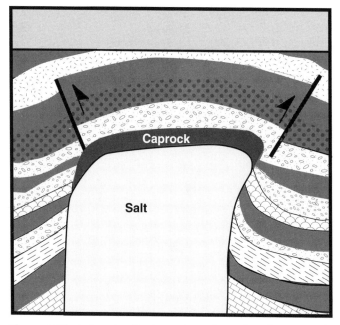

Figure 159. A longitudinal section of a salt column, with faults its growth has caused in the formations above it. If the curved formations nearest the surface are resistant enough, they will form a salt dome at the surface.

above the original salt layer, with its tip 200 feet (60 m) below the ground surface. In some cases the salt is close enough to the surface to form salt lakes and be directly mined from the surface; in others, the tops of the salt bodies are 6,000 feet (1,830 m) or more below the surface. Deep salt bodies are mined by drilling into them and pumping fresh water down into the salt. The water dissolves the salt, and the resulting brine is pumped back to the surface. A typical salt well extends 2,500 feet (760 m) below the surface.

A typical salt body is 95 percent halite, or rock salt (sodium chloride) and 5 percent anhydrite (calcium sulfate). Many of the bodies terminate in "caprock," thick masses of other minerals that have accumulated between the salt and the bedrock above (Figure 159). The caprock develops when groundwater dissolves the salt at the crest of the body, leaving the insoluble anhydrite behind. If conditions for bacterial life are right, chemical reactions promoted by bacteria convert some of the anhydrite to the mineral calcite and the gas hydrogen sulfide. Further reactions may break the gas down into its elements—hydrogen (a gas) and sulphur—thus

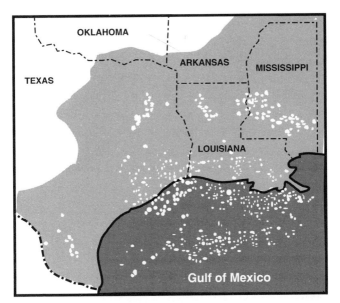

Figure 160. Salt domes and columns in the Gulf Coast region.

building a deposit of sulphur. A typical caprock has three zones; the lower one is rich in anhydrite, the middle zone has a mixture of anhydrite, calcite, sulphur, and gypsum (anhydrite with attached water molecules), and the upper zone is rich in calcite.

As the salt forces its way through the sedimentary layers, it bends their edges upward and causes some faults to develop. The bending and faulting creates places where oil accumulates, which is one reason why so much oil is produced in the Gulf coastal area (Figure 161). Many of the salt bodies have overhangs in their upper margins, caused by slight changes in the upward direction of the salt (Figure 159).

Avery Island, Louisiana, is a well-known salt dome southwest of New Iberia, an "island" of Ice Age sand and gravel in

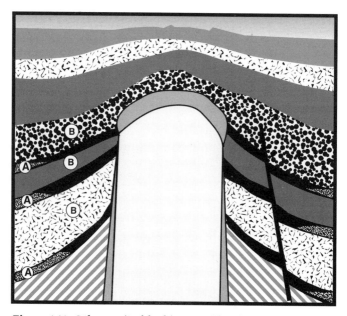

Figure 161. Oil traps (in black) created by the growth of a salt column. The traps are in three permeable sandstone layers (A) interbedded with impermeable sedimentary rocks (B). Because the layers were tilted by the column, oil has migrated up against the impermeable edge of the salt. On the right, a fault has offset the beds. Oil has migrated up to the fault and was stopped there by impermeable rocks.

sea-level marshes with a maximum elevation of 152 feet (46 m). Rock salt was discovered there, 13 feet (4 m) below the surface, in 1862. The base of the salt is estimated to be at about 35,000 feet (10,700 m) below the surface. Weeks Island and Côte Blanche Island are two other salt domes southeast of Avery Island.

The tip of the salt column at the Palestine Dome, near Palestine, Texas, is 120 feet (37 m) or more below the surface, depending upon where the measurement is made. The rock layers above and surrounding the salt mass are greatly faulted, and they dip steeply away from the dome. Their broken edges have been eroded into indistinct ridges surrounding two-thirds of the dome, although Ice Age sand and gravel cover most of it. The diameter of the salt mass at 250 feet (76 m) below sea level is about 1 mile (1.6 km). Palestine is on U.S. 84, east of I-45. Another Texas salt dome is at Grand Saline, U.S. 80, east of Dallas.

The salt deposits originated in the Jurassic Period, when offshore block faulting resulted in troughs and basins more or less parallel to the shore, with underwater ridges in between. According to one theory, some of the ridges extended up close to the ocean surface, and shallow basins formed between nearshore ridges and the shore. The ridges, rising to near the surface, kept open-ocean water from circulating freely into and out of the basins. The water in the basins stagnated, and salts in the water accumulated as the water evaporated in the arid climate of the Jurassic Coastal Plain region. After hundreds of years, thick and extensive deposits of salt formed. Eventually, conditions changed in Earth's crust as the breakup of Pangaea continued: the larger basins under the continental shelf deepened, the salt deposits tilted downward, water circulation was restored, and sand and silt covered them up.

Geologists presume that a typical upward-moving salt body evolved in the following way: As a major basin, such as the Gulf of Mexico Basin, developed, the bedrock and ocean deposits tilted downward from the shore into the basin. Rock salt (halite) flows slowly under pressure, so as the weight of deposits above the Jurassic salt increased, the salt started to flow down the slope. Some irregularities in the surface of the salt layer may have become larger bumps and ridges as the salt crept. While the deposits above the salt were accumulating, they varied in thickness and density from place to place. One can assume that the sediment load above a bump was relatively lighter than the load over a low

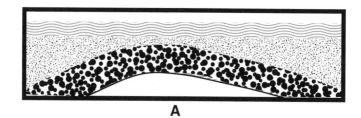

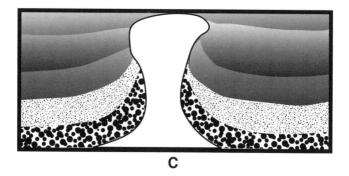

Figure 162. The development of a salt column. A, sediments surrounding an irregularity on the salt surface begin to build up and press down around the irregularity. B, salt flows upward in a column as the weight of sediments increases. C, the salt column continues to flow upward as more and more sediments are deposited around it.

area. Where the load above the salt was heavier, the weight forced the salt to flow upward into the area where the load was lighter (Figure 162). This process was greatly helped by faults that occurred before the upward flow started or while it was going on. In many cases, the salt body apparently started to grow as soon as deposits began to form over and around it, so that as the sediments thickened, it rose upward, and its top was always near the ocean bottom or land surface. In other cases, the process started after a great thickness of sediments had accumulated.

◆Beaches and Other Sandy Phenomena

The large proportion of Coastal Plain sediments that are sandstone and unconsolidated sand furnish an unlimited supply of sand to be eroded away, carried to the shore and the ocean, and moved around by the ocean waves and currents. Figure 163 shows the cross section of a beach. In the strict sense, a beach is the zone washed by waves between the lowest low tide level and the highest high tide level, including storm waves at the highest tide. Inland of the wave-washed beach is a zone of sand dunes formed by wind. The sloping part of the beach, alternately covered and uncovered by waves, is called the "swash and backwash" zone. There, the upslope-downslope motion of the water moves the sand back and forth or along the shore. Offshore bars develop beyond the swash and backwash zone as long, low ridges of sand more or less parallel to the shore within the zone of surf. Most offshore bars are underwater.

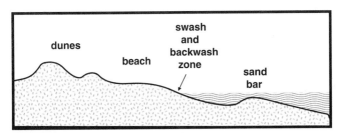

Figure 163. A cross section of a complete beach.

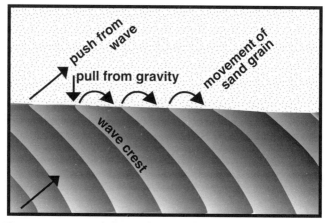

Figure 164. As seen from above, waves striking the shore at an angle move sand grains along the beach.

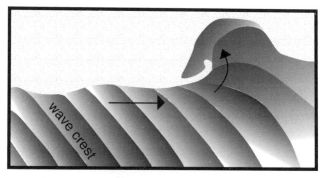

Figure 165. Waves that move sand along a beach will tend to keep the sand moving in the same direction at an indentation in the coast and will move the sand outward across the bay or inlet to form a spit.

A wave striking the beach at an angle (Figure 164) pushes sand grains diagonally along the beach, then slides back and carries the grains perpendicular to the shoreline. Each wave does the same, moving the sand grains farther along the beach. If there is an indentation in the shoreline, the wave action will tend to move the sand in the same direction in which it had been moving down the beach. The waves move the sand out into the indentation, little by little, to form a curving extension of the beach called a spit (Figure 165). Sandy Hook in New Jersey is a very large spit.

Waves and currents pile up sand to form the long, narrow barrier islands that parallel the coast from Long Island to the Rio Grande. Some of them may be former spits that have been cut off from the mainland. Others have been built up by some wave and current action that is not clearly understood.

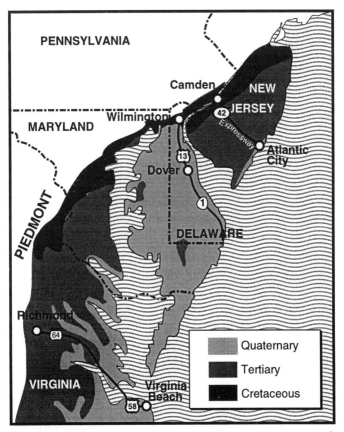

Figure 166. Deposits and highways of the northern Coastal Plain.

COASTAL PLAIN HIGHWAY SECTIONS

Camden to Atlantic City, New Jersey: Route 42, Atlantic City Expressway (Figure 166)

Between Camden and the Route 534–Blackwood exit, Cretaceous sand, sandstone, silt, and clay underlie the topsoil. The overlying Tertiary sediments, which are similar to the Cretaceous ones, start about at the Blackwood exit. At the coast, in the Pleasantville area, is a zone of Quaternary Ice Age gravel, sand, and silt, with recent beach and marsh deposits. Atlantic City occupies the north end of a typical although highly populated barrier island.

Wilmington to Dover, and the Delaware south shore: U.S. 13, Route 1 (Figure 166) *This route crosses typical Coastal Plain deposits to the coast, where seashore evolution is evident.*

Wilmington lies on Lower Cretaceous sand and silt layers. South of the city, Ice Age gravel, sand, and silt cover the Cretaceous sediments. In the vicinity of the U.S. 301 exit, streams have eroded through the Ice Age material to expose Tertiary silt, sand, and gravel between there and Blackbird. The rest of the route traverses Ice Age deposits, which cover almost all of the state. There is a spit at Cape Henlopen State Park. South of the park, along Route 1 and in the Delaware Seashore State Park, there is good evidence of how shorelines change. Waves and currents have transported enough sand to close the Rehoboth and River bays. The Army Corps of Engineers dug the present opening to the ocean in 1941, after an earlier inlet was closed during a 1940 storm. Storm waves still occasionally wash over the highway. Should they erode a new opening where one of the spitlike ridges of sand joins the mainland, they would create a barrier island.

Richmond to Virginia Beach, Virginia: I-64, U.S. 58 (Figure 166)

Richmond is built on Tertiary sand, gravel, silt, and clay, deposits that continue almost all the way to Hampton. The Quaternary sediments (Ice Age sand and silt) begin just beyond the Yorktown exits. At Virginia Beach, the deposits are recent beach material.

Rocky Mount to Manteo, North Carolina: U.S. 64 (Figure 167) *This is about the most extensive section of Coastal Plain sediments that can be traversed in North Carolina, although it includes very little Cretaceous material.*

Rocky Mount sits right at the junction of Piedmont granite

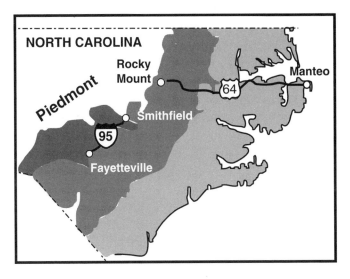

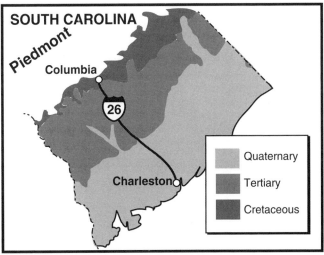

Figure 167. Coastal Plain deposits and highways of North and South Carolina.

and Coastal Plain Tertiary sandstone, limestone, sand, and silt. The Tertiary sediments continue to near Robersonville. However, in between, the Tar River at Tarboro has eroded down into Cretaceous sand and gravel, which outcrops in the valley walls and bottom. Quaternary Ice Age sands and silts extend from Robersonville to just beyond Plymouth; from there to the coast the material consists of recent river, marsh, and beach sediments. Manteo is in the barrier island zone. There is a much longer stretch of the Cretaceous sediments on I-95 between Smithfield and Fayetteville.

Columbia to Charleston, South Carolina: I-26 (Figure 167)

Columbia lies at the edge of the Piedmont, with Upper Cretaceous sand and silt to the northeast and southwest. Some Tertiary formations occur above the Cretaceous along I-26 in the southwest side of the Congaree River valley. After the river curves eastward away from the highway, Tertiary sediments alone extend to just beyond the U.S. 15 intersection. The sediments consist of sand, silt, and limestone. Quaternary Ice Age sand and silt continue to Charleston.

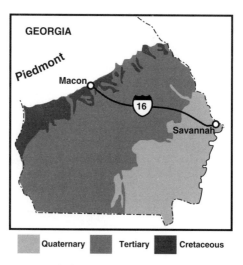

Figure 168. Coastal Plain deposits and Interstate 16 in Georgia.

Macon to Savannah, Georgia: I-16 (Figure 168)

Northeast to southwest of Macon are Cretaceous sands, gravels, clays, and silts. About 12 miles (19 km) southeast of where I-16 leaves I-75, the Tertiary sandstone, sand, silt, and clay overlap the Cretaceous sediments. The highway traverses a wide swath of the Tertiary formations, to the Denmark-Nevils area. From there to Savannah, the highway crosses Quaternary sand, silt, and gravel.

Gainesville to Tampa, Florida: I-75, I-275 (Figure 169)
This route follows the crest of the Ocala Arch and goes down its west side. It crosses successively younger Tertiary sediments and passes into Quaternary ones.

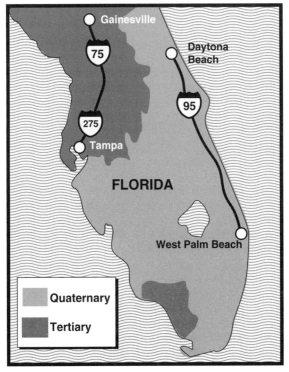

Figure 169. Coastal Plain deposits and highways in the Florida peninsula.

From Gainesville to the Bushnell exit, the rocks are mostly Early Tertiary Ocala Limestone, with large erosion remnants of overlying later Tertiary limestone from the Levy Lake area about to the Route 484 (Belleview) exit. Since the Ocala Arch dips to the south, the limestone at its crest dips underneath the younger limestone as the highway continues southward, just beyond the Bushnell exit. At that point, the highway is going diagonally down the west flank of the arch. The Tertiary rocks extend almost to the Route 52 exit, where Quaternary sandy and clayey sediments overlap the Tertiary ones. Beyond the beginning of I-275, erosion at the head of Tampa Bay has cut down into the underlying Tertiary limestone, on which the city of Tampa was built.

Daytona Beach to West Palm Beach, Florida: I-95 or U.S. 1 (Figure 169)

These coastal routes are built on a zone of Ice Age limestone and shale and recent shore and river deposits, including sand and silt from beaches, lagoons, and marshes. U.S. 1 passes close to the beaches and barrier islands along the coast. The coastline southward has become rather too civilized and crowded for serious geologizing, but beyond Miami, U.S. 1 makes available the interesting limestones of the Keys.

Montgomery to Mobile, Alabama: I-65 (Figure 170)

Cretaceous sand, silt, clay, and gravel mostly surround Montgomery. To the southwest, Cretaceous marl and chalk show up, and just beyond the exit to Routes 4 and 45, the highway enters the overlying Early Tertiary formations of sand, silt, clay, marl, and limestone. About at the U.S. 31/U.S. 84 interchange, it crosses onto a strip of later Tertiary sands, silts, clay, marl, and gravel. Those layers continue all the way to Mobile, except for the valley of the Mobile River, which is floored with recent river deposits.

Tupelo, Mississippi to Hammond, Louisiana: Natchez Trace Parkway, I-55 (Figure 171)

Since the state of Mississippi occupies the vast Mississippi Embayment of the Coastal Plain, one must drive all across the state to experience a cross section of the Coastal Plain there. Tupelo lies west of center in a wide band of Cretaceous sandstone, shale, chalk, sand, clay, and marl. Just beyond the Route 8 exit, Early Tertiary shale, clay, silt, and sand appear; the beds continue to the vicinity of Jackson, where later Tertiary gravel, sand, silt, and clay overlie them. Going south on I-55, the Quaternary deposits occur at the

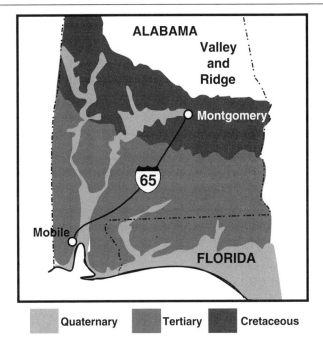

Figure 170. Coastal Plain deposits and highway I-65 in Alabama.

Louisiana state line. The sediments from there to Hammond are Ice Age gravel, sand, and clay.

U.S. 641 in Tennessee to Little Rock, Arkansas: I-40

From the U.S. 641 interchange on I-40 to the Tennessee Route 22 exit, the highway follows Cretaceous sand and silt. From Route 22 to U.S. 45 north of Jackson, Tertiary sand and silt occur, and from there to Memphis the route crosses Quaternary sand, along with wind-blown deposits (loess). At the Mississippi River, the valley is filled with Quaternary (Ice Age and recent) sand, silt, and gravel deposited by the great river in its incessant flow to the sea. From West Memphis to Little Rock, Arkansas, the highway traverses Quaternary alluvium of the Mississippi and Arkansas rivers.

For a side trip over Tertiary sand and silt in Arkansas, take U.S. 79, which samples the Tertiary between Pine Bluff and Fordyce.

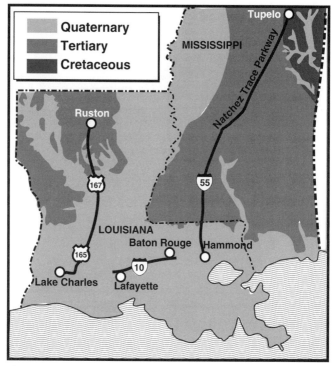

Figure 171. Coastal Plain deposits and highways in Louisiana and Mississippi.

Lake City to West Paducah, Kentucky: I-24

Lake City sits on an island of Cretaceous gravel surrounded by younger sediments. Across the Tennessee River are Quaternary sand and wind-blown dust (loess), with two small bodies of Late Cretaceous/Early Tertiary sand and silt. At the U.S. 68 intersection, the road crosses into the recent river deposits. U.S. 641 between Murray and Benton traverses Kentucky Tertiary sand and silt.

The southern tip of Illinois: I-57, I-24

The Coastal Plain's northernmost extension in the central United States consists of a mixture of Cretaceous gravel, sand, and clay with Quaternary Ice Age and recent river de-

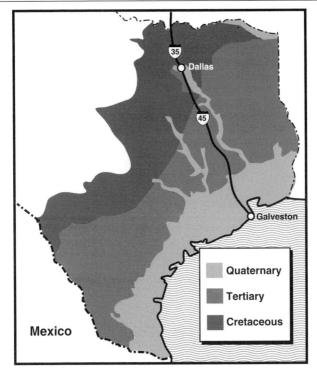

Figure 172. Coastal Plain deposits and highways in Texas.

posits. It is crossed by I-24 from the Ohio River to Ganntown and by I-57 from the Mississippi River to Dongola.

The southeast corner of Missouri: I-55

The Coastal Plain corner of the state consists almost entirely of Quaternary Ice Age and recent river deposits. Early Tertiary clay and silt occur about 10 miles (16 km) south of Cape Girardeau, crossed by I-55.

Ruston to Lake Charles, Louisiana: U.S. 167, U.S. 165, I-10 (Figure 171)

Ruston lies in the middle of a relative highland of Tertiary sediments between the Red River Valley and the bayous east of Monroe, in which there are recent river deposits. The sediments of the higher land are Early Tertiary gravel, sand, silt, and clay with some limestone, shale, and marl. In the Winn-field area, Ice Age sand, silt, clay, and gravel occupy the local river valleys. Just before Williana, the road enters the main body of the Quaternary zone, which extends to beyond Lake Charles. South of Lake Charles, along the coast, marshlands have recent stream and shore deposits.

Lafayette to Baton Rouge, Louisiana: I-10 (Figure 171)

This trip gives an idea of the scope of the Mississippi Valley. Lafayette and Baton Rouge lie on the uplands on either side of the valley, underlain by Ice Age sand, gravel, and clay. The western valley wall crosses the highway about 5 miles (8 km) east of the U.S. 167 intersection with I-10 north of Lafayette. The eastern valley wall is just beyond the river, at Baton Rouge. In between, the valley contains a bit more than 50 miles (80 km) of recent river deposits. Figure 146 shows how the river has moved from one side of the valley to the other since the end of the Ice Age. From I-10 to the ocean, most of the Mississippi River sediments belong to its delta.

Ardmore, Oklahoma to Galveston, Texas: I-35, I-45 (Figure 172)

This highway section starts in Oklahoma, near the Texas border, where the Coastal Plain (Cretaceous) rock layers sit-ting upon the Paleozoic rocks of the Central Lowland (in this case, Pennsylvanian) can be seen. Otherwise, the route crosses a typical section of Coastal Plain formations.

The escarpment formed by the Cretaceous formations crosses the highway about halfway between Ardmore and Marietta, Oklahoma. The rocks are clay, marl, and limestone in the Marietta area. Similar rocks continue down to Denton; beyond, Cretaceous shale and sandstone extend from Denton to Dallas. Chalk and shale outcrop in the Dallas area (Plate 35). From Dallas to about 5 miles (8 km) south of Corsicana, the bedrock is Cretaceous shale, marl, and limestone.

Between Corsicana and Streetman, the highway passes over three major faults of the Mexia fault system, as it enters Early Tertiary sand, sandstone, silt, clay, and shale. The con-tact with Middle and Late Tertiary sand-sandstone and silt-shale formations is about halfway between Madisonville and Huntsville; they continue to the vicinity of Conroe. From there to the shore beyond Houston, Quaternary Ice Age sand and clay cover the country. Along the shore are recent river and beach deposits of sand, gravel, and silt, including a bar-rier island, Galveston Island.

APPENDICES
REFERENCES
INDEX

APPENDIX A:
THE ORIGINS OF NAMES
IN THE GEOLOGIC TIME CHART

The geologic time chart (see front endpapers) started with the observations of Abraham Gottlob Werner (1750–1817) in his native Saxony. In studying local minerals, he became well aquainted with the formations in his area and realized that he could separate the rocks into three major categories. First came a basement series of crystalline rocks: granite, schist, gneiss, and other metamorphics, some of which he could see had been folded. He called them the Primitive Series (the *Urgebirge*), which later became the Primary Series. On top of the Primitive Series, Werner recognized layers of more or less horizontal sedimentary rocks, which he named the Flat-lying Series (the *Flötzgebirge*), later called the Secondary Series. Surmounting the Flat-lying Series were unconsolidated sands and gravels that Werner called Alluvium (*Angeschwemmtgebirge*). Our North American equivalent of Werner's series would be the edge of the Canadian Shield, an area such as that illustrated in Figure 52, where folded, metamorphosed, and intruded Precambrian rocks underlie nearly horizontal Cambrian and Ordovician sedimentary rocks, with unconsolidated glacial till and outwash on top.

As time went by, early geologists identified a thick body of sedimentary rocks in Europe between the Primary and Secondary series. Werner called those rocks the Transition Series (*Übergangsgebirge*). Later, other geologists recognized sedimentary rocks that belonged between the Secondary Series and the Alluvium. As the rocks were studied more and more carefully by more and more people, the list of geological time categories grew and was subdivided. Students agreed on eras and periods, and as they gathered more information, they found they could subdivide periods into epochs. Further research has enabled geologists to subdivide epochs into ages. Thus, an earliest Miocene rock or fossil belongs to the Aquitanian Age (from the historic Aquitaine region of southwest France) of the Miocene Epoch of the Tertiary Period of the Cenozoic Era of the Phanerozoic Eon. The work continues today, as geologists seek to know more about what has happened during the evolution of Earth's crust.

The geologic time chart comprises a miscellaneous assortment of names. The names reflect the variety of individuals involved in creating the time chart and the time span over which they worked.

The Eons: The name of the *Archean* Eon comes from the Greek for "ancient." It was proposed in 1872 and referred to the oldest known rocks at that time. *Proterozoic* (1888) is from the Greek for "before animals." At the time it was first used, Precambrian fossils had not yet been discovered; now we know there is a long history of Proterozoic organisms. *Phanerozoic* (1896) is from the Greek for "evident animals." That eon encompasses the rocks that contain a good fossil record, starting in the Cambrian Period.

The Eras: *Paleozoic* comes from the Greek for "ancient animals," *Mesozoic* means "middle animals," and *Cenozoic* is a corruption of the original name, Cainozoic, meaning "recent animals."

The Periods: *Cambrian* comes from Cambria, the Roman name for Wales. *Ordovician* reflects the name of an ancient British tribe, the Ordovices. *Silurian* comes from the name of an ancient Welsh tribe, the Silures. *Devonian* is from Devon, England. *Carboniferous* refers to the coal deposits of the period; coal consists largely of the element carbon. The *Permian* Period is named after the city of Perm, Russia, which lies in a large basin that contains rocks of that age on the west flank of the Ural Mountains. *Triassic* refers to the fact that European rocks of the period can be conveniently divided into three subdivisions. *Jurassic* is from the Jura Alps of southeastern France. *Cretaceous* is derived from the Latin word for "chalky"; the famous chalk deposits of southern England are from this period. *Tertiary* refers to an early third major subdivision of rocks (after Werner's Primary and Secondary series). *Quaternary* is an early fourth subdivision.

The Epochs: The *Mississippian* is named after the central part of the Mississippi River valley, where rocks of that age are prominent. *Pennsylvanian* memorializes the rocks of that age that abound in Pennsylvania. *Paleocene* is from the Greek for "ancient new"; the Paleocene Epoch is the oldest part of the newest era, or the earliest epoch of modern animals. *Eocene* is from the Greek for "dawn new." *Oligocene* comes from the Greek for "few new"; in Europe, where the name was coined, few fossils of that age weather out, although there are many Oligocene fossils in other regions. *Miocene* and *Pliocene* mean, respectively, "less new" and "more new" in Greek, presumably referring to the fossils becoming more modern in the latter epoch. Pleistocene and Holocene are also from Greek, meaning "most new" and "whole new," respectively.

Precambrian Orogenies: *Uivakian* refers to the Uivak gneisses, which were formed during the orogeny; Uivak is a local geographic name from the coast of Labrador. *Laurentian* was used originally for Precambrian granites in the Laurentian Highlands of eastern Canada. *Kenoran* refers to the last widespread folding, metamorphism, and intrusion in the Superior Province of the Canadian Shield; the name comes from rocks near the town of Kenora, in southwestern Ontario, at the north end of the Lake of the Woods. *Hudsonian* stands for rocks in the Hudson Bay region that were affected by the last important folding event there. The *Grenvillian* Orogeny was named for rocks along the Ottawa River in the vicinity of Grenville, Québec.

APPENDIX B: RADIOACTIVE DATING

Rocks that contain certain minerals, made up of certain elements, can reveal their approximate ages. The secret is in the fact that some elemental atoms are radioactive. The molecules that make up everything we see are composed of atoms. For instance, two atoms of the element hydrogen and one atom of the element oxygen constitute a molecule of water; a molecule of sugar is actually 12 carbon atoms, 22 hydrogen atoms, and 11 oxygen atoms bonded together.

Every atom contains a variety of particles, the description of whose discoveries, properties, and activities would fill a book. Students of geologic history care mostly about the three particles called protons, neutrons, and electrons. Each proton has a positive charge; some electrons have negative charges, others have positive charges. Protons have significant mass (for an atomic particle, which is unbelievably small) while electrons have essentially no mass. Mass is the measure of the amount of matter in something; the force of gravity attracts an object according to its mass to give it what we call weight. Neutrons also have significant mass but, as their name implies, they have no charge.

A specific number of protons characterizes the atoms of each element and gives the element its basic properties; that number is the element's atomic number. The number of protons in elemental atoms ranges from 1 to 105; the atomic number for hydrogen is 1, that for carbon is 6, that for iron is 26, and that for gold is 79.

Scientists long ago assigned a mass number to the atoms of each element, which represents the total number of protons plus neutrons in every typical atom of that element. The mass numbers for the elements listed above are: hydrogen, 1; carbon, 12, iron, 56; and gold, 197. Thus a typical carbon atom has six protons and six neutrons, while a gold atom has 79 protons and 118 neutrons.

While each atom of each element has a certain number of

protons, atoms can vary in the numbers of neutrons. The different varieties are called isotopes. For example, hydrogen has three isotopes. The basic atom has one proton and no neutrons, but some hydrogen atoms have one proton and one neutron, and others have one proton and two neutrons. Scientists designate the three isotopes as hydrogen-1, hydrogen-2, and hydrogen-3. The abbreviated versions of each are ^{1}H, ^{2}H, and ^{3}H. Since neutrons have significant mass, hydrogen-3 weighs more than hydrogen-2, which weighs more than hydrogen-1. The term "heavy water" refers to water with heavier isotopes of hydrogen in its molecules. Since the heavier isotopes 2 and 3 are rare, the mass number (1) reflects the common variety of hydrogen. On the Periodic Table of the Elements, the assigned relative atomic weight of hydrogen is 1.00797. The atomic weight is an average of individual isotope weights, and the mass number is the atomic weight, rounded off.

Some elements have isotopes that are radioactive, in that their nuclei actively change by emitting alpha or beta particles or capturing their own electrons. When these changes happen, the elements become different elements. An alpha particle consists of two protons and two neutrons. When a nucleus emits an alpha particle, it loses two protons, and therefore turns into another element. A beta particle is an electron emitted from the nucleus, which originates when a neutron changes into a proton plus an electron. The electron leaves the nucleus and the proton stays behind. The new atom therefore has one more proton than its precursor and is a different element. In the so-called electron capture transitions, a proton in the nucleus absorbs one of the electrons from outside the nucleus and therefore changes to a neutron. So there is one less proton in the resulting atom, which makes it a different element.

Three of the better-known radioactive atom types are isotopes of the elements potassium, rubidium, and uranium. Uranium offers two radioactive isotopes, ^{235}U and ^{238}U, which today exist in the proportion of 1 ^{235}U to 137.8 ^{238}U, so the heavier atom is much more common. Each of the isotopes changes through a series of intermediate steps to a stable isotope of lead. Uranium-235 becomes lead-207, and uranium-238 becomes lead-206. The uranium decay involves several steps in which alpha and beta particles leave atoms. For instance, uranium-238 emits an alpha particle; that is, it loses two protons and two neutrons. In losing two protons, it automatically becomes another element, thorium, as its atomic number changes from 92 to 90. The thorium atom then emits a beta particle. Since the formation of the beta particle involves the changing of a neutron into a proton, the atomic number increases to 91, which identifies the element protactinium. And so it goes, until the stable isotope of lead evolves.

The radioactive atoms of the element rubidium emit beta particles. Thus each adds a proton to change from an atomic number of 37 to 38, which is the element strontium. The radioactive isotope of potassium changes by electron capture, so that a proton becomes a neutron, thus reducing the atomic number from 19 to 18 and forming an argon atom.

Scientists have been able to measure the rates of radioactive decay, and they measure them in half-lives. A half-life is the time required for half of a given mass of an element to change into a stable form. Starting out with a given sample, after one half-life there will be one-half of the original left; after two half-lives there will be one-quarter; after three half-lives there will be one-eighth, and so forth. The radioactive isotopes of uranium decay at greatly different rates; uranium-235 has a half-life of 0.71 billion years, and uranium-238 has a half-life of 4.47 billion years.

Radioactive uranium is found in the mineral uraninite, or uranium oxide (UO_2), which occurs in some igneous rocks. Nowadays, geologists have found the mineral zircon to be very useful in helping them to determine ages by the uranium method. Zircon is zirconium silicate ($ZrSiO_4$), which forms in igneous and metamorphic rocks, and whose crystals, through weathering and erosion, can become included in sedimentary rocks. Zircon crystals contain uranium atoms that enter the crystal structure while the mineral is forming. Importantly for the purpose of radioactive dating, lead atoms apparently cannot enter the zircon crystal structure. Therefore, any lead found in a zircon crystal must have come from the decay of uranium.

Generally speaking, once geologists know the rate of radioactive decay, they can compare the amount of newer lead with the amount of older uranium and determine approximately how many years ago a radioactive mineral (uraninite) or a zircon crystal formed. Where two radioactive isotopes of an element exist, as in uranium, they serve to double-check the date. Since the uranium isotopes decay at very different rates, their proportions in a sample will change over time. A billion years ago, the proportion of uranium-235 must have been much greater than it is today, because it decays so much faster. Therefore, the relation between the proportions of the two uranium isotopes to their lead end-products in a zircon crystal indicates the age of the crystal.

The isotope rubidium-87, which occurs in biotite and muscovite micas, some feldspars, and some clay minerals, decays to strontium-87, with a half-life of 48.8 billion years. Strontium has two isotopes, -86 and -87. The age of a rock sample is calculated from the ratios of strontium-87 and rubidium-87 to strontium-86. The isotope potassium-40 occurs in micas, some feldspars, amphiboles, and clay minerals. It decays to argon-40 with a half-life of 1.31 billion years. The amounts of the two isotopes in a sample indicate the age of the rock.

Geologists use a sophisticated instrument called a mass spectrometer to measure the masses and abundances of isotopes from a sample. The instrument separates the isotopes by their differing masses and sends signals to a plotter. The plotter then graphs the masses of the isotopes and their proportions in the original mixture.

REFERENCES

Many books and maps offer more information on various areas. What follows is a sampling to give you an idea of the books, pamphlets, and maps that are available in bookstores and libraries. Professional field trip guides to specific areas can be found in college and university geology libraries.

Maps

The American Association of Petroleum Geologists (AAPG) publishes a series of 12 geological highway maps that cover the entire United States. Each map shows the surface geology of one or more states along with many highways. The maps showing the area covered in this book are those of the northern great plains region, the mid-continent region, Texas, the Great Lakes region, the northeastern region, the mid-Atlantic region, and the southeastern region. Each map is accompanied by plenty of useful geologic information about the region it covers. Write, call, or fax:

AAPG Bookstore
P.O. Box 979
Tulsa, OK 74101-0979
1-800-364-2274
Fax (918) 560-2652

The maps described above show surface geology, which is appropriate because you drive over the surface. Thus, in glaciated regions, the maps show the glacial deposits along with whatever bedrock is exposed; bedrock covered by glacial material is not shown. If you wish a more complete picture, you might want to supplement the AAPG map with a bedrock geology map. State geological surveys provide bedrock maps. Addresses and telephone numbers of state geological surveys may be found in your library in the publication *State Administrative Officials Classified by Function*, published by the Council of State Governments.

Canadian geological highway maps are available from the following sources:

Manitoba
Manitoba Energy and Mines
Marketing Publications
Unit 360
1395 Ellice Avenue
Winnipeg, Manitoba R3G 3P2
(204) 945-4154
Fax (204) 945-0586
Available in English and French editions.

New Brunswick
New Brunswick Department of Forests, Mines, and Energy;
    Mineral Resources Division
P.O. Box 6000
Fredericton, New Brunswick E3B 5H1
(506) 453-2206
Fax (506) 453-3671
Available in English and French editions.

Newfoundland
Geological Association of Canada
Publications Department, Net 95
c/o Department of Earth Sciences
Memorial University of Newfoundland
St. John's, Newfoundland A1B 3X5
(709) 737-7660
Fax (709) 737-2532
Map and guidebook together available in English and French editions.

Nova Scotia
Nova Scotia Department of Natural Resources Library
P.O. Box 698
Halifax, Nova Scotia B3J 2T9
(902) 424-8633
Fax (902) 424-7735

Ontario

A bilingual map for northern Ontario (Map 2506) is available from:

Ontario Geological Survey
Publication Sales
933 Ramsey Lake Road
Level B2
Sudbury, Ontario P3E 6B5
(705) 670-5691
Fax (705) 670-5770

The accompanying map for southern Ontario (Map 2441) is out of print but can be found in libraries. The reference is: Freeman, E.B., 1979: Geological Highway Map, southern Ontario, Ontario Geological Survey Color Map 2441.

Québec

Ministère des Ressources Naturelles
311 Rue Marquette
Sherbrooke, Québec 1M2 J1H
(819) 820-3122
Fax (819) 820-3948

There are three maps available, in French; one of the north shore of the Gulf of St. Lawrence, one of southeastern Québec to the Gaspé Peninsula, and one of the Eastern Townships region.

Books

The following books can be found in bookstores and libraries or ordered from the publishers:

General

Bolt, Bruce A. 1993. *Earthquakes.* New York: W. H. Freeman and Co.

Fortey, Richard. 1991. *Fossils: A Key to the Past.* Cambridge, Massachusetts: Harvard University Press.

Gurnee, Howard N. and Russell H. 1966. *Visiting American Caves.* New York: Crown Publishers.

Hamburg, Michael, and Jurg Alean. 1992. *Glaciers.* Cambridge, UK: Cambridge University Press.

Heppenheimer, T. A. 1990. *The Coming Quake.* New York: Paragon House.

Imbrie, John and Katherine P. 1986. *Ice Ages.* Cambridge, Massachusetts: Harvard University Press.

Lambert, David, and the Diagram Group. 1988. *The Field Guide to Geology.* New York: Facts On File Publications.

———. 1985. *The Field Guide to Prehistoric Life.* New York: Facts On File Publications.

Manley, Seon and Robert. 1968. *Beaches, Their Lives, Legends, and Lore.* Philadelphia: Chilton Book Company.

Rhodes, Frank H. T., Herbert S. Zim, and Paul R. Shaffer. 1962. *Fossils, a Golden Guide.* New York: Golden Press.

Silverman, Sharon Hernes. 1991. *Going Underground: Your Guide to Caves in the Mid-Atlantic States.* Philadelphia: Camino Books.

Thomson, Betty F. 1977. *The Changing Face of New England.* Boston: Houghton Mifflin Co.

Weisel, Dorian. 1994. *Fire on the Mountain: The Nature of Volcanoes.* San Francisco: Chronicle Books.

Iowa

Anderson, Wayne I. 1983. *Geology of Iowa.* Ames: Iowa State University Press.

Troeger, Jack C. 1983. *From Rift to Drift: Iowa's Story in Stone.* Ames: Iowa State University Press.

Kansas

Buchanan, Rex C., and James R. McCauley. 1987. *Roadside Kansas.* Lawrence: University of Kansas Press.

Massachusetts

Little, Richard D. 1986. *Dinosaurs, Dunes, and Drifting Continents.* Greenfield, Massachusetts: Valley Geology Publications.

Maine

Kendall, David L. 1987. *Glaciers & Granite: A Guide to Maine's Landscape and Geology.* Camden, Maine: Down East Books.

Michigan

Dorr, John A. Jr., and Donald F. Eschman. 1970. *Geology of Michigan.* Ann Arbor: University of Michigan Press.

Feldman, Rodney M., Alan H. Coogan, and Richard A. Heimlich. 1977. *Southern Great Lakes, Field Guide.* Dubuque: Kendall/Hunt Publishing Company.

Paull, Rachel K. and Richard L. 1977. *Geology of Wisconsin and Upper Michigan.* Dubuque: Kendall/Hunt Publishing Co.

Minnesota

Ojakangas, Richard W., and Charles L. Matsch. 1982. *Minnesota's Geology.* Minneapolis: University of Minnesota Press.

Schwartz, George M., and George A. Thiel. 1954. *Minnesota's Rocks and Waters.* Minneapolis: University of Minnesota Press.

Missouri

Beveridge, Thomas R. 1978. *Geologic Wonders and Curiosities of Missouri.* Rolla, Missouri: Division of Geology and Land Survey Educational Series No. 4.

Unklesbay, A. G., and Jerry D. Vineyard. 1992. *Missouri Geology.* Columbia: University of Missouri Press.

New Hampshire

Van Diver, Bradford B. 1987. *Roadside Geology of Vermont and New Hampshire.* Missoula, Montana: Mountain Press Publishing Co.

New Jersey

Wolfe, Peter E. 1977. *The Geology and Landscapes of New Jersey.* New York: Crane Russak and Co.

New York

Isachsen, Yngvar W. 1980. *Continental Collisions and Ancient Volcanoes* (pamphlet). Albany: New York State Geological Survey Educational Leaflet 24.

Isachsen, Y. W., et al, editors. 1991. *Geology of New York: A Simplified Account.* Albany: New York State Museum Education Leaflet No. 28.

Jaffe, Elizabeth B. and Howard W. 1986. *Geology of the Adirondack High Peaks.* Glens Falls, New York: The Adirondack Mountain Club.

Titus, Robert. 1993. *The Catskills: A Geological Guide.* Fleischmanns, New York: Purple Mountain Press.

Van Diver, Bradford B. 1985. *Roadside Geology of New York.* Missoula, Montana: Mountain Press Publishing Co.

Ohio

Feldman, Rodney M., Alan H. Coogan, and Richard A. Heimlich. 1977. *Southern Great Lakes, Field Guide.* Dubuque: Kendall/Hunt Publishing Company.

Ontario

The Ontario Geological Survey has published a series of geological guidebooks to the geology and scenery, fossils, and minerals of certain areas. For information, contact the office listed in the section on maps above. Two are out of print but can be found in libraries.

ROCK ONtario. The Ontario Geological Survey (English and French editions). 1994.

Pennsylvania

Van Diver, Bradford B. 1990. *Roadside Geology of Pennsylvania.* Missoula, Montana: Mountain Press Publishing Co.

Texas

Spearing, Darwin. 1991. *Roadside Geology of Texas.* Missoula, Montana: Mountain Press Publishing Co.

Virginia

Frye, Keith. 1986. *Roadside Geology of Virginia.* Missoula, Montana: Mountain Press Publishing Co.

Vermont

Van Diver, Bradford B. 1987. *Roadside Geology of Vermont and New Hampshire.* Missoula, Montana: Mountain Press Publishing Co.

Wisconsin

Paull, Rachel K. and Richard L. 1977. *Geology of Wisconsin and Upper Michigan.* Dubuque, Iowa: Kendall/Hunt Publishing Co.

Schultz, Gwen M. 1986. *Wisconsin's Foundations.* Dubuque, Iowa: Kendall/Hunt Publishing Co.

INDEX

Acadian Orogeny, 53, 121, 125, 127, 129, 135, 136, 163, 174, 180, 182, 225, 231, 234, 240, 242, 243, 252, 257, 280, 281, 284, 285, 287–99, 293, 297, 300, 302, 305, 306, 307, 310, 311, 314, 315, 318, 320

Acadian Zone, 280–81, 284–90, 297, 299–01, 303–4, 305, 307, 310–13, 318–19, 320–22, 323–24, 326–27

actinolite, 31

Adirondack Mountains, 100, 101–2

Alabama
geology, 140–43, 240, 264, 333, 336, 341, 364
highways, 264, 364

Alabaster Cavern, Oklahoma, 199

Alabaster State Park, Oklahoma, 199

Algonquin Arch, 146, 147–49, 173, 178–79

Alleghenian Orogeny, 53, 121–22, 125, 127, 129, 134, 141, 143, 163, 173, 181, 182, 183, 199, 200, 211–12, 226–27, 231, 234, 239, 240, 241–43, 252, 256, 266–67, 271, 273, 288, 289, 293, 305, 307, 312,

Allegheny Front, 126

Allegheny Plateau, 125–40

Alley Spring, Missouri, 191

alpha particle, 376

Amnicon Falls State Park, Wisconsin, 87

amphibole, 8, 10 (defined), 11, 12, 29, 31

amphibolite, 11, 12, 31 (defined), **Pl. 44**

Anadarko Basin, 199

andesite, 8 (defined), 14, **Pl. 28**

anhydrite, 25 (defined), 211, 216, 312, 320, 353–54

anorthosite, 11–12 (defined), 64, 85, 101, 102, 103, 107, 109, 243, 256, 272, 309, **Pl. 7**

anticline, 33–34 (defined)

anticlinorium, 33–34 (defined)

Anticosti Basin, 146, 147, 150–52

Appalachian Mountains, 233–40, 253–254, 257–258, 259, 261, 262, 264, **Pl. 17**

Appalachian Orogeny, 54, 220

Appalachian Plateaus, 123, 124–44

Appalachian Province structure, 220–21
  rock types, 221–22
  history, 219–20, 222–29
Arbuckle Mountains, 270–71, 274–76
Arch, Algonquin, 146, 147–49, 173, 178–79
  Cape Fear, 349
  Cincinnati, 114, 118, 161–71
  Ocala, 334, 364
  Transcontinental, 113, 203, 214
argillite, 25–26 (defined)
Arkansas
  geology, 193, 196–97, 266–69, 336, 365
  highways, 196–97, 268–69, 365
Arkansas Novaculite, 265
arkose, 18 (defined)
asbestos, 309, 317–18
ash, volcanic, 15
Atlanta, Georgia, 263
Atlantic City, New Jersey, 360
Atlantic Provinces and southern Québec, 306–27
atom, 375
Augusta, Georgia, 263, 334
Ausable Chasm, New York, 160
Avalonia, 225–26, 246–47, 281, 284, 288, 289–93, 297, 301, 303, 305–307, 314, 324, **Pls.** 28, 29, 48
Avalonian Zone, 281, 209–95, 297, 300, 301, 304–5, 307, 313–16, 320, 323, 324–26
Avery Island, Louisiana, 354–55

Baie Comeau Segment, 98
Balcones Escarpment, Texas, 339
Balcones Fault Zone, Texas, 349
Bald Eagle Mountain, Pennsylvania, 233, 235, 236
Baltimore, Maryland, 242, 257
Bangor, Maine, 306
barite, 183, 191, 192
Barre Quarries, Vermont, 287
barrier islands, 332, 335, 359, 368
basalt, 8 (defined), 14, 31, **Pls.** 8, 24, 28
basement complex, 57, 60
basin, Anadarko, 199
  Anticosti, 146, 147, 150–52
  Black Warrior, 264
  Boston, 297
  Cahoba, 264
  Coosa, 264
  Dunkard, 113, 118, 119
  Gulf of Mexico, 343, 355
  Illinois, 114, 118, 180–88
  Lake Erie, 132
  Lake Superior, 81–82, 86, 92
  Michigan, 70, 171–80
  Mississippi, 335, 336–339
  Narragansett, 293, 297
  Norfolk, 293
  Québec, 146–47, 149–50
  Sudbury, 64
  Triassic (see Triassic Basin)
batholith, 13 (defined), 68, 76, 91, 98, 271, 287–88, 289, 304, 310, 314, 315, 321, 322, 326
Baton Rouge, Louisiana, 368
Bay of Fundy Triassic Basin,

308, 316–17, 322, 323
beach, 357
bedding plane, 16, 21, 26, 37
Bedford Limestone, 22
Bennett Spring, Missouri, 195
beta particle, 376, 377
Big Spring, Missouri, 191
biotite, 8, 9 (defined), 11, 12, 27, **Pl. 41**
biotite schist, 27
Birmingham, Alabama, 264
Black Warrior Coal Basin, Alabama, 264
Blow-Me-Down Mountain, Newfoundland, 309
Blue Mountain, Pennsylvania, 232, 251–52, 253–54
Blue Ridge Mountains, 240–42, 257, 258, 259, 260–61, 262
Bordas Escarpment, Texas, 339
Boston Basin, 297
Boston, Massachusetts, 291, 297, 301
Boston Mountains, 193, 196
boulder, 16–17 (defined)
Boulevard Lake Park, Ontario, 92
Boundary Waters Canoe Area, Minnesota, 75
Break of the Plains, 200
breccia, 18 (defined), **Pl. 43**
Bringhurst Woods Park, Delaware, 256–57
Bristol Caverns, Tennessee, 232
Buhrstone Escarpment, Alabama, 336
Bull Run Mountains, 257, 258
Burlington, Vermont, 302

Cahoba Coal Basin, Alabama, 264
calcite, 16 (defined), 18, 22, 24, 28, 64, 353–54
Cameron Cave, Missouri, 203
Campbellsport Drumlins, Wisconsin, 205
Canadian Shield, rock ages and history, 57, 60–62
  rock types, 57–60
  topography, 58
Cape Cod, 331, 340
Cape Fear Arch, 349
Cape Henlopen State Park, Delaware, 360
carbon dioxide, 35
carbonic acid, 37
Cascade River State Park, Minnesota, 86
Catamaran Brook Provincial Park, Newfoundland, 326
Catoctin Mountain, Maryland, 257
Catskill Mountains, 115, 135–36
caves, formation of, 35, 37
  Florida, 334
  Indiana, 183, 188
  Iowa, 204
  Kentucky, 163
  Minnesota, 207–8
  Missouri, 191, 192, 195, 203
  New York, 132
  North Carolina, 242
  Ohio, 165
  Oklahoma, 199
  Pennsylvania, 240
  Tennessee, 165, 232, 242
  Virginia, 232
  West Virginia, 132
  Wisconsin, 208

Cave of the Mounds, Wisconsin, 208
cement, mineral, 16, 18, 22
Central Appalachians, 229–64
Central Gneiss Belt, 99, 104, 105, 107
Central Granulite Terrane, 101–2, 107, 109
Central Lowland, 115–16, 123, 124, 144–217
Central Metasedimentary Belt, 99–100, 102, 104, 105
Central Park, New York, 285
chalcopyrite, 321
chalk, 23 (defined), 24, 340, **Pl. 35**
channel, stream/river, 18, 19, **Pls. 15, 34**
Charlotte, North Carolina, 245, 261
Charlottesville, Virginia, 259
Chattanooga, Tennessee, 263
chemical limestone, 23 (defined)
chert, 24 (defined), **Pls. 6, 13, 22, 45**
chert nodules, 24, 191, **Pl. 13**
Chester Escarpment, 183
Chicago, Illinois, 186
Chippewa Falls Park, Ontario, 70
Chippewa Moraine, Wisconsin, 205
chlorite, 26, 31
chlorite schist, 28
Chocorua, Mt., New Hampshire, 304
Chutes Provincial Park, Ontario, 95
Cincinnati Arch, 114, 118, 161–71

Cincinnati, Ohio, 118, 167, 168
cirque, 40 (defined), 103, 228–29, 280, 304, 308, **Pl. 40**
clastic limestone, 22 (defined)
clay, 17, 20, 21
clay minerals, 21, 27
Clearwater Lake Provincial Park, Manitoba, 216
cleavage, 9 (defined)
  in argillite, 27, **Pl. 27**
  in minerals, 9, 10, 11, 22, 28
  in slate, 26
Cleveland, Ohio, 137, 168
coal, 127, 130, 134, 142–43, 165, 173, 184, 198, 200, 212, 252, 313, 324
Coastal Plain structure, 329–30
  history, 329, 341–48
  rocks, 339–41
  topography, 330–31
cobalt, 96, 191
cobble, 16, 17 (defined), 18
Cobequid Mountains, Nova Scotia, 315, 323
Columbia, South Carolina, 242, 261, 334, 362
Columbus, Georgia, 334
Columbus, Ohio, 167
Concord, New Hampshire, 301
conglomerate, 18 (defined), **Pls. 28, 32, 42**
Connecticut
  geology, 284–85, 289–90, 297, 299
  highways, 297, 299
Connecticut Valley Triassic Basin, 281, 295–96, 297, 299–300
continental drift, 49–54

continental glacier, 40 (defined)

convection, 51

Coosa Coal Basin, Alabama, 264

copper, 63, 64, 68, 70, 72, 87, 92, 94, 96, 208, 309, 310, **Pl. 45**

coral reefs, Devonian, 211, 217

   Pleistocene, 335

   Silurian, 175–76, 188, 204, **Pl. 16**

Cormorant Provincial Forest, Manitoba, 216

Corner Brook, Newfoundland, 327

cross bedding, 19–20, **Pls.** 12, 34

Cross Plains, Wisconsin, 205

crust, Earth's, 11, 12–14, 25, 33, 35, 46–54

   oceanic, 11, 46–54

Crystal Cave, Ohio, 165

Crystal Cave, Wisconsin, 208

Cumberland Overthrust Block, 141, 142, 143

Cumberland Plateau, 140–44

current cross bedding, 19

current ripple marks, 19, **Pl. 46**

Dallas, Texas, 368, **Pl. 35**

Delaware

   geology, 256–57, 332, 360

   highways, 256–57, 360

Delaware Seashore State Park, 360

delta, 19 (defined), 337, 338–39

delta cross bedding, 19, **Pls.** 12, 34

Devil's Lake State Park, Wisconsin, 205

diabase, 12 (defined), 31

dike, 13 (defined), 57, 61, 62, 64, 70, 71, 75, 82, 86, 90, 91, 92, 93, 95, 104, 107, 108, 110, 190, 232, 295–6

Dinosaur State Park, Connecticut, 296, 299

dinosaurs, 123, 248, 296, **Pl. 30**

diorite, 8, 10 (defined), 28

dolomite, mineral, 22, 24, 28

   rock , 22–4 (defined), 28, 35, **Pl. 14**

dolostone, 24

Dome, 113, 117, 119

   Lexington, 162

   Nashville, 162

   Ozark, 188–98

Driftless Area, 184, 206–7, 214–15

Dripping Springs Escarpment, 164

drumlin, 42–43 (defined), 131, 155, 205, 214, 228, 280, 308

Dubuque, Iowa, 214

Duck Cove Park, New Brunswick, 320

Duluth, Minnesota, 85, 86

Dunbar Cave, Tennessee, 165

Dunkard Basin, 113, 118, 119, 124, 127, 130, 140

Durham, North Carolina, 260

earthquake, 120, 213

Eastern Grenville Province, 98

Eastern Gulf Coastal Plain, 335–36

electron, 375, 376, 371

Endless Caverns, Virginia, 232

epidote, 31
erratic, glacial, 41 (defined),
    Pl. 40
escarpment, 116, 123–24,
    137, 198, 203, 211, 214,
    215, 240, 330, 331, 336,
    339, 368, **Pl. 33**
  Allegheny Front, 126
  Balcones, 339
  Bordas, 339
  Break of the Plains, 200
  Buhrstone, 336
  Chester, 183
  Dripping Springs, 164
  Eureka Springs, 191
  Highland Rim, 162
  Knobstone, 169, 170
  Muldraughs Hill, 169, 170
  Nacogdoches, 339
  Niagara, 148, 174
  Onondaga Limestone,
    134, **Pl. 13**
  Red Rock, 93
  Ripley, 336
  White Rock, 339
esker, 42 (defined), 103, 131,
    205, 228, 280, 308, **Pls.
    38, 39**
Eureka Springs Escarpment,
    191
evaporite, 25 (defined)
Evits Mountain, Pennsylva-
    nia, 253
exposures of rock, occur-
    rences of, 31–32
extrusive igneous rocks, 14–
    16 (defined)

Fairmount Park, Pennsylva-
    nia, 255
Fall Line Hills, 336
Fall Zone, Fall Line, 242,
    333–34
Falling Waters State Recre-
    ation Area, Florida, 334

fault, 12, 35, 36, 46, 119,
    238–40, 247–50, 261,
    262, 280, 282, 285, 309,
    310, 349–50, **Pl. 32**
feldspar, 8, 9, 10, 18, 20, 21,
    Pl. 41
Finger Lakes, New York, 131
First Mountain, New Jersey,
    250
flint, 24, **Pl. 45**
Flint, Michigan, 178
flood plain, 18, 19, 21
Florida
  geology, 334–35, 363–64
  highways, 363–64
Florida Caverns State Park,
    Florida, 334
Florida Keys, 335
fluorite, fluorspar, 183
Fluorspar District, Illinois,
    182–83
flute casts, **Pl. 46**
folds, 33–34, 46, **Pls. 17, 21–
    23, 26**
formation, 32 (defined)
fossil, 21, 32, 45, 47, 48, 49,
    116–17, 123, 125, 127,
    129, 130, 132, 135, 137–
    38, 139, 167, 175, 212,
    213, 291–92, 311, 316,
    318, 320, 324, 341, **Pls.
    35, 47, 48**
franklinite, 256
Fredericton, New Bruns-
    wick, 320
fresh water limestone, 23
Ft. Ridgely State Park, Min-
    nesota, 77
Fusulinid *(Fusulina)*, 212,
    213

gabbro, 8, 11 (defined), 12,
    28, 31
gabbrodiorite, 11 (defined)
Gainesville, Florida, 364

galena, 184, 191, 322
garnet, 28, 29, 104, **Pls.** 25, 44
Genessee River Gorge, New York, 134, 159
geologic time, 45–6
Georgia
geology, 240, 242–43, 245–46, 263, 333, 363
highways, 263, 363
Giant Mountain, New York, 103
Gibraltar Island, Ohio, 116, 119
glacial epoch (pre-Pleistocene) Proterozoic, 80, 241
Pennsylvanian, 48, 49
Glacial Lake Agassiz, 208–9
Duluth, 86, 87
Hitchcock, 296, 300
Wisconsin, 205, 214–15
glacial striae, 40, 48, 49, 80, 166, 168, **Pl. 46**
glacier, 40–45
glauconite, 340
gneiss, 29 (defined), 31, 5, 7, 25, 44
granite gneiss, 29
gold, 63, 64, 68, 69, 71, 72, 96, 261
Gondwana, 125, 223–26, 234, 242, 247, 264, 267, 271, 274, 289, 293, **Pl. 2**
Gooseberry Falls State Park, Minnesota, 86
Grand Saline Dome, Texas, 355
granite, 8, 9 (defined), 12, 29, 31, **Pls.** 4, 31
granite gneiss, 29
granodiorite, 11 (defined), 12
granule, 17 (defined), 20
granulite, 29 (defined), 101, 108, 109, 285

gravel, 18 (defined), **Pls. 36, 37**
graywacke, 18 (defined), 20, 28
Great Falls District Park, New Jersey, 255
Great Plains, 122–23
Great Smoky Mountains, 241–42, 260–61
Great Smoky Mountains National Park, 260–61
Great Valley, 231–32, 250, 251, 254, 256, 257, 258, 259, 263, 264
Green Mountains, 278, 281, 282, 303
greenschist, 31 (defined)
greenstone, 31 (defined), **Pl. 44**
Grenville Front Tectonic Zone, 98, 105
Grenville Province, 64, 98–110, 273, 280, 300, 307
Grenvillian Orogeny, 61, 98, 99, 100, 102, 103, 107, 108, 222, 282, 303, 326
Gros Morne National Park, Newfoundland, 98
Gulf of Mexico Basin, 343, 344, 355
gypsum, 17, 25 (defined), 35, 199, 201, 211, 312, 320, 354
Gypsy Hill Park, Virginia, 259

Hackberry Grove State Preserve, Iowa, 203
Halifax, Nova Scotia, 322
halite, 25 (defined), 353, 355
Hamilton, Ontario, 178, 179
Hartford, Connecticut, 299
Hawks Nest State Park, West Virginia, 130

hematite, 16 (defined), 19, 22, 26, 59, 76, 93, 103, 191, 264
Herkimer diamonds, 159
Highland Rim, 162
highways, see state references
Hillcrest Park, Ontario, 92
Horicon Marsh, Wisconsin, 205
hornblende, 28, 31
hornblende schist, 28
Hot Springs, Arkansas, 268
Hot Springs Mountain, Arkansas, 268–69
Hot Springs National Park, 268
Howe Caverns, New York, 132
Hudsonian Orogeny, 61, 78, 80

Ice Age, 43–45 (defined)
  deposits, etc. 43, 115, 116, 127, 131–32, 153–55, 166, 172, 175, 178, 184, 186, 190, 201, 203, 204–5, 208–9, 214, 228–29, 280, 287, 293–96, 304, 308, 320, 331, 336–38, 340
Ice Age National Scientific Reserve, Wisconsin, 205
Ichetucknee Springs, Florida, 334
igneous rocks, 7–16
Illinois
  geology, 174–75, 176, 180–83, 184–87, 366
  highways, 184–87, 366
Illinois Basin, 114, 118, 180–88
Indian Caverns, Pennsylvania, 240

Indiana
  geology, 162, 174, 179–80, 182, 183–84, 187–88
  highways, 179–80, 187–88
Indianapolis, Indiana, 188
Inks Lake State Park, Texas, 277
Interstate Park, Minnesota, 84
Interstate State Park, Wisconsin, 84, 205
intrusive igneous rocks, 8 (defined) –14
Iowa
  geology, 202–4, 213–14
  highways, 213–14
iron, 67, 96, 103, 191, 192, 208, 264, 309, 310
iron formation, 59 (defined), 64, 67, 68, 69, 71, 72, 76, 79, 84–85, 87, 89, 90, 92, 321, **Pl. 6**
isotope, 376

Jay Cooke State Park, Minnesota, 84
Jewel Cave, Tennessee, 165
joints, 9 (defined), 37, **Pls.** 16, 31
Judge C. R. Magney State Park, Minnesota, 86

Kakabeka Falls Provincial Park, Ontario, 92
Kansas
  geology, 198, 200–1, 212–13
  highways, 198, 212–13
Kansas City, Kansas, 212
Kansas City, Missouri, 194–95, 213
karst topography, 35–37 (defined), 164, 183, 207, 232, 334

Katahdin, Mt., Maine, 278, **Pl. 31**

Kearsarge, Mt., New Hampshire, 288

Kenoran Orogeny, 60–61, 65–66, 70, 91

Kentucky
geology, 141–44, 162, 163–64, 165, 168–70, 366
highways, 143–44, 168–70, 366

Kepler State Park, Iowa, 203

kettle, 42 (defined), 43, 214, 280

Kettle Moraine State Forest, Wisconsin, 205

Kittatinny Mountain, New Jersey, 137, 229–30, 232
Pennsylvania, 253–54

Knobstone Escarpment, 169–70

Knoxville, Tennessee, 239–40

laccolith, 13 (defined)

Lafayette, Louisiana, 368

lake deposits, 18, 21

Lake Erie Basin, 132

Lake Superior Basin, 81–82, 86, 92

Lake Superior Provincial Park, Ontario, 69

Lake Superior sill, 82, 85, 86

Lansing, Michigan, 171

Laurentia, 81, 125, 223–26, 240, 246, 247, 264, 271, 272, 274, 279, 281–82, 284, 288, 289, 290, 297, 301, 303, 305–7, 314

Laurentian Mountains, 58, 102, **Pl. 3**

Laurentides Provincial Reserve, Québec, 109–10

lava, 14 (defined), 16, 21

lead, 68, 183, 184, 191–93, 208, 310, 322

Letchworth Gorge (State Park), New York, 134–35

Lexington Dome, 162, 168–170

Lexington, Kentucky, 168, 169–170

lignite, 341

limestone, 22 (defined) –24, 28, 35, **Pls.** 12–16, 23, 32, 46, 47

Lincoln Caverns, Pennsylvania, 240

Linville Caverns, North Carolina, 242

Little Mt. Sheridan, Oklahoma, 276

Llano Uplift, 273, 276–77

Long Range Mountains, Newfoundland, 98

Lost River Cave, Wisconsin, 208

Louisiana
geology, 336–39, 352–57, 367–68
highways, 367–68

Louisville, Kentucky, 169

Luray Caverns, Virginia, 232

Mackinac Breccia, 176–78

Macon, Georgia, 334, 363

Madison, Wisconsin, 180

magnetite, 59, 69, 103

Maine
geology, 285, 288–89, 294–95, 305–6
highways, 305–6

Mammoth Cave, 163

Manchester, New Hampshire, 301

Manitoba
geology, 208–11, 216–17
highways, 216–17

mantle, 12
Marathon Region, 273–74, 277–78
marble, 28 (defined), **Pls.** 5, 8, 26
Mark Twain Cave, Missouri, 203
marl, 340
Maryland
  geology, 244, 257–58, 332
  highways, 257–58
Massachusetts
  geology, 282–85, 289–94, 297, 299–300, 301, 304–5, 331
  highways, 297, 299–300, 301, 304–5
Meramec Caverns, Missouri, 195
Meramec Spring, Missouri, 192, 195
Mesabi Range, Minnesota, 84–85
metagraywacke, 28
metamorphic rocks, 7 (defined), 25–31
meteoroid crater, 96–97, 136, 195
Mexia Fault Zone, Texas, 349
Miami, Florida, 335
mica, 9, 21, 27
Michigan
  geology, 82, 87–90, 171–73, 176–78
  highways, 87–90, 176–78
Michigan Basin, 70, 171–80
migmatite, 31 (defined), 69, 75, 101, 104–5, **Pl. 23**
Mill Bluff State Park, Wisconsin, 205
Minnesota
  geology, 64–67, 73, 75–77, 78–80, 82–86, 204–8, 215

highways, 73, 75–77, 82–86, 215
Mississippi
  geology, 333, 336–37, 341, 364
  highways, 364
Mississippi Basin (Embayment), 335–39
Missouri
  geology, 189–96, 203, 213, 367
  highways, 193–96, 213, 367
Mitchell, Mt., North Carolina, 242
Moat Mountain, New Hampshire, 288
Mobile, Alabama, 364
Monadnock, Mt., New Hampshire, 278
Montgomery, Alabama, 364
Montpelier, Vermont, 302
Montréal, Québec, 149
monzonite, 11 (defined), 12
moraine, 41 (defined), 42, 73, 131, 132, 135, 153, 155, 166, 172, 184, 203, 205, 214, 228, 255, 280, 287, 295, 308, 340, **Pl. 38**
Moraine State Park, Pennsylvania, 131
mountains
  Arkansas
    Hot Springs Mountain, 268–69
    North Mountain, 268–69
    Sugarloaf Mountain, 269
  Georgia
    Stone Mountain, 242–43
  Maine
    Mt. Katahdin, 278, **Pl. 31**
  Maryland
    Catoctin Mountain, 257
    South Mountain, 240–41, 257

Newfoundland
  Blow-Me-Down Mountain, 309
  North Arm Mountain, 309
  Table Mountain, 309
New Hampshire
  Moat Mountain, 288
  Mt. Chocorua, 304
  Mt. Kearsarge, 288
  Mt. Monadnock, 278
  Mt. Washington, 304, **Pls.** 20, 40
New Jersey
  First Mountain, 250
  Kittatinny Mountain, 137, 229–30
  Second Mountain, 250
New York
  Giant Mountain, 103
North Carolina
  Mt. Mitchell, 242
Oklahoma
  Little Mt. Sheridan, 276
  Mt. Scott, 276
  Mt. Sheridan, 272
  Raggedy Mountain, 272
Pennsylvania
  Bald Eagle Mountain, 233, 235, 236
  Blue Mountain, 232, 251–52, 253–54
  Evits Mountain, 253
  Kittatinny Mountain, 253–54
  Nittany Mountain, 233, 235–36, 237
  Second Mountain, 252
  South Mountain, 240, 241, 257
  Tuscarora Mountain, 253–54
  Tussey Mountain, 253
  Wills Mountain, 253
mountain ranges

Adirondack, 58, **Pl. 3**
Appalachian, 233–40, 253–54, 257–58, 259, 261, 262, **Pl. 17**
Arbuckle, 270–71, 274–76, **Pl. 19**
Blue Ridge, 240–42, 257, 258, 259, 260–61, 262, **Pl. 18**
Boston, 193, 196
Bull Run, 257, 258
Catskill, 135–36
Cobequid, 315, 323
Great Smoky, 241–42, 260–61
Green, 278, 303
Laurentian, 58, **Pl. 3**
Long Range, 98
Notre Dame, 309, 317
Ouachita, 264–69, **Pl. 18**
Schickschock, 309
Shawangunk, 135, 160–61, 230, 232
St. Francois, 189–91
Sutton, 309
Taconic, 278
White, 278, 288, 304, **Pl. 20**
Wichita, 272–73, 276, **Pl. 19**
mudstone, 21 (defined)
Muldraughs Hill, 169, 170
muscovite, 9 (defined), 26, 31
mylonite, 35 (defined), 98, 100, 102, 105, 106, 241, 263, 305, **Pl. 43**
Mystery Cave, Minnesota, 207

Nacogdoches Escarpment, 339
Narragansett Basin, 293, 297
Nashville Dome, 162, 171
Natural Bridge, Virginia, 232

Nebraska
 geology, 200–201, 213
 highways, 213
Nels Rasmussen Park, Wisconsin, 91
Nemaha Uplift, 212, 213
neutron, 375, 376, 377
New Brunswick
 geology, 310–14, 315, 316–17, 320–22
 highways, 320–22
New England, 278–306
New Hampshire
 geology, 284–85, 287–88, 301, 303–4, 305
 highways, 301, 303–4, 305
New Haven, Connecticut, 297
New Jersey
 geology, 243, 248–50, 255–56, 332, 359, 360
 highways, 255–56, 360
New River Gorge, West Virginia, 130
New York
 geology, 101–3, 125, 131, 134–36, 152–53, 155, 158–161, 243, 247–51, 282–83, 285–87, 297, 299, 331
 highways, 102–3, 134–36, 158–61, 250–51, 297, 299
New York City, 285–287
New York–Virginia Triassic Basin, 248–50, 254, 255, 258
Newfoundland
 geology, 98, 150–52, 157–58, 309–11, 314–15, 324–27
 highway,s 157–58, 324–27
Niagara Cave, Minnesota, 208

Niagara Escarpment, 148, 174
Niagara Falls, 149,
nickel, 64, 95
Nittany Mountain, Pennsylvania, 233, 235–36, 237
Norfolk Basin, 293, 297
normal fault, 35–36 (defined)
North Arm Mountain, Newfoundland, 309
North Carolina
 geology, 241–42, 244–46, 259–60, 261–62, 263, 332, 360–62
 highways, 259–60, 261–62, 263, 360–62
North Mountain, Arkansas, 268–69
Northeast Corridor, 145–61
Northern End, Coastal Plain, 331–32
Notre Dame Mountains, 309, 317
Nova Scotia
 geology, 310–14, 315–17, 322–24
 highways, 322–324
novaculite, 265 (defined), 268, 269, 274, 277

obsidian, 8, 14 (defined)
Ocala Arch, Florida, 334, 364
Ocala Caverns, Florida, 334
ocean deposits, 18, 21, 22–25
offshore bar, 357
Ohio
 geology, 127–29, 132, 134, 137–39, 152, 154–55, 158, 162, 165, 167–68, 174, 179
 highways, 137–39, 158, 167–68, 179
Ohio Caverns, 165
oil, 354

Oklahoma
  geology, 193, 197–99, 211–
    12, 266–67, 269, 270–76
  highways, 197–98, 211–
    12, 269, 274–76
Olentangy Caverns, Ohio,
  165
olivine, 8, 11 (defined), 12,
  17
Onondaga Cave, Missouri,
  195
Ontario
  geology, 64–72, 78–80, 82,
    92–97, 98–100, 103–7,
    147–49, 153, 155–57,
    173–74, 178–79
  highways, 67–72, 92–97,
    103–7, 155–57, 178–79
Ontario-Hudson Region,
  147, 152–53
Onyx Mountain Caverns,
  Missouri, 195
Organ Cave, West Virginia,
  132
organic limestone, 23 (defined)
orogeny, 46 (defined)
orthoclase feldspar, 8, 9 (defined), 11, 12, 27, 29,
  31, **Pl. 41**
Osage Plains, 198–99
oscillation ripple marks, 19
Ouachita Mountains, 264–
  69
outcrops, occurrence o,f 31–
  32
outwash, 42 (defined), 73,
  127, 131, 153, 166, 172,
  178, 184, 201, 203, 204,
  208, 228, 280, 293
Ozark Dome, 188–98

Palestine Dome, Texas, 355
Palisades State Park, New
  Jersey, 255

Palisades-Kepler State Park,
  Iowa, 203
Palo Duro Canyon State
  Park, Texas, 200
Pangaea, 47–49, 52, 53–54,
  120, 121, 125, 134, 143,
  149, 156, 173, 181, 182,
  187, 195, 198, 203, 212,
  226, 227, 247–48, 264,
  281, 288–90, 292, 293,
  303, 308, 316, 324, 329,
  341–42, 355
pebble, 17 (defined), 18
Pennsylvania
  geology, 126–27, 131, 134,
    136–37, 152, 233–37,
    243–44, 248–50, 251–55
  highways, 136–37, 251–55
Pennsylvanian glacial epoch,
  48, 49
peridotite, 8, 11 (defined), 12
phenocryst, 16 (defined)
Philadelphia, Pennsylvania,
  244, 255
phyllite, 27 (defined), **Pl. 27**
Pictured Rocks National
  Lakeshore, Michigan,
  176
Piedmont, 242–46, 254–60,
  262–63
Pikes Peak State Park, Iowa,
  204
pillow lava, 59, 69, 75, 76,
  89, 90, 105, **Pl. 6**
Pine Creek Gorge, Pennsylvania, 137
pipestone, 83, **Pl. 8**
Pipestone National Monu-
  ment, Minnesota, 82–
  83
Pittsburgh, Pennsylvania,
  137
plagioclase feldspar, 8, 10
  (defined), 11, 12, 29, 31
plate, crustal, 50, 51

plate tectonic cycle, 53,54
plate tectonics theory, 51–54
platinum, 96
Point Pleasant Park, Nova
    Scotia, 322
Porcupine Mountains Wil-
    derness State Park,
    Michigan, 87
porphyry, 15–16 (defined)
Portland, Maine, 305–6
Powers Bluff County Park,
    Wisconsin, 91
Precambrian, 46 (defined)
Presque Isle, Pennsylvania,
    152
Prince Edward Island, 307,
    308, 313
proton, 375, 376, 377
Providence, Rhode Island,
    297
pumice, 8–14
pyrite, 59, 321
pyroxene, 8, 11 (defined), 12,
    29
pyrrhotite, 59, 321

Quarry Lake State Park, Wis-
    consin, 176
quartz, 8, 9 (defined), 12, 14,
    16, 18, 20, 27, 28, 29,
    31, 64, 69, 94, **Pl. 41**
quartzite, 28 (defined), **Pls.** 8,
    26
Québec
    geology, 63, 64–67, 72–73,
      74, 98–102, 107–10,
      149–50, 153–54, 156–
      57, 309–10, 311, 317–20
    highways, 72–73, 107–10,
      156–57, 317–20
Québec Basin, 146–47, 149–
    50, 156–57
Québec City, 149, 157, 309,
    317

radioactive dating, 375–378
Raggedy Mountain, Okla-
    homa, 272
Raleigh, North Carolina, 334
Red Hills, 333, 336
Red Rock Escarpment, On-
    tario, 93
redbeds, 22 (defined)
reef, Devonian, 211–17
    Pleistocene, 335
    Silurian, 175–76, 188, 204,
      Pl. 16
regression, ocean shore, 346
    (defined) –348
reverse fault, 35–36 (defined)
Rhode Island
    geology, 291, 293, 297
    highway, 297
rhyolite, 8 (defined), 14
Rib Mountain State Park,
    Wisconsin, 91
Richmond Triassic Basin,
    259
Richmond, Virginia, 242,
    259, 334, 360
ridge, sea floor, 49
rifting, continental, 50, 51,
    54, 61, 62, 81, 227–28,
    247–50, 281, 288, 291,
    292, 303, 304, 308, 316,
    324, 329, 355
Ripley Escarpment, 336
ripple marks, 19 (defined),
    Pl. 46
river channel, 18, 19, **Pls.** 15,
    34
river deposits, 18–22
Rochester, New York, 159
rock salt, 25, 35, 176–77,
    202, 341, 352–57

Saint John, New Brunswick,
    320
Salem Limestone, 184
Salem Plateau, 190, 191

salt, 25, 35, 176–77, 202, 341, 352–57
salt dome (column), 352–57
sand, 17 (defined), 18, 19, 20
sand dune, 178
sandstone, 18 (defined) –21, 22, 28, **Pls.** 12, 15, 16, 21, 22, 30, 32, 34, 36, 42, 46
Schickschock Mountains, 309
schist, 27–28, 31, **Pl. 24**
scoria, 8 (defined), 14
Scott, Mt., Oklahoma,a 276
Sea Island Region, Coastal Plain, 332–33
sea-floor spreading, 50
Second Mountain, New Jersey, 250
    Pennsylvania, 252
Secret Caverns, New York, 132
sediment, 16–25
sedimentary rocks, 7–16 (defined), 16–25
serpentinite, 286, 317–18
shale, 21 (defined), 22, **Pls.** 12, 14–16, 22, 28, 34, 36, 42, 47
Shawangunk Mountains, 135, 160–61, 230, 232
Shenandoah Caverns, Virginia, 232
Sheridan, Mt., Oklahoma, 272
Shoreline and Inland Bays Region, Coastal Plain, 332
Sibley Provincial Park, Ontario, 92
siderite, 59
silica, 24
sill, 13 (defined), 57, 61, 64, 82, 85, 86, 92, 93, 94, 95, 96, 190, 232, 249, 254, 255, 272, 289, 295
silt, 17 (defined), 20, 21
siltstone, 21 (defined)
silver, 63, 64, 72, 93, 96, 191
Silver Springs, Florida, 334
sinkhole, 37 (defined)
slate, 25–26 (defined)
slaty cleavage, 26
Sleeping Bear Dunes National Lake Shore, Michigan, 178
Soudan Underground Mine State Park, Minnesota, 76
South Bend, Indiana, 188
South Carolina
    geology, 245–46, 261, 262–63, 332–33, 362
    highways, 261, 262–63, 362
South Mountain, Pennsylvania/Maryland, 240, 241, 257
Southern Province, 64, 77–97
sphalerite, 322
spit, 358–59, 360
Split Rock Lighthouse State Park, Minnesota, 86
Spook Cave, Iowa, 204
spring, 191–92 (defined), 232
    Alley Spring, Missouri, 191
    Bennett Spring, Missouri, 195
    Big Spring, Missouri, 191
    Ichetucknee Springs, Florida, 334
    Meramec Spring, Missouri, 192–95
    Silver Springs, Florida, 334
    Wakuklla Springs, Florida, 334
Springfield, Illinois, 186
Springfield Plateau, 190, 191

Stable Interior, history, 119–23
  rock ages, 114–15
  rock types, 114, 116–17
  structure, 113
  topography, 114–16
State College, Pennsylvania, 233–37
St. Francois Mountains, 189–91
St. John's, Newfoundland, 324
St. Louis, Missouri, 193–94, 195
stock, 13 (defined), 62, 64, 272, 287, 289, 310
Stone Mountain, Georgia, 242–43
striae, glacial, 40, 48, 49, 80, 166, 168, 280, 308, **Pl. 46**
strike-slip fault, 35–36 (defined)
submarine trenches, 50
Sudbury, Ontario, 64, 95, 96
Sudbury Basin, 64, 95–97
Sugarloaf Mountain, Arkansas, 269
sulphur, 353–54
Superior Province, 63–64, 65–77
Sutton Mountains, 309
swash and backwash zone, 357
syenite, 11 (defined), 12
syncline, 33–34 (defined)
synclinorium, 33–34 (defined)
Syracuse, New York, 160

Table Mountain, Newfoundland, 309
Taconian Zone, 280, 281–84, 299–300, 302, 303, 307, 308–10, 317–18, 319–20, 327
Taconic Mountains, 278, 281, 282
Taconic Orogeny, 53, 121, 147, 150, 152, 153, 156, 158, 163, 204, 225, 234, 240, 242, 243, 246, 252, 255, 256, 257, 262, 280, 282, 285, 287, 289, 290, 299, 302, 308–10, 317, 327
Tampa, Florida, 364
Temperance River State Park, Minnesota, 86
Tennessee
  geology, 140–43, 144, 165, 171, 238, 240, 241–42, 260–61, 263, 365
  highways, 144, 171, 260–61, 263, 365
terrace, wave-cut, 154
Texas
  geology, 199–200, 211, 212, 273–74, 276–78, 339, 341, 352–57, 368
  highways, 211, 212, 276–78, 368
thrust fault, 35–36 (defined)
Thunder Mountain State Park, Wisconsin, 91
till, 41 (defined), 73, 127, 131, 153, 166, 172, 184, 190, 201, 203, 204, 208, 228, 280, 293, 308, **Pl. 37**
tillite, 45 (defined), 49, 79, 80, 94, 241, 291, **Pl. 43**
titanium, 64
Toronto, Ontario, 156
trachyte, 288
Transcontinental Arch, 113, 203, 214
transgression, ocean shoreline, 346–48

Triassic Basin, 247–250, **Pls.** 20, 30
  Bay of Fundy, 308, 316–17, 322, 323
  Connecticut Valley, 281, 295–96, 297, 299, 300
  New York–Virginia, 248–50, 254, 255, 258
  Richmond, 259
  Wadesboro-Deep River-Durham, 260
Tri-State Mineral District, 192, 193, 195, 198
Trowbridge Falls Park, Ontario, 92
Tuckaleeche Caverns, Tennessee, 242
tuff, 15 (defined), 21
Tulsa, Oklahoma, 197
Turner Falls Park, Oklahoma, 276
Tuscarora Mountain, Pennsylvania, 253–54
Tussey Mountain, Pennsylvania, 253
Twin Cities, Minnesota, 207, 215
Two Creeks Buried Forest, Wisconsin, 205
Two Harbors city park, Minnesota, 86

unconformity, 37–39 (defined), 120, 156, 180, 181, 186, 187, 190
uraninite, 377
uranium, 64, 70, 94, 376–77

Valley and Ridge mountains, 233–40, 253–54, 257–58, 259, 261, 262, 264
Valley and Ridge Zone, 229–40, 253–54

valley glacier, 40 (defined)
Vermont
  geology, 282–84, 287, 302–3
  highways, 302–3
Virginia
  geology, 141, 142, 241, 244, 245, 248–50, 258–59, 262, 332, 360
  highways, 258–59, 262, 360
volcanic ash, 15
volcanic breccia, 18 (defined)
volcanic rock, 14–16
Voyageurs National Park, Minnesot,a 73

Wadesboro-Deep Rover-Durham Triassic Basin, 260
Wakulla Springs, Florida, 334
Warm Springs, Virginia, 232
Washington, D.C., 258
Washington, Mt., New Hampshire, 304, **Pls.** 20, 40
weathering, 15, 18, 21, 14, 35, 37
West Virginia
  geology, 130, 134, 137, 139–40
  highways, 137, 139–40
Western Gulf Coastal Plain, 339
Western Tier, 188–89, 198–217
Western Windows, 269–78
White Lake Provincial Park, Ontario, 69
White Mountains, 278, 288, 304
White Rock Escarpment, 339

Wichita Mountains, 272–73, 276
willemite, 256
Wills Mountain, Pennsylvania, 253
Wilmington, Delaware, 256–57
Winnipeg, Manitoba, 216
Wisconsin
  geology, 78–80, 82, 84, 86–87, 90–92, 175, 176, 203, 205–8, 214–15
  highways, 86–87, 90–92, 214–15
Wisconsin Dells, 214
Wonder Cave, Iowa, 204
Wyandotte Cave, Indiana, 188

zinc, (zincite) 63, 68, 72, 183, 192, 193, 208, 256, 310, 322
zircon, 377

Far Western Mountains

Columbia
Plateau

Rocky
Mountains

Basin and
Range
Province

Colorado
Plateau

Stable Inter

Areas covered by this book